# Waste-based Zeolite

Synthesis and Environmental Applications

# Waste-based Zeolite

## Synthesis and Environmental Applications

**Mihir Kumar Purkait**
Department of Chemical Engineering, Indian Institute of Technology
Guwahati, Guwahati, Assam, India

**Piyal Mondal**
Department of Chemical Engineering, Indian Institute of Technology
Guwahati, Guwahati, Assam, India

**Niladri Shekhar Samanta**
Department of Chemical Engineering, Calcutta Institute of Technology,
Uluberia, Howrah, West Bengal, India

**Pranjal Pratim Das**
Department of Chemical Engineering, Indian Institute of Technology
Guwahati, Guwahati, Assam, India

ELSEVIER

Elsevier
Radarweg 29, PO Box 211, 1000 AE Amsterdam, Netherlands
125 London Wall, London EC2Y 5AS, United Kingdom
50 Hampshire Street, 5th Floor, Cambridge, MA 02139, United States

**Notices**
Knowledge and best practice in this field are constantly changing. As new research and experience broaden our understanding, changes in research methods, professional practices, or medical treatment may become necessary.

Practitioners and researchers must always rely on their own experience and knowledge in evaluating and using any information, methods, compounds, or experiments described herein. In using such information or methods they should be mindful of their own safety and the safety of others, including parties for whom they have a professional responsibility.

To the fullest extent of the law, neither the Publisher nor the authors, contributors, or editors, assume any liability for any injury and/or damage to persons or property as a matter of products liability, negligence or otherwise, or from any use or operation of any methods, products, instructions, or ideas contained in the material herein.

ISBN: 978-0-443-22316-7

For Information on all Elsevier publications
visit our website at https://www.elsevier.com/books-and-journals

*Publisher:* Matthew Deans
*Acquisitions Editor:* Stephen Jones
*Editorial Project Manager:* Deepak Vohra
*Production Project Manager:* Anitha Sivaraj
*Cover Designer:* Christian Bilbow

Typeset by MPS Limited, Chennai, India

Working together
to grow libraries in
developing countries

www.elsevier.com • www.bookaid.org

# Contents

# Chapter 1

# Introduction to zeolite

## 1.1 Background of zeolite

In the 18th century, a Swedish chemist and creator of modern mineralogy, Baron Axel Fredrik Cronstedt, revealed that when mineral "stilbite" is heated, moisture forms on its surface (Margeta & Farkaš, 2020). His finding sparked interest in the physical, mineralogical, and chemical properties of natural aluminosilicate minerals.

Despite the restricted capabilities of zeolite structural studies at the time of their discovery (compared to advanced methodologies), scientists laid the groundwork for today's applications of synthetic and natural zeolites. The first synthetic zeolite-levinite was created in the middle of the 19th century, and a synthetic analog zeolite-mordenite was created in the first part of the 20th century, according to the literature that is now accessible (after the first structural analysis of aluminosilicate compounds) (Bajpal et al., 1978).

Zeolites are crystalline microporous solids made by $TO_4$ tetrahedral (where T represents Al, Si, Ge, P, B, and other elements), the structures of which surround cavities of molecular dimensions ($0.3-1.5$ nm dia.). These silicate-based compounds can be discovered in the environment, particularly in volcanic regions, soils, and sediments from alkaline desert lakes, or in marine deposits. There are roughly 45 different types of natural zeolites, and while some of them are found in large quantities, their restricted compositions and structures, combined with the presence of impurities, limit their commercial use to animal feed and water filtration (Chakraborty, Gautam, Das, & Hazarika, 2019; Das, Anweshan, Mondal, & Purkait, 2021; Das, Anweshan, & Purkait, 2021; Das & Mondal, 2021; Das, Sharma, & Purkait, 2022; Sontakke & Das, 2021), though other applications have been mentioned as well (Misaelides, 2011). As a result, the majority of zeolites used in industrial processes today that include ion exchange, catalysis, and adsorption are very pure synthetic crystals with distinct compositions. There are 253 distinct structures contained in the International Zeolite Association's database as of now (N.S. Samanta et al., 2021). Zeolites are used often as adsorbents (N.S. Samanta et al., 2021; N.S. Samanta, Das, Mondal, Bora, et al., 2022; N.S. Samanta, Das, Mondal, Changmai, et al., 2022; Shekhar Samanta et al., 2023), ion exchangers for water softening, or as industrial catalysts for upgrading liquid fuels and intermediates for the petrochemical,

Waste-based Zeolite. DOI: https://doi.org/10.1016/B978-0-443-22316-7.00001-2

pharmaceutical, or chemical industries. Zeolites are also used in many products that we use on a daily basis (A.D. Sontakke et al., 2023). The fact that zeolites are safe, environmentally friendly, and sustainable, in addition to their exceptional physicochemical qualities and high functionality, is a major factor in the growing adoption of them as "green" substitutes for things like polyphosphates in detergents, chlorine in swimming pools, mineral acids such as hydrofluoric, hydrochloric, and sulfuric acid in different catalytic industrial applications.

The synthesis of zeolites via different techniques including conventional and advanced methods, as well as their primary characteristics and uses of some zeolite material are also discussed in this chapter.

## 1.2 Classification of zeolite

Zeolites are three-dimensional crystalline minerals with a good arrangement of silicon and aluminum atoms. They are classed based on the arrangement of aluminum, silicon, and oxygen atoms in the crystal lattice, which defines their framework structure. The primary categorization distinguishes between natural and synthesized zeolites. Natural zeolites are found in geological deposits and are given different names according to the main mineral species, such as clinoptilolite, chabazite, and heulandite. On the other hand, synthetic zeolites are created artificially in labs and industrial settings for specialized uses. The arrangement of pores and channels inside the crystal structure influences their ion exchange capabilities and adsorption capacity. Zeolites are divided into four categories: three-dimensional (3D) framework zeolites, two-dimensional (2D) layer zeolites, one-dimensional (1D) channel zeolites, and zero-dimensional (0D) cluster zeolites. Because each group has specific properties, zeolites are adaptable materials that are widely utilized for a broad range of environmental, industrial, and catalytic purposes (Pranjal P. Das et al., 2023).

## 1.3 Structure of zeolite

As mentioned earlier, the zeolite material can be composed of several constituents such as Si, Ge, Al, Ti, B, and Zn. The neighboring tetrahedrals are linked by shared bridging oxygens, resulting in an overall O/T ratio of 2. Thus zeolites and zeolite-like compounds are classified as tectosilicates and can be separated from thicker phases by their framework density (FD), which is characterized by the number of T-atoms per 1000 Å. For dense formations like quartz or feldspars, the minimum FD ranges from 20 to 22, but zeolites have FD values that are often less than 21, and occasionally even less than 19. The FD is closely connected to the pore volume, as we shall learn in greater detail later, although it does not always represent the pore opening size. Subsequently, a preliminary investigation of 70 network structures conducted in 1989 by Brunner and Meier revealed that the minimum FDSi for each group decreases with the smallest ring when the frameworks are grouped as a function of the smallest ring contained. A similar association was discovered when taking into account the 191 structures that the International Zeolite Association (IZA) acknowledged in 2010 (McCusker et al., 2007).

From a strictly structural perspective, aluminate $[AlO_4]^-$ and silicate $[SiO_4]^-$ tetrahedral create the linked zeolite framework. $Si^{4+}$ is substituted for $Al^{3+}$ in network locations, creating a negative charge that must be balanced by charge-balancing cations in extra-network positions, which are found inside the structure's cages and/or pores. Pure silicate networks are neutral, however, the replacement of $Si^{4+}$ with $Al^{3+}$ creates a positive charge. These cations may be both organic and inorganic, and the initial ones present in the sample when it was produced can be swapped by other, more practical cations, giving the zeolite cation-exchange capability and other unique qualities that will be discussed further. The micropore may also include neutral guest molecules, like water, which in certain situations hydrate the cations discussed before. In zeolites, the $Al-O-Al$ functional group has not been seen. This observation also referred to as the Lowenstein rule is supported by the fact that negatively charged clusters are less stable than isolated negative charges (N.S. Samanta et al., 2021). Si/Al ratios will thus always be greater than 1. A zeolite can be defined by the empirical formula provided as follows:

$$M^{n+}_{x/n}Al_xSi_{1-x}O_2 \cdot yX \tag{1}$$

$M^{n+}$ is the organic or inorganic charge-balancing cation, and X is neutral guest molecules ($H_2O$) that may become blocked within the pores, with $x$ ranging from 0 to 0.5. Si and Al are not the only elements that may be included in the framework's tetrahedral locations. The resulting substances, identified as zeotypes, comprise silicoaluminophosphates (SAPOs), crystalline microporous aluminophosphates (AlPOs), and titanosilicates (McCusker et al., 2003). The fundamental building blocks are the framework tetrahedra, which may be arranged into secondary building units (SBUs) that can hold up to 16 T-atoms. A finite group of T-atoms must exist in at least two distinct tetrahedral network topologies to be regarded as an SBU. Fig. 1.1 shows some of the SBUs that are the most common.

These SBUs are assembled one on top of the other to create cages or channels that will ultimately create the 3D structure typical of a particular zeolite. Fig. 1.2 shows three potential configurations that house the sodalite cage as an illustration.

A majority of frameworks have channels characterized by at least eight T-atoms (8R); however, some, like the pure silica clathrasil, only have cages

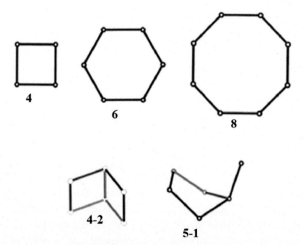

**FIGURE 1.1  Secondary building units of zeolite**. Secondary building units and their symbolic representation. *From Martínez, C., & Corma, A. (2013). Zeolites. In Comprehensive inorganic chemistry II (second edition): From elements to applications (Vol. 5, pp. 103–131). Elsevier Ltd. https://doi.org/10.1016/B978-0-08-097774-4.00506-4*

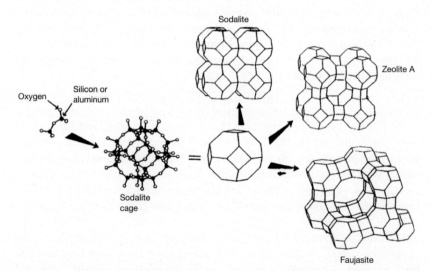

**FIGURE 1.2  Formation of zeolite network**. Zeolite network and formation of ball-structured sodalite cage and creation of another zeolite network by the arrangement of sodalite cage. *From Martínez, C., & Corma, A. (2013). Zeolites. In Comprehensive inorganic chemistry II (second edition): From elements to applications (Vol. 5, pp. 103–131). Elsevier Ltd. https://doi.org/ 10.1016/B978-0-08-097774-4.00506-4*

with six T-atoms or less in their maximum ring count (6Mr). As a result, zeolites may be divided into four categories: small (8R), extra-large (14R or larger), large (12R), and medium (10R) depending on the channel size, which is determined by the number of T-atoms constituting the channel ring or window. These channels will be a part of a channel system, which can be sinusoidal, straight, or zigzag, and which can be mono- (1D), di-dimensional (2D), or tridimensional. The channel system may be interlinked to a second channel system or it may be autonomous, with no direct access from one channel system to another. The Atlas of Zeolite Networks is once more recommended to those who want additional in-depth knowledge, terminology, and particular examples (McCusker et al., 2007). Despite the comparatively few frameworks that have been synthesized to date, Foster and Treacy's database has more than 1 million thermodynamically viable aluminosilicate cages. The extra-large 18-R pore ITQ-33 is one of the structures that were anticipated by computational modeling to be present there. There are other databases that contain fictitious zeolite or zeotype frameworks (Yan et al., 2009). The existence of channel intersections (3D or 2D type channels) is especially beneficial in a variety of catalytic processes involving zeolite-like compounds because it encourages the dispersion of reactive particles and products associated with the chemical reaction. As a result, 3D microporous zeolites like Beta and Na-Y have been used extensively and are available commercially. The pore size in these two samples, for instance, 12R, is distinct despite the tridimensionality of the channel system. Zeolites are thought of as multidimensional when the interconnecting pores have varying sizes, and in this situation, the idea of "molecular traffic control" (MTC) may be used. Derouane is credited with first describing this particular type of shape selectivity. He defined it as "a situation where the reactant molecules more frequently enter the catalyst's pores through a given channel framework, while the products diffuse out by the other, thus avoiding the development of significant counter diffusion limitations in the catalytic transformations." Therefore molecules of various sizes might diffuse differently into one channel or the other in zeolites with linked pores of varying diameters, as shown in Fig. 1.3A and B, respectively.

## 1.4   Zeolite formation mechanism and effect of synthesis parameters on the obtained products

### 1.4.1   Zeolitization

When aluminate and silicate precursor solutions are combined and heated to a high temperature (probably 40°C−190°C) for various treatment durations, a supersaturated solution is produced that triggers impulsive nucleation and crystallization processes. This is how zeolite is typically synthesized. Fig. 1.4 illustrates the crystallization processes engaged in creating zeolite from waste material containing Al and Si. The processes that lead to the creation of zeolite include (1) the dissipation of Al and Si materials, (2) the development of aluminate-silicate gel by the condensation of Si and Al

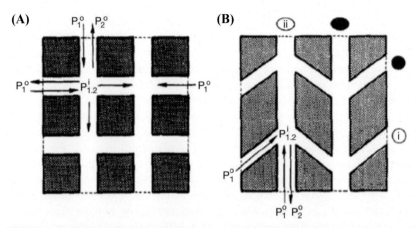

**FIGURE 1.3** **Various types of zeolite pores or channels**. (A) Single-type pore channel and (B) void space in ZSM-5-like zeolite material, where (i) represents sinusoidal and nonlinear channels and (ii) represents an elliptical-type straight channel. *From Martínez, C., & Corma, A. (2013). Zeolites. In Comprehensive inorganic chemistry II (second edition): From elements to applications (Vol. 5, pp. 103−131). Elsevier Ltd. https://doi.org/10.1016/B978-0-08-097774-4.00506-4*

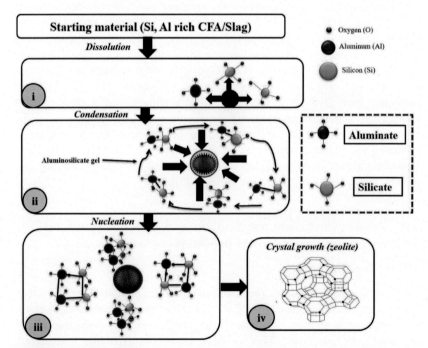

**FIGURE 1.4** **Zeolite crystallization step**. Schematic representation revealed the steps involved in the zeolite crystallization. *From Samanta, N. S., Das, P. P., Mondal, P., Changmai, M., & Purkait, M. K. (2022). Critical review on the synthesis and advancement of industrial and biomass waste-based zeolites and their applications in gas adsorption and biomedical studies.* Journal of the Indian Chemical Society, 99(11). https://doi.org/10.1016/j.jics.2022.100761

constituents, (3) the nucleation of crystals, and (4) the crystals growth and cage creation. The crystallization of zeolite during its synthesis is thus influenced by a variety of conditions. However, several synthesis methods need to be taken into account to obtain the correct zeolite form and crystallinity.

## 1.5    Effect of various parameters on zeolite preparation

### 1.5.1    Effect of crystalline time

In zeolite production, crystalline time refers to the time necessary for the development and development of aluminosilicate minerals during the synthesis time. It is a crucial parameter that can have a considerable effect on the qualities and ion exchange or catalytic performance of the finished sample (zeolite). Some literature reviews are presented next to provide a full explanation of the aforementioned issue.

In a study, X. Zhang et al. (2013) used the composite sample made from 3.5Na$_2$O: Al$_2$O$_3$: 2.9SiO$_2$: 150H$_2$O to examine the impact of crystallization duration. Their research indicates that all of the samples crystallized at a temperature of 90°C. After 180 minutes of hydrothermal synthesis, the X-ray diffraction (XRD) examination of the prepared sample verified that no crystalline pattern had developed (Fig. 1.5A). After 7 hours, only a few weaker peaks that resembled the Faujasite (FAU)-type pattern were seen (Fig. 1.5B), indicating that the sample still included a significant quantity of amorphous phases. The hydrothermal treatment also showed that the rate of crystallization went up quickly between 7 and 15 hours (Fig. 1.5B, C, and D), and after 15 hours, a prominent crystalline faujasite-type zeolite mineral was seen. They further confirmed from the XRD analysis that no distinct peak was developed even after increasing the crystallization period past 15 hours (Fig. 1.5E and F). Additionally, after 15 hours of crystallization, the silicon/aluminum molar ratio (1.29) was discovered to be within the range of zeolite (Gollakota et al., 2020). However, Wang et al. (2008) noted that because of the extremely complicated characteristic of the polymeric SiO$_4$ units, the creation of an X-type zeolite (Na-X) structure required lengthy crystallization time. Zeolite type A (Na-A) was therefore the foremost crystalline stage during the reaction period. Even after 180 minutes of hydrothermal treatment, the Fourier transform infrared (FTIR) spectroscopy shows that there was no FAU-type functional group, indicating that the material is still amorphous (Vassilev & Vassileva, 2005) (Fig. 1.5A). However, the IR spectra obtained after 7−34 hours and at a digestion temperature of 90°C show the existence of faujasite adsorption bands. Fig. 1.5B−F illustrates the Si−O−Si, Si−O−Na, and Si−O−Al species that have been found in IR spectra at the band range of 1200−450 cm$^{-1}$. The adsorption bands within the range of 1365 and 1445 cm$^{-1}$ gradually weakened over time, which is what caused the progressive decline in the amount of amorphous chemicals (Gollakota

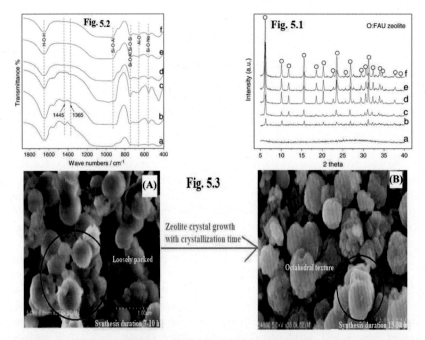

**FIGURE 1.5 XRD and FESEM analysis of synthetic zeolite Na-X**. 1.5.1(A)–(F) shows the XRD analysis of Na-X zeolite crystallized from 3 to 34 hours. 1.5.2(A)–(F) FTIR analysis of as-synthesized zeolite nanopowder and 1.5.3(A) and (B) indicate the development of zeolite cages varying with different crystallization times. *From Samanta, N. S., Das, P. P., Mondal, P., Changmai, M., & Purkait, M. K. (2022). Critical review on the synthesis and advancement of industrial and biomass waste-based zeolites and their applications in gas adsorption and biomedical studies. Journal of the Indian Chemical Society, 99(11). https://doi.org/10.1016/j.jics.2022.100761*

et al., 2020). Only crystalline faujasite adsorption bands remained after 15 hours of crystallization, with the bands for the amorphous phase disappearing. The scanning electron microscopy (SEM) pictures showed that a sufficient amount of loosely packed zeolite minerals were formed, some of them showed an octahedral shape after 7–10 hours of crystallization (Fig. 1.5A). The product growth greatly improved when the crystallization time was increased to 15, 20, or 34 hours (Fig. 1.5B), and both IR and XRD data showed that this improvement was sustained. The structure and inherent features of zeolite, namely network deformation, may be negatively impacted by longer crystallization times during zeolite synthesis. Additionally, a shorter nucleation time during zeolite crystallization may have an impact on the dissolution factor of Si and Al, which might have a big impact on the crystal structure and phase creation of the zeolite. Therefore it is recommended that crystallization time optimization be done to achieve a zeolite structure with better zeolite morphology.

### 1.5.2    Effect of crystallization temperature

To create zeolites, which are crystalline aluminosilicate materials with distinct nanoporous architectures, the crystallization temperature is an important factor to consider. It has an impact on several steps in the zeolite production as well as the characteristics of the finished sample.

Fig. 1.5A−F illustrates how XRD results validated the effect of crystallization temperature on the development of zeolite cages. The crystal formation at 70°C−90°C was verified by the XRD study (Yang et al., 2019). According to Kim and Ahn (Kim & Lee, 2009), the crystallization temperature has a substantial impact on both nucleation and crystal growth. They found that higher temperatures promoted crystallization and a strengthening of chemical group concentration in the mixture. Additionally, several investigations have discovered that the production of hydroxysodalite occurred when the hydrothermal reaction temperature was set at 130°C. Hydroxysodalite (HSOD) is produced when metastable Na-X is changed into thermodynamically balanced aluminosilicate minerals (Cardoso et al., 2015). SEM investigation validated the octahedral morphology of Na-X zeolite within the temperature of 70°C−90°C (Fig. 1.5B). At higher synthesis temperatures, however, enormous crystal sizes and marginal crystal size distributions were achieved. The octahedral crystal shape vanishes as the crystallization temperature hits 130°C, leaving only hemispherical hydroxysodalite pieces. The crystallinity of zeolite was impacted by the temperature ranging from 70 to 130°C. At 130°C, the size distribution of aluminosilicate crystals is dramatically altered; nevertheless, a further rise in temperature may cause the zeolite framework to distort structurally. However, the zeolite crystallinity was somewhat enhanced in the 70°C−90°C temperature range, where very little hydrothermal (crystallization) time may impede the binding of alumina and silica, resulting in the production of amorphous zeolite and potentially affecting the ability to adsorb metals. As a result, it is strongly advised to fabricate zeolite at a moderate temperature as opposed to a higher temperature.

### 1.5.3    Aging time

Aging is a transitional process that aids in the dissolution of the reaction solution to create the crystal-like phase of the aluminosilicate mineral. According to Yoo et al. (2009), the impact of aging time and temperature reveals an impactful part during the crystallization time. They attempted to demonstrate that whereas sodalite formed after 24 hours of aging at a temperature of 50°C, zeolite Na-A, and integrated faujasite-type zeolite required 48 and 72 hours, respectively, to form at the same condition. The research made sure the Na-A zeolite crystals were properly constructed for aging for 48 and 72 hours. Additionally, it was discovered that the XRD pattern of

FAU-type zeolite exhibited an increase in peak intensity with an aging duration of 48−72 hours. Three steps are involved in zeolite preparation: gel precipitation, nucleation, and crystal formation. Zeolite structure creation occurs as a result of the production of the nucleus on the gel surface during the aging phase. Additionally, as crystals age, the nucleation rate rises noticeably while the rate of crystal growth decreases, leading to the production of diverse sodalite cages. The synthesis of different sodalite cage zeolite structures (4R, 6R) with regard to aging time was also a major focus of the previous section. As a result, it was found that the kinds and properties of the zeolites produced at the various aging durations and high sintering temperatures varied. When producing various crystal structures, such as sodalite and faujasite, researchers used comparable temperature ranges but varied aging durations. Longer aging times lead to greater alumina and silica dissolution coupled with alkaline source formation, which results in zeolite crystal development and improved particle size. On the other hand, short aging times and high temperatures (90°C−130°C) might delay nucleation and produce structural variations in zeolite minerals. Therefore it is advised to develop many zeolite structures rather than sodalite and faujasite by altering the temperature while maintaining the same aging time.

### 1.5.4 Variation of aging temperature

Similar to aging time, aging temperature plays an important part in crystal growth. According to several scientific reports, zeolite's crystal size grew proportionately with the aging temperature. Additionally, the aging temperature during zeolite production regulates the quality of the crystalline phase. According to Yoo et al. (2009), a highly crystalline zeolite material was produced at a temperature of 70°C as opposed to 50°C for the corresponding zeolite. It is believed that because participating reactant molecules dissolve more quickly at higher temperatures, the aging temperature improves the crystallization kinetics, resulting in a faster zeolitization rate. In a different experiment, zeolite Y (Na-Y) was created by Bo and Hongzhu (1998) and the reaction mixture was aged at temperatures ranging from 30°C to 70°C to identify the changes in crystal size and form. They hypothesized that the size of the crystal grew with age at temperatures between 30°C and 40°C. According to J. Park et al. (2001), clean zeolite crystals may form at temperatures between 25°C and 35°C, but when crystals are aged over 45°C, considerable imperfections can be detected. The reactant molecules in the mixed liquor interact more quickly at the higher temperature. In the process of synthesizing zeolite, such a phenomenon ensures a quicker crystallization rate. The high aging temperature had a greater impact than the low aging temperature on the well-defined crystal form and zeolite crystallinity. When the hydrothermal process is complete, smaller crystals will develop since the low aging temperature (25°C−75°C) generates a greater number of nuclei.

The nanoparticles are very useful in applications like composite materials and as adsorbents. Therefore, it is proposed that zeolite preparation at a moderate temperature (less than 100°C) may considerably generate zeolite nanopowder that can then be used for heavy metal removal or the water softening process, leading to a high level of economic and environmental sustainability.

### 1.5.5   Silica/alumina molar ratio

The primary components required for zeolite-like mineral preparation are $SiO_2$ and $Al_2O_3$, which exist as an aluminosilicate composite substance in the generated zeolite. For the numerous forms of zeolite production, for instance, zeolite Na-X, Na-A, mordenite, type-Y zeolite, and so on, the silica/alumina molar ratio is a crucial factor. The impact of the silica/alumina molar ratio on the development of cages and aluminosilicate crystal growth has been hypothesized in several studies. To examine the finished product's structural properties, X. Zhang et al. (2013) synthesized FAU zeolite by altering various silica/alumina molar ratios of 0.5, 1.5, 1.0, 2.9, 4.0, 3.5, and 4.3 and examined the structural characteristics of the finished sample. They also made an effort to explain how analytical techniques such as Brunauer−Emmett−Teller (BET), XRD, and SEM could be employed to show how the silica and alumina ratio may alter the size of the crystal and surface area. The results also indicated that the surface area of the final sample increased from 464 to 604 $m^2/g$ due to the lowering trend of the silica/alumina molar ratio in the reaction solution. Although the Si/Al molar ratio reduced from 4.0 to 1.5, the micropore volume size of FAU-type zeolite rose from 0.14 to 0.30 $m^3/g$. Using greater silica/alumina molar concentrations, ranging from 4 to 3.5, may have caused the production of poorer crystallinity. However, FAU zeolite has large relative pore volumes (0.32 $cm^3/g$). The chemical makeup of the reactant solution is undoubtedly essential for the zeolite production. Additionally, it is mentioned that a high sodium aluminate concentration might cause the synthesis gel's silica/alumina molar ratio to fall. In addition, there are two distinct forms of the FAU-type zeolite. Zeolite Na-X has a Si/Al molar ratio of 1−1.5, while Na-Y zeolite has a Si/Al atomic ratio of 1.5−3.0 (Li et al., 2016). Additionally, according to Fotovat et al. (2009), zeolite is referred to as Na-A, Na-X, and Na-Y depending on whether its Si/Al ratio is lower than 2.0, between 2.0 and 2.4, or between 2.4 and 3.0, respectively. Additionally, they claimed to have found densely packed FAU zeolite crystals with diameters ranging from 100 to 1200 nm at various silica/alumina molar ratios. However, a cubic crystal with a molar ratio of $SiO_2/Al_2O_3$ = 1.0 was produced. Only well-crystallized cubic-shaped Na-A zeolite is present in the sample generated at a silica/alumina molar ratio of 0.5. Small silicate crystals like the double-membered four-ring (D4R) formed when the solution's concentration of $Si^{4+}$ decreased (Valtchev & Bozhilov, 2004). Therefore, zeolite A is most likely generated at a lower $SiO_2/Al_2O_3$ ratio. Tanaka et al. (2002) also achieved comparable outcomes. This may be because hydrogel has a high amount of Al, which produces fewer prenuclei and leads to the creation of bigger particles. Zeolites' surface characteristics and particle size are dependent on the molar ratio of $SiO_2/Al_2O_3$. Due to the presence of a high concentration of silica species will make more nuclei, which in turn form tiny particles, a larger ratio of $SiO_2$ to $Al_2O_3$ causes the creation of small minerals. Low $SiO_2/Al_2O_3$ ratios cause the crystallinity to decay, which may result in the creation of microscopic pores. This increases the particle's specific surface area, which may have a major impact on the ability to absorb elements during the treatment of wastewater or industrial effluent (P.P. Das, Dhara, et al., 2023; S. Dhara et al., 2023; Simons Dhara, Das, et al., 2023; M. Sharma et al., 2023; Mukesh Sharma et al., 2023). However, this ratio might have a negative impact on crystal development and structure if it is extremely low or extremely high (Bharti, Das, & Purkait, 2023; Das, Dhara, & Purkait, 2023, 2024; Das, Duarah, & Purkait, 2023; Das & Mondal, 2023; Das, Sontakke, & Purkait, 2023; Sharma & Das, 2021, 2022a, 2022b, 2023). Therefore a moderate silica−alumina ratio is more likely to be used to produce zeolites with high crystalline content and unique shapes.

### 1.5.6 Concentration of alkaline source

Sodium hydroxide (NaOH), is an alkaline substance. It is the material that is preferentially utilized as a mineralizing agent in the production of zeolite. Several researchers have suggested that the use of greater zeolite concentration affects the crystallization and particle size distribution of zeolite. The impact of NaOH concentration on the crystal structure of hydrothermally produced zeolite was investigated by Fukui et al. (2006). The phillipsite selectivity toward zeolite production is increased after the hydrothermal reaction employing a lesser concentration of NaOH, whereas the creation of hydroxysodalite is reduced. Higher NaOH concentrations have been reported to speed up reactions and shorten crystallization times. They also showed that during zeolite formation, the rate of $SiO_2$ and $Al_2O_3$ ion dissolution did not raise the concentration of NaOH. The NaOH concentration promotes the selectivity of phillipsite synthesis by incorporating additive quartz powder, which in turn reduces the crystallization time. The zeolitization process is determined by the NaOH concentration, as was mentioned in the previous section. It is also obvious that low NaOH content improves the phillipsite's selectiveness for zeolite production, whereas high NaOH concentration speeds up the reaction and shortens the crystallization process. It was also considered that NaOH serves as a binding agent throughout the zeolite crystallization process, and it is advised to have an adjustable alkali/ash ratio to help the final product become more crystalline and thermally stable.

## 1.6    Various techniques for zeolite synthesis and their limitations

### 1.6.1    Hydrothermal technique

During the 19th and 20th centuries, several findings on the preparation of zeolite minerals from fly ash were conducted. A typical flow diagram for the traditional hydrothermal method is presented in Fig. 1.6A. Three phases make up the process methodology: (1) blending of Si and Al-contained ash with various alkaline materials, such as KOH and NaOH; (2) blending the powdered mixture with determined DI; allowing the mixture to age (8−12 hours) or cure; followed by centrifugation and washing; (3) drying the resulting thicker product at 100°C to achieve the final product. Mondragon et al. (1990) initially published the essential research on the alkaline pretreatment of coal combustion ash under a variety of experimental circumstances, including temperature (90°C−100°C), NaOH concentration (2−13M), and working duration (8−48 hours). The outcome demonstrated that the coal combustion ash's amorphous component was predominantly transformed

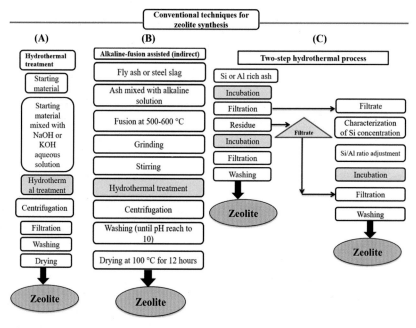

FIGURE 1.6    **Various techniques for zeolite synthesis**. Schematic illustration of (A) traditional hydrothermal treatment, (B) alkaline fusion−boosted hydrothermal process, and (C) two-step hydrothermal method. *From Samanta, N. S., Das, P. P., Mondal, P., Changmai, M., & Purkait, M. K. (2022). Critical review on the synthesis and advancement of industrial and biomass waste-based zeolites and their applications in gas adsorption and biomedical studies.* Journal of the Indian Chemical Society, 99(11). https://doi.org/10.1016/j.jics.2022.100761

into crystalline phases, taking on the shapes of Na-P, Na-X, and HSOD zeolite. The direct hydrothermal procedure for converting fly ash to zeolite entails dispersing ash in a basic solution, like NaOH or KOH solution, which leads to the expulsion of $SiO_2$ and $Al_2O_3$ elements with a following heating step, culminating in the formation of zeolite crystals from the reaction solution. By adding silicate or aluminate compounds, the ratio of Si/Al may be successfully changed to create the desired kind of zeolites. Through hydrothermal treatment, Flores et al. (2021) produced zeolite Na-P from rice straw ash, which comprised zeolite feedstock. Various treatment periods (8−24 hours) and temperatures (varying from 125°C to 175°C) were used for the hydrothermal curing process. The results demonstrated that several zeolite phases, including merlinoite, chabazite-k, and analcime, formed during zeolitization. Furthermore, the findings showed that a longer temperature or synthesis time caused a decrease in a certain surface area. A notable scientific report on the production of zeolite minerals via alkaline hydrothermal from fly ash was prepared by Moreno et al. (2001). They looked at how the solution property, hydrothermal duration, and solid/liquid ratio affected the synthesis of zeolite. The solid-to-liquid ratio, raw material mineralogy and chemical composition, and reaction time, according to the scientists, all had a significant bearing on the volume, crystal size, and newly formed minerals (Mondragon et al., 1990). The study also revealed that the hydrothermal process of volcanic glass in an alkaline solution may effectively produce zeolite. The author found that using raw materials with a high amount of $SiO_2$ and $Al_2O_3$ is adequate for the synthesis of aluminosilicate by varying the working temperatures, namely 250°C, 200°C, and 180°C and producing chabazite, phillipsite, and Na-A. The results also indicated that temperature, pH, and reaction time had a dominating influence on the preparation of several kinds of aluminosilicate minerals.

The aforementioned section shows how several kinds of aluminosilicate complexes were produced using fly ash or rice husk ash as zeolite feedstock via a hydrothermal process that was successfully regulated by operating conditions such as temperature, solution pH, alkaline concentration, and silica/alumina ratio. A very high temperature was employed to create the zeolite material that resembles analcime, chabazite, during the crystallization process, which was discovered to significantly raise the operating cost. As a result, it is important to examine the implications of the less operating thermal treatment for zeolite synthesis on operational costs and topological characterization for future environmental applications.

### 1.6.2 The alkaline fusion−facilitated hydrothermal technique

This chapter also pays close attention to fusion-assisted hydrothermal treatment for the production of zeolites or zeolitic-based materials that have greater durability and crystallinity under high temperatures and various pH conditions in water and wastewater (Changmai et al., 2018; S. Dhara, Shekhar Samanta, et al., 2023; Taghizadeh et al., 2013). The hydrothermal process with alkaline fusion assistance is demonstrated in Fig. 1.6B. The traditional hydrothermal treatment mentioned in the previous section is likely to be used in conjunction with this treatment approach. To improve the shape and stability of zeolite, only fusion of the combination of alkaline source and ash material is necessary before hydrothermal reaction (Dere Ozdemir & Piskin, 2019).

According to research (Volli & Purkait, 2015), zeolite-like minerals were synthesized using fly ash as feedstock material by adding a fusing phase before the hydrothermal procedure. NaOH was heated with the solid waste at elevated temperatures during the solid phase. After that, this fusion was mixed with an alkaline solution and let to mature. Due to the breakdown of its structural components, the mineral components of coal fly ash (CFA), namely mullite and quartz, were dissociated in the alkaline precursor solution after the conclusion of both the fusing and aging phases. A significant quantity of $Al_2SiO_5$ was absorbed into the solution by the dissolved mineral phases, which increased the rate of CFA conversion. Additionally, it was claimed that the zeolite minerals produced by the fusion stage have a higher surface area, well-defined crystallinity, and a better ion exchange capacity than the unfused crystalline sample (Purnomo et al., 2012). The research also suggested that in contrast to the nonfusion stage, which results in a mixture of HSOD, Na-P, and Na-X, the creation of Na-A or Na-X zeolite was mainly favored by the fusion step.

In comparison to the conventional hydrothermal process, the crystallization time and temperature needed for zeolite preparation in this method are substantially lower. However, in fusion-assisted hydrothermal treatment, high temperature is necessary for fusion, which in turn greatly raises the operational cost. Therefore the main objective should be to produce sorbent compounds that take less time, money, and energy to operate. Therefore future research should concentrate on high-throughput processes with quicker rates for aluminosilicate compound production. In this regard, the high-throughput ionothermal method is a hybrid methodology that may be utilized to build zeolites as well as inorganic−organic hybrid materials like metal−organic frameworks. It is recommended that future studies make use of this technology to create zeotype frameworks from other waste sources.

### 1.6.3  Zeolite preparation via two-stage hydrothermal treatment

In a study, Bukhari et al., (2015) reported that a direct hydrothermal process may be developed into a two-stage process that involves incubating a combination of 500 g of fly ash and 1.25 $dm^3$ of 2 M alkaline (NaOH) solution at a reaction duration and temperature of 6 hours and 90°C. The solid residue is dried in the second step, followed by filtering whereas, the filtrate is changed to have a molar ratio of 0.8−2. The subsequent reaction solution was then cleaned, and the resulting thicker slurry was evaporated after the calibrated filtrate had been incubated for 48 hours at 90°C. It was combined with the fresh filtrate and the incubated fly ash residue. Finally, the entire combination was incubated for 24 hours at the aforementioned temperature before the recovered residues were filtered and dried. The identical reaction mixture is prepared for one-stage hydrothermal treatment, which requires holding it there for 96 hours at 90°C (Bukhari et al., 2015). The two-stage hydrothermal technique for aluminosilicate production is shown in Fig. 1.6C.

The original hydrothermal procedure was modified to include a second stage of thermal treatment and subsequent filtering following the initial treatment with an alkaline source like KOH or NaOH. This is known as a two-stage hydrothermal treatment. The simplified procedure lowers the response time by different hours, using lower electrical energy and requiring less setup money. By adjusting the NaOH and solid waste ash concentrations to get the desired ratio, such a process may be improved. Additionally, it is recommended that by adjusting the temperature range, any change in surface morphology and crystalline characteristics may be explored. However, utilizing solid waste as the starting material, the fusion-pretreatment hydrothermal technique generates zeolite mineral which is characterized by a cubical shape. The high energy and extended time needed during the reaction are the main problems with this type of zeolitization procedure. Additionally, fusion-assisted hydrothermal and two-stage hydrothermal treatment procedures use a lot more heat energy than more sophisticated synthesis processes such as ultrasound (US), ball milling, and microwave (MW). Consequently, the methods may not be economically viable. Therefore, importance should be given to the synthesis of zeolites from different waste sources with minimal energy consumption to lower the cost of manufacturing.

## 1.6.4 Molten-salt method

Mineral wastes were transformed into zeolitic materials while in a molten state, without the addition of water. Normally, powdered combinations of 0.3 g base, 0.7 g fly ash, and 1 g salt were used. For different treatment times, the obtained mixture was melted at $350 \pm 5$°C. The obtained lump was grounded and seven times rinsed with 50 mL of DI water to expel exterior salts and bases after cooling to room temperature. As bases, KOH, NaOH, or $NH_4F$ were used, and as salts, $KNO_3$, $NaNO_3$, or $NH_4NO_3$. Using the same technique, the zeolitization of fly ash and other minerals such as montmorillonite, kaolinite, and zeolites was also investigated. In another study, M. Park et al. (2000) produced zeolite-like minerals using sewage waste as precursor material. Before synthesis, the collected sludge was thermally activated at an elevated temperature (550°C) and transformed into granular form by the ball milling process. Externally, sodium sulfate and

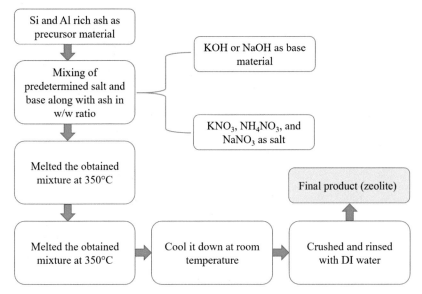

**FIGURE 1.7    Molten-salt method for zeolite synthesis**. Molten-salt method for the synthesis of zeolite-like material from silica- and alumina-contained solid waste.

sodium hydroxide were used as salt sources and mixed up with the powdered sludge to obtain the specimens. The resultant mixture was melted up in a muffle furnace at a very high temperature $\sim 950°C$ and aged at an ambient temperature for 3 days to crystallize. The resulting thicker solution was filtered and rinsed with distilled water sequentially and evaporated the resulting sample overnight at $105°C$ to get the final product. In addition, variation of alkaline source namely NaOH was examined in this study and observed that zeolite crystal growth was regulated by alkaline concentration. The reason is presumably due to reactant solubility and rate of crystallization reaction in the alkaline solution. The steps involved in preparing waste-based zeolite via the molten-salt technique have been shown in Fig. 1.7.

### 1.6.5    Green synthesis (ionothermal) technique

Ionothermal is a green technology for zeolite preparation. Nowadays, researchers are focusing on zeolite synthesis through this empirical method. In this context, Azim and Mohsin (2020) synthesized zeolite A (Na-A) via an ionothermal technique, which is discussed next:

A $10 \, cm^3$ glass tank was filled with $Al(OiPr)_3$ (0.93 mmol; Across Organics), $[C_4Py]Cl$ (6.85 mmol) and $H_3PO_4$ (1.04 mmol; 85 wt.-% in $H_2O$; Sigma Aldrich). In a fume closet, HF (0.70 mmol; 47−51 wt.-% in water; Fluka) was introduced to the reaction solution. As a result, the reaction mixture's initial molar ratio was $[C_4Py].Al(OiPr)_3$: Cl: $H_3PO_4$: HF: $H_2O$ = 7.37:

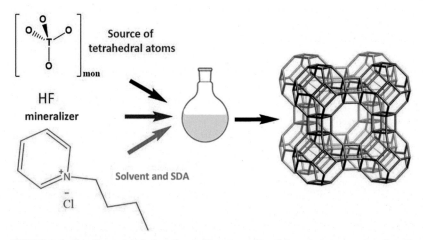

**FIGURE 1.8   Ionothermal technique for zeolite preparation.** Schematic representation of zeolite synthesis via ionothermal technique. *From Azim, M. M., & Mohsin, U. (2020). An efficient method for the ionothermal synthesis of aluminophosphate with the LTA framework type.* Microporous and Mesoporous Materials, *295, 109957. https://doi.org/10.1016/j.micromeso.2019.109957*

1.00: 1.12: 0.75: 1.95. The crystallization process took place at 160°C in an oil bath while being stirred for varied reaction durations. After being taken out of the oil bath, the reaction vessel was cooled at ambient temperature. The prepared solution mixture was supplemented with 5 cm$^3$ of deionized water. The product was recovered by filtering, thoroughly cleaned in 50 cm$^3$ of deionized water, followed by 50 cm$^3$ of acetone, dried in a 120°C oven for 4 hours, and then heated at 550°C in an oven for 4 hours. The supplemental material is where you can get a thorough description of the characterization techniques used for the AlPO-LTA. Fig. 1.8 depicts the zeolite synthesis via the ionothermal method.

## 1.6.6   Microwave-irradiation technique

On the electromagnetic spectrum, microwave energy has a frequency that is between infrared and radio waves. The range of the MW's working frequency is 915 MHz to 2450 MHz. Water absorbs the majority of the energy at 2450 MHz, making it an advantageous frequency for utilization, and commercial microwaves have easy access to magnetrons (Li & Yang, 2008). In comparison to traditional synthesis techniques, MW has a clear advantage since its energy transfer procedure differs from that of ordinary heating. The contact between MW and the dielectric molecules is what causes the energy transfer. Dielectric heating happens when the dipoles of polar molecules move in the opposite direction of an alternating electric field. The oscillations of the magnetic field are followed by the molecules, which have a

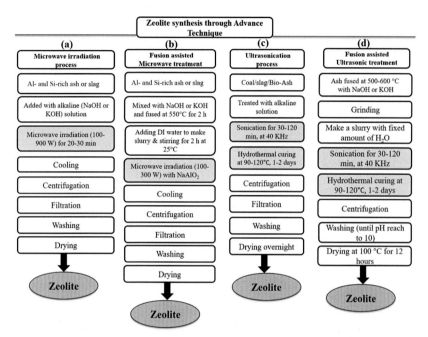

**FIGURE 1.9 Zeolite synthesis via advanced techniques.** Schematic diagram of (A) microwave irradiation, (B) fusion pretreatment followed by microwave irradiation, (C) ultrasonication, and (D) fusion-facilitated ultrasonic techniques. *From Samanta, N. S., Das, P. P., Mondal, P., Changmai, M., & Purkait, M. K. (2022). Critical review on the synthesis and advancement of industrial and biomass waste-based zeolites and their applications in gas adsorption and biomedical studies. Journal of the Indian Chemical Society, 99(11). https://doi.org/10.1016/j.jics.2022.100761*

tendency to align the dipole molecules in the irradiation field. Their inability to move is due to particle interactions and electric resistance. In contrast to traditional heating, the created energy scatters as heat as a consequence, resulting in a homogenous spreading of heat. Additionally, it was mentioned that MW irradiation increases the solvent's boiling point. When the nucleation of a gas condenses under MW irradiation, this is referred to as the "superheating" effect (Li & Yang, 2008). In spite of this, nucleation retardation can be decreased by agitating the reaction mixture (Perreux & Loupy, 2001). Nonpolar compounds can also benefit from the addition of susceptors to MW irradiation since these materials absorb MW well and convey that energy as heat to neighboring molecules that do not absorb MW well (Leadbeater & Torenius, 2002). Fig. 1.9A illustrates the straightforward processes involved in the MW alkaline hydrothermal process.

This section will discuss the zeolite synthesis through this empirical method using solid waste as a silica and alumina source. In this method, sodium aluminate and Si-Al-rich ash were thoroughly ground and combined to create a homogenous mixture. The predefined combination is allowed to be suspended in a specific volume of KOH or NaOH solution with a given concentration. To obtain the homogeneous solution of solid-alkaline, the resulting liquid is agitated for 30 minutes. The resulting suspension was subjected to fixed microwave radiation for $20-30$ minutes at a power level of $100-900$ W at a frequency of 2.5 GHz. To remove the remaining particles, the obtained sample was segregated by filtration and sequentially rinsed with DI water. In a zeolite synthesis experiment, Querol et al. (1997) have prepared Na-A zeolite using a noncrystalline phase (62%) containing fly ash (FA) (Si/Al = 1.85), and a crystalline phase containing $SiO_2$ (7%), magnetite (13%), mullite (13%), anhydrite (0.7%), and anorthite (0.3%). Coal FA was changed into zeolite-like minerals using both KOH and NaOH. Under the working conditions of $1-5M$ NaOH and $150°C-225°C$ temperature, CFA was converted to aluminosilicate minerals using both conventional energy and microwave. The effectiveness of conventional hydrothermal treatment and hydrothermal treatment aided by microwaves was then evaluated in terms of zeolite synthesis. It was found that, even though the kinds and yield of zeolites generated were quite comparable, the activation time in microwave-facilitated hydrothermal process was found to be much shorter-decay from 24 hours to just 30 minutes. Additionally, Tanaka et al. (2008) proposed a method incorporating the application of a standard MW for the quick crystallization of zeolite Na-A from FA. The aforementioned technique is combined by two stages: the first stage includes dissolving amorphous aluminosilicate from milled ash to extract $Si^{4+}$ and $Al^{3+}$ under the microwave heat treatment, and the second stage entails accomplishing the creation of zeolite minerals by MW treatment for one hour. Additionally, Behin et al. (2014) conducted a thorough investigation on the preparation of Linde-type zeolite A using microwave irradiation. The findings of this experiment demonstrate that microwave heating at a relatively low power of $100-300$ W and a reaction period of $10-30$ minutes may be used to create zeolite Na-A with a very high adsorption capacity and CEC. Additionally, Ng et al. (2019) reported that the effective integration of microwave irradiation and conventional heating led to the synthesis of zeolite-4A with a cation-exchange capacity of 5.5 meq/g. To keep the silica/alumina ratio at 2.07, they used an industrial-grade MW having a frequency of 2.45 GHz and with an operating power range of $0-2$ kW. Next, they dispersed preestimated sodium aluminate into a solution of CFA and NaOH to keep the Si/Al ratio of 2.07.

Several studies concentrate on fusion pretreatment followed by the MW treatment approach for producing high-quality zeolite minerals from Al and Si-rich solid waste. In addition, it takes a short period to prepare zeolite. However, extremely high temperatures are applied throughout the process to fuse the solid waste compounds and other alkaline sources such as NaOH/KOH and $NaAlO_2$. After cooling, the obtained mixture was crushed and the fine powder was mixed with DI water in a 1:10 weight-to-volume ratio, and the resulting slurry was agitated all night. To crystallize the slurry, the resulting mixture was then exposed to microwave treatment for 1−2 hours at 100 W. The final solution's pH was adjusted to 10 by filtering and rinsing the resultant product multiple times. To get the finished product, the residual mass was evaporated overnight at 100°C. The method is elaborately depicted in Fig. 1.9B.

Microwave energy has been widely employed for zeolite-like mineral preparation from FA due to its great efficiency. Additionally, it has been noted that the fusion-pretreated microwave approach is more effective as compared to the microwave-pretreated hydrothermal method in terms of the zeolite crystallinity, which is derived from bagasse ash (Oluyinka et al., 2020). Additionally, the BET surface area during the fusion-facilitated microwave technique was reported to be better than the traditional microwave-hydrothermal method. It can be concluded that using the microwave approach will significantly shorten the activation time required for the production of zeolite. The molarity of NaOH and the silica/alumina ratio are also determined to be within the necessary range. In addition, the fusion-assisted MW technique may improve a number of factors, including specific surface area and crystallinity level. As a result, this method may be widely used to produce unique zeolites.

In recent years, the MW-facilitated hydrothermal approach has gained popularity as a zeolite-creation technique. This method may be used to create a variety of zeolites utilizing solid waste that contains Si and/or Al as a precursor material. The procedure is more efficient as well as appealing to the research community due to the distinctive zeolite network and shorter nucleation time. However, because of its high energy consumption, the industrial-scale manufacturing of aluminosilicate via MW-energy-induced treatment has significant drawbacks. Therefore, more research is necessary to lower the synthesis cost of the aluminosilicate minerals employing a microwave-irradiation approach. It is stated that the problem of energy consumption can be solved by using green synthesis techniques.

### 1.6.7  Ultrasound irradiation technique

US-energy irradiation may considerably accelerate the synthesis of zeolite since it speeds up chemical reactions more quickly than conventional methods. The fast development and collapse of microbubbles in an aqueous medium is known as cavitation, and it is a phenomenon that originated in the US (Belviso et al., 2011). The oscillations of the gas nuclei in an aqueous media, such as their compression and expansion, result in cavitations, and the generation of a vacuum during the development of the ultrasonic wave frequencies causes gas diffusion (Ankush D. Sontakke & Purkait, 2020). Collisions of high-speed particles are brought on by these cavitations. The straightforward ash zeolitization process, shown in Fig. 1.9C, involves treating silica- and alumina-contained FA with the presence of an alkaline solution. The resulting precursor mixture was subjected to a 40 kHz ultrasonic bath treatment for 30−120 minutes, followed by 90°C−120°C hydrothermal treatment for 1−2 days. To obtain the required product, the prepared solution was centrifuged, washed with DI water, and then evaporated at 80°C−120°C for 12 hours (Ng et al., 2019). Woolard et al. (2002) produced hydroxysodalite from CFA using a hybrid US-hydrothermal process. In a zeolite synthesis experiment, Andaç et al. (2005) used a solid-liquid ratio of 0.5 g/cm$^3$ and concentrations of NaOH ranging from 1 to 5 M. The solid FA used in the production of the zeolite had a silica/alumina ratio of 1.9. The mixture was then sonicated for 30−120 minutes at room temperature using a 300 W ultrasonic bath. The slurry combination showed a much higher surface area (35.4 m$^2$/g) than the raw FA (5.6 m$^2$/g$^1$) even though no phase shift was seen from the XRD analysis of the alkaline CFA mixture (sonicated or unsonicated). This might be because unprocessed CFA contains contaminants that prevent cage formation and interfere with zeolite nucleation. To create "diamond" shaped zeolite P, Vaičiukynienė et al. (2021) used fertilizer plant by-products that included Si that had been treated with aluminum fluoride in a hydrothermal process. This study showed that zeolite type-P1 pseudospherical crystal structure was created when the making of aluminosilicate crystal was only conducted by hydrothermal processes. When zeolite-like minerals preparation was conducted using ultrasonic assistance as opposed to the traditional hydrothermal method, distinct crystal edge zeolite morphology was seen. Small plate-type zeolites were generated during conventional hydrothermal treatment. Due to the high nucleation rate, it was found that the zeolites generated by the ultrasonic-assisted hydrothermal method had smaller zeolite crystal sizes than those produced by traditional hydrothermal synthesis. The outcome also showed that ultrasonic pretreatment (20 minutes) causes the hydrothermal synthesis time to be cut in half, from 24 to 12 hours. The effectiveness of the fusion-assisted ultrasonic synthesis process has also been demonstrated in several research. This method produces more stable and crystalline zeolites or zeolite-like compounds by fusing Al and Si-rich ash at a high temperature before hydrothermal treatment. To generate zeolites (as shown in Fig. 1.9D), Ojumu et al. (2016) investigated the viability of employing a high-power ultrasonic technique as opposed to a fusion step. According to their findings, the optimal amount of time for the creation of zeolite A at 90°C was 10 minutes of sonication, followed by 2 hours of hydrothermal process. It was determined that the high-power ultrasonic procedure (10 minutes) could successfully replace the standard energy-assisted fusion step (90 minutes). In a different work, Belviso et al. (2011) successfully prepared zeolite Na-X utilizing FA by first treating it with alkali and then sonicating it at a range of temperatures (25°C−60°C). The outcomes were contrasted with the most widely utilized hydrothermal treatments. The results showed that the hydrothermal technique produced zeolites with acceptable crystalline shapes. Moreover, following ultrasonic irradiation at 60°C, a reduced crystal size was achieved. The high nucleation rate, which grew faster as compared with the crystal development rate following ultrasonic treatment, caused the crystals to become smaller (Qiu et al., 1998). The outcomes also showed that ultrasonic energy had no impact on the crystal structure. At lower temperatures, sonication was shown to have a stronger effect on the crystal size. When compared to zeolite synthesized at 35°C, Na-X zeolite produced at 25°C had bigger crystals. To counteract this impact, it may be assumed that higher temperatures enhance supersaturation, which would increase the rate of nucleation rather than the rate of growth. In fact, the rate of growth ratio and nucleation have an impact on the crystal size during batch crystallization. Both rates rise with supersaturation, although the exponential law of nucleation grows more dramatically than the low-order power law of growth rate (Sung et al., 2007).

Zeolite may be made in a variety of forms by US irradiation, which ultimately leads to its many uses. By creating a pure zeolite network by US-energy treatment during the crystallization, which delivers high energy and speeds up nucleation formation, the procedure replicates the fusion process. Future research can therefore produce STA-7, type-L-like amorphous zeolite structures for possible uses by using the US approach or by using surface directing agents.

### 1.6.8    Ball milling energy approach (solvent-free and solvent-aided synthesis route)

Ball milling is a method for combining different materials or dissolving larger particles into smaller ones using mechanical energy. For procedures such as grinding, mixing, and mechanical alloying, it has been extensively employed in a variety of disciplines, including chemistry and material science. Ball milling can be used to increase the output of zeolites or to change their characteristics when it comes to zeolite synthesis. The following section contains some evidence of zeolite synthesis via the ball milling approach.

In a study, Ahmedzeki et al. (2018) used the ball mill energy methodology, a template-free and solvent-free method, to manufacture mordenite and ZSM-5-like zeolite crystals without adding any crystal grain. The graphical depiction of ball mill energy for the production of zeolite is shown in Fig. 1.10A. $Al_2(SO_4)_3$ and $Na_2SiO_3$ reactants were combined in this procedure to form a solid-state reaction that produced crystalline, evenly mixed aluminosilicate and $Na_2SO_4$ precursors. Without the use of solvents, the resulting precursor is thermally activated at 180°C to produce crystalline ZSM-5 and mordenite. The study also shows that the parent mixture's Si/Al ratio may be altered and that these two factors can help the crystalline ZSM-5 or mordenite crystal grow significantly. Using the ball mill process, it was reported that the maximal BET area of prepared zeolite was 300 m$^2$/g$^1$ (Lin et al., 2015). The solvent-free treatment technique for zeolite also has a number of benefits, such as the creation of meso-, micro-, and

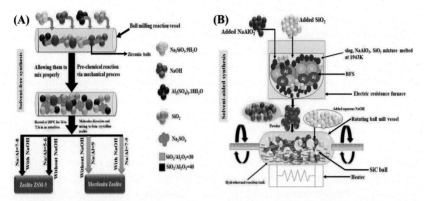

**FIGURE 1.10   Ball milling approach for zeolite synthesis**. Ball milling approach for zeolite synthesis (A) solvent-free technique and (B) ball milling-facilitated hydrothermal treatment. *From Samanta, N. S., Das, P. P., Mondal, P., Changmai, M., & Purkait, M. K. (2022). Critical review on the synthesis and advancement of industrial and biomass waste-based zeolites and their applications in gas adsorption and biomedical studies.* Journal of the Indian Chemical Society, 99(11). *https://doi.org/10.1016/j.jics.2022.100761*

macrostructures that affect mass transfer during reactions, faster reaction rates at higher temperatures, and an effective reduction in autogenous pressure during crystallization due to the absence of solvent. Sugano et al. (2005) recently demonstrated type A zeolite synthesis utilizing synthetic slag (BFS) containing Si and Al as the starting material, followed by an alkali hydrothermal method in a ball mill reactor vessel. To preserve the silica/alumina ratio in aluminosilicate-like minerals, they externally combined BF-slag with $SiO_2$ and $NaAlO_2$ and powder (as silica and alumina feedstock). The resulting combination was heated at 1670°C, cooled with cold water, and evaporated before being put into the ball mill vessel as a sample. The precursor alkaline solution and the obtained solidified mixture were digested in a Teflon-lid hydrothermal digestor. Later, 30 SiC balls with a 10-mm diameter were added to the revolving tank to quicken the hydrothermal reaction rate. With temperatures ranging from 343 to 473 K and a set autoclave rotating speed of 75 rpm, ball milling hydrothermal aging was performed. The obtaining slurry was removed and sequentially cleaned with DI water and then dried. The water vapor adsorption experiment made use of the manufactured zeolite A powder. Fig. 1.10B shows the ball milling hydrothermal technique discussed earlier for making zeolite.

## 1.7   Characterization of zeolite

### 1.7.1   Energy dispersive X-ray spectroscopy

Energy dispersive X-ray spectroscopy (EDX), sometimes known as energy dispersive X-ray analysis (EDXA) or EDS, is a method for determining the

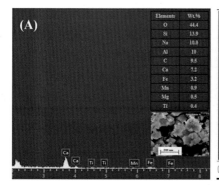
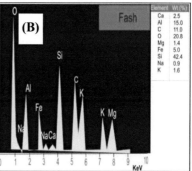

**FIGURE 1.11** **EDX spectra of zeolite mineral**. EDX spectroscopic analysis of (A) zeolite A (N.S. Samanta et al., 2021) and (B) zeolite Y (Oyebanji et al., 2020). *From Oyebanji, J. A., Okekunle, P. O., & Fayomi, O. S. I. (2020). Synthesis and characterization of zeolite-Y using* Ficus exasperata *leaf: A preliminary study.* Case Studies in Chemical and Environmental Engineering, 2, 100063. https://doi.org/10.1016/j.cscee.2020.100063

elements in a substance or its composition. It frequently works in combination with transmission electron microscopy (TEM) or SEM to examine details about a sample's chemical constituents. Here are a few examples of various types of zeolite synthesized from silica and alumina-contained waste sources.

### 1.7.1.1   EDX analysis of zeolite Na-A

N.S. Samanta et al. (2021) synthesized A-type zeolite using steel industry-processed LD-slag as precursor material through a preheat treatment followed by a hydrothermal process. The EDX analysis of the as-synthesized aluminosilicate material has been discussed next.

Fig. 1.11A depicts the EDX analysis of zeolite type A. The EDX spectrum of the sample was taken from the particular portion which is shown in the inset (Fig. 1.11A). It revealed the presence of several compounds such as Fe, Al, Na, Ca, and Si. Wherein, the zeolite composition namely Si and Al was found to be 13.9 and 10 wt.%, respectively.

### 1.7.1.2   EDX-ray spectroscopy of Na-Y zeolite

Oyebanji et al. (2020) prepared Na-Y zeolite using Ficus exasperata leaves as an abundant source of silica and alumina. The zeolite was prepared via gel formation followed by aging and crystallization steps. The EDX analysis of the obtained zeolite sample is shown in Fig. 1.11C. The existence of zeolite compounds is confirmed by EDX analysis, which indicates the cumulative percentage weight of major zeolite components to be 78.2% (O = 20.8%, Si = 42.4%, and Al = 15.0%), whereas other constituents supplied a percentage weight of 21.8%.

## 1.7.2 Electron microscopy analysis for imaging (FESEM)

The effective method of FESEM, or field emission scanning electron microscopy, is employed to investigate the surface structure and microstructure of materials at high resolution. FESEM may be used to analyze zeolite minerals to learn more about their crystal structure, grain boundaries, particle sizes, and surface characteristics. The FESEM analysis of several kinds of zeolites has been emphasized in the following section.

Madhu et al. (2022) have prepared zeolite Na-X and Na-A from rice husk ash-derived sodium silicate as an alternative source of silica for zeolite preparation. Fig. 1.12A and B represents the FESEM image of rice husk-derived zeolite A, and Fig. 1.12C and D shows the X-type zeolite, which was derived from the same silica source. The images of both the zeolite types were captured at different magnification ranges. The zeolite A micrographs as shown in Fig. 1.12A and B imply that the end sample has an "Ice cube" shape with chamfered (rounded-off) edges. The sodalite cage's (β-cage) secondary building block unit (SBU) is joined to D4R to create a 3D micropore cubic

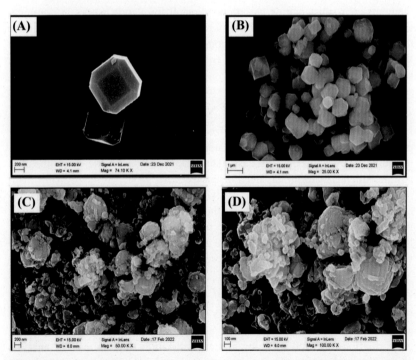

**FIGURE 1.12 Morphological analysis of Na-X zeolite.** (A, B) FESEM images of rice husk-based zeolite A and (C, D) FESEM micrographs of Na-X zeolite at two magnification ranges. *From Madhu, J., Santhanam, A., Natarajan, M., & Velauthapillai, D. (2022). CO$_2$ adsorption performance of template free zeolite A and X synthesized from rice husk ash as silicon source.* RSC Advances, 12(36), 23221−23239. https://doi.org/10.1039/d2ra04052b

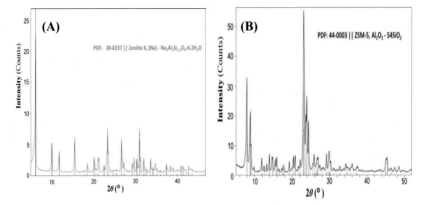

**FIGURE 1.13  XRD analysis of zeolite 13X and ZSM-5**. XRD graph of (A) zeolite 13X and (B) zeolite ZSM-5. *From Liu, H. (2022). Conversion of harmful fly ash residue to zeolites: Innovative processes focusing on maximum activation, extraction, and utilization of aluminosilicate. ACS Omega, 7(23), 20347−20356. https://doi.org/10.1021/acsomega.2c02388*

structure (α-cage). Zeolite A has been reported to have particles between the sizes of 0.6 and 0.9 μm, with an average diameter of 0.8 μm. Images of zeolite type X captured at two different magnifications are shown in Fig. 1.12C and D. The micrographs demonstrate that the particles have an identical size, octahedral shape, and an average diameter of between 1 and 2 μm. Similar to this, the octahedral cage is formed by joining the repeating unit of the SBU sodalite cage or β-cage with a double six-membered ring.

### 1.7.3   X-ray diffraction analysis

In a zeolite preparation experiment, H. Liu (2022) synthesized blended zeolite, which was a mixture of ZSM-5 and Na13X from coal FA. Fig. 1.13A, B, and C represents the XRD analysis of as-synthesized zeolite samples. From XRD data, they reported that the precursor's gel pH has a significant impact on hydrothermal growth as well as zeolite crystallinity (Fig. 1.13B). From the XRD, it is evident that the prepared zeolite 13X was in crystalline form with a diameter of < 15 μm. The prominent peaks that appeared for zeolite ZSM-5, as shown in Fig. 1.13C, also reveal the existence of a crystalline zeolite phase in the obtained sample (H. Liu, 2022).

### 1.7.4   Fourier transform infrared spectroscopy

A common strategy for determining the functional groups and chemical makeup of materials, including zeolites, is FTIR Spectroscopy. FTIR spectra of several types of zeolite have been provided in the following:

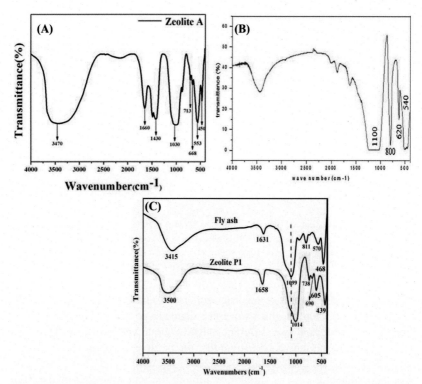

**FIGURE 1.14 FTIR analysis of zeolite**. FTIR analysis of (A) zeolite A (N.S. Samanta et al., 2021), (B) ZSM-5 (Omar et al., 2018), and (C) zeolite Na-P1 (Y. Liu et al., 2018). *From Liu, Y., Yan, C., Zhao, J., Zhang, Z., Wang, H., Zhou, S., & Wu, L. (2018). Synthesis of zeolite P1 from fly ash under solvent-free conditions for ammonium removal from water.* Journal of Cleaner Production, 202, *11−22. https://doi.org/10.1016/j.jclepro.2018.08.128*

Fig. 1.14A depicts the LD-slag-derived zeolite type A, which was synthesized via fusion pretreatment followed by hydrothermal technique. The functional group included in the synthesized Na-A zeolite was identified using FTIR employing infrared light with a wavelength between 4000 and 400 $cm^{-1}$. The OH functional group of adsorbed water molecules is shown in the FTIR spectrum (Fig. 1.14A) as a band at 1660 $cm^{-1}$. $SiO_4$ and $AlO_4$ both include tetrahedral units, and the stretching vibrations associated with those units were reported to be at around 710 and 1030 $cm^{-1}$, respectively (N.S. Samanta, Anweshan, Mondal, et al., 2023). The existence of Al−O and Si−O bonds in the zeolite A framework was said to be responsible for the absorption band at 450 $cm^{-1}$. The double-membered four-ring framework (D4R) for Na-A zeolite was ascribed to the peaks that were seen at 553 and 668 $cm^{-1}$. The bending vibrational peak at 1430 $cm^{-1}$ is ascribed to the existence of hydrous metal oxides and hydroxides. The spectrum also showed that the −OH groups in zeolite A had a stretching vibration that falls between 3300 and 3700 cm$^{-1}$ (Simons Dhara, Samanta, et al., 2023).

The FTIR analysis of zeolite ZSM-5 has been shown in Fig. 1.14B. As shown in Fig. 1.14B, the zeolite lattice vibration modes between 1100 and 540 cm$^{-1}$, confirm the presence of the zeolite ZSM-5 group in the sample (Omar et al., 2018). Both structure-sensitive exterior tetrahedron and structure-insensitive internal tetrahedron symmetric stretching vibrations are responsible for the band at about 800 cm$^{-1}$ (Omar et al., 2018). In the meanwhile, all of the samples showed the distinctive band of the double five-ring structure of mobil-type five (MFI)-type zeolites, which may be utilized to determine the degree of crystallinity in the samples. Despite the weak band of the subnanocrystals at 620 cm$^{-1}$, it may be proven that the subnanocrystals might have zeolites of the MFI type as their fundamental structure.

Zeolite Na-P1 is another class of aluminosilicate or zeolite mineral that can be synthesized by traditional as well as advanced methods. Fig. 1.14C shows the FTIR analysis of the P1-type zeolite which was synthesized by Liu and their research team using FA as silica and alumina feedstock (Y. Liu et al., 2018). Fig. 1.14C also demonstrates the FTIR spectrum of FA. For the FA, the characteristic bands between 1631 and 3415 cm$^{-1}$ are attributed to vibrations of absorbed water's hydroxyl groups. In contrast, the hydroxyl group vibrations or the solid phase hydrate water in zeolite channels are responsible for the bands about 1658 and 3500 cm$^{-1}$ in the zeolite P1. The bands at 811, 570, and 1099 cm$^{-1}$ in the FA sample are suggestive of mullite (Z. Zhang et al., 2012). The Si(Al)−O−Si antisymmetric stretch's band with a center at 1099 cm$^{-1}$ is characteristic of silicate glass spectra. In zeolite P1, the band is ascribed to the TO$_4$ (T = Al or Si) tetrahedral asymmetric stretch vibration. Due to the polycondensation of alternating Al−O and Si−O bonds, it changes to lower wavenumbers (from 1099 to 1014 cm$^{-1}$).

## 1.8    Application of zeolite

### 1.8.1    Catalysis

A crucial instrument for the sustainability of chemicals is catalysis. Improved catalytic processes will utilize natural resources more effectively, produce fewer by-products, remove contaminated effluents, and need less energy overall (Duarah et al., 2022). Catalytic methods are therefore preferred since they achieve both economic and environmental goals. In actuality, catalytic reactions create more than 90% of all industrial chemicals (A. D. Sontakke et al., 2023). With an emphasis on heterogeneous catalysts, they offer further benefits by facilitating separation and creating waste and less salt. The most often utilized components in commercial applications for heterogeneous catalysts are molecular sieves. Despite a growing growth tendency, only 20% of the world's zeolite output is used for catalytic purposes, with the remainder going to detergents (70%), adsorbents (10%), and other applications. However, catalytic applications, particularly in the oil-refining industry, represent the biggest market value. The majority of large-scale traditional processes currently employing catalysts based on zeolite are in the petrochemical and petroleum refining sectors (Vermeiren & Gilson, 2009), but their use in the chemical sector, environmental remediation, unconventional processes for converting oil, gas, coal, and biomass conversion (P.P. Das, Samanta, et al., 2023; P.P. Das, Sontakke, & Purkait, 2023; Pranjal P. Das et al., 2022), and car exhaust treatment is growing quickly.

### 1.8.2 Adsorption

Although zeolites' microporous structure and the name "molecular sieve" could lead us to believe that size exclusion is the primary method of separation in these materials, the majority of commercial separation methods actually make use of these materials' high adsorption capacities. Therefore, the majority of commercial gas separation is based on the preferential sorption of polar molecules by Al-enriched zeolites. Midway through the 1960s and the 1970s made an evident the development of the first industrial adsorption techniques for the creation of $O_2$, the purification of $H_2$ and air, and the drying of gaseous streams. Natural zeolites have historically been suggested for various uses due to their availability and affordability. However, the usage of natural zeolite-based adsorbents may be constrained by the existence of contaminants, a lack of consistency, and the requirement for processing; as a result, synthetic zeolites are currently mostly used (P.P. Das, Sontakke, Samanta, et al., 2023; Niladri Shekhar Samanta et al., 2023; N.S. Samanta, Anweshan, & Purkait, 2023; N.S. Samanta, Mondal, et al., 2023).

### 1.9 Summary

This chapter is basically focused on various aspects of zeolite chemistry and includes a brief introduction to zeolite, the type of zeolites, and synthesis techniques available for zeolite synthesis using different waste feedstock. Even though a wide variety of zeolite networks and constituents have been created, it is still difficult to synthesize custom zeolites due to their porous nature as well as the distribution and placement of their active sites. Another important consideration is the hydrothermal stability of large- and extra-large-pore zeolites, particularly in light of the possibility of their commercial use in processes such as catalytic cracking or hydrocracking. The application of zeolites as heterogeneous catalysts is the sector that is expanding the fastest and having the most effects on the economy and environment. This is true of all zeolite structures, including small, medium, large, and extra-large zeolites.

# References

Ahmedzeki, N. S., Abbas, M. N., Joodee, A. M., & Jaed, Y. M. (2018). Waste resources utilization for zeolite a synthesis. *Journal of Chemical Technology and Metallurgy, 53*(2), 239−244, rom. Available from http://dl.uctm.edu/journal/node/j2018-2/10_17_128_Ahmedzeki_p_239_244.pdf.

Andaç, O., Tatlier, M., Sirkecioğlu, A., Ece, I., & Erdem-Şenatalar, A. (2005). Effects of ultrasound on zeolite A synthesis. *Microporous and Mesoporous Materials, 79*(1−3), 225−233. Available from https://doi.org/10.1016/j.micromeso.2004.11.007.

Bajpal, P. K., Rao, M. S., & Gokhale, K. V. G. K. (1978). Synthesis of mordenite type zeolites. *Industrial and Engineering Chemistry Product Research and Development, 17*(3), 223−227. Available from https://doi.org/10.1021/i360067a009.

Behin, J., Bukhari, S. S., Dehnavi, V., Kazemian, H., & Rohani, S. (2014). Using coal fly ash and wastewater for microwave synthesis of LTA zeolite. *Chemical Engineering and Technology, 37*(9), 1532−1540. Available from https://doi.org/10.1002/ceat.201400225, http://www3.interscience.wiley.com/journal/10008333/home.

Belviso, C., Cavalcante, F., Lettino, A., & Fiore, S. (2011). Effects of ultrasonic treatment on zeolite synthesized from coal fly ash. *Ultrasonics Sonochemistry, 18*(2), 661−668. Available from https://doi.org/10.1016/j.ultsonch.2010.08.011, http://www.elsevier.com/inca/publications/store/5/2/5/4/5/1.

Bharti, M., Das, P. P., & Purkait, M. K. (2023). A review on the treatment of water and wastewater by electrocoagulation process: Advances and emerging applications. *Journal of Environmental Chemical Engineering, 11*, 111558. Available from https://doi.org/10.1016/j.jece.2023.111558.

Bo, W., & Hongzhu, M. (1998). Factors affecting the synthesis of microsized NaY zeolite. *Microporous and Mesoporous Materials, 25*(1−3), 131−136. Available from https://doi.org/10.1016/S1387-1811(98)00195-4, http://www.elsevier.com/inca/publications/store/6/0/0/7/6/0.

Bukhari, S. S., Behin, J., Kazemian, H., & Rohani, S. (2015). Conversion of coal fly ash to zeolite utilizing microwave and ultrasound energies: A review. *Fuel, 140*, 250−266. Available from https://doi.org/10.1016/j.fuel.2014.09.077, http://www.journals.elsevier.com/fuel/.

Cardoso, A. M., Paprocki, A., Ferret, L. S., Azevedo, C. M. N., & Pires, M. (2015). Synthesis of zeolite Na-P1 under mild conditions using Brazilian coal fly ash and its application in wastewater treatment. *Fuel, 139*, 59−67. Available from https://doi.org/10.1016/j.fuel.2014.08.016, http://www.journals.elsevier.com/fuel/.

Chakraborty, S., Gautam, S. P., Das, P. P., & Hazarika, Manuj K. (2019). Instant Controlled Pressure Drop (DIC) treatment for improving process performance and milled rice quality. *Journal of The Institution of Engineers (India): Series A, 100*, 683−695. Available from https://doi.org/10.1007/s40030-019-00403-w.

Changmai, M., Banerjee, P., Nahar, K., & Purkait, M. K. (2018). A novel adsorbent from carrot, tomato and polyethylene terephthalate waste as a potential adsorbent for Co (II) from aqueous solution: Kinetic and equilibrium studies. *Journal of Environmental Chemical Engineering, 6*(1), 246−257. Available from https://doi.org/10.1016/j.jece.2017.12.009, http://www.journals.elsevier.com/journal-of-environmental-chemical-engineering/.

Das, P. P., Dhara, S., & Purkait, M. K. (2023). *Hybrid electrocoagulation and ozonation techniques for industrial wastewater treatment. Sustainable industrial wastewater treatment and pollution control* (pp. 107−128). India: Springer Nature. Available from https://link.springer.com/book/10.1007/978-981-99-2560-5.

Das, P. P., Samanta, N. S., Dhara, S., & Purkait, M. K. (2023). *Biofuel production from algal biomass. Green approach to alternative fuel for a sustainable future* (pp. 167−179). India: Elsevier. Available from https://www.sciencedirect.com/book/9780128243183.

Das, P. P., Sontakke, A. D., & Purkait, M. K. (2023). *Rice straw for biofuel production. Green approach to alternative fuel for a sustainable future* (pp. 153−166). India: Elsevier. Available from https://www.sciencedirect.com/book/9780128243183.

Das, P. P., Sontakke, A. D., Samanta, N. S., & Purkait, M. K. (2023). *Emerging contaminants in wastewater: Eco-toxicity and sustainability assessment. Industrial wastewater reuse: Applications, prospects and challenges* (pp. 63−87). India: Springer Nature. Available from https://link.springer.com/book/10.1007/978-981-99-2489-9.

Das, P. P., Anweshan, A., Mondal, P., & Purkait, M. K. (2021). Integrated ozonation assisted electro-coagulation process for the removal of cyanide from steel industry wastewater. *Chemosphere*, *263*, 128370. Available from https://doi.org/10.1016/j.chemosphere.2020.128370.

Das, P. P., Anweshan, A., & Purkait, M. K. (2021). Treatment of cold rolling mill (CRM) effluent of steel industry. *Separation and Purification Technology*, *274*, 119083. Available from https://doi.org/10.1016/j.seppur.2021.119083.

Das, P. P., Deepti., & Purkait, M. K. (2023). *Industrial wastewater to biohydrogen production via potential bio-refinery route* (pp. 159−179). Springer Science and Business Media LLC. Available from http://doi.org/10.1007/978-3-031-20822-5_8.

Das, P. P., Dhara, S., & Purkait, M. K. (2023). The anaerobic ammonium oxidation process: Inhibition, challenges and opportunities. In Maulin P. Shah (Ed.), *Ammonia oxidizing bacteria: Applications in industrial wastewater treatment* (pp. 56−82). Royal Society of Chemistry. Available from https://doi.org/10.1039/BK9781837671960-00056.

Das, P. P., Dhara, S., & Purkait, M. K. (2024). Ozone-based oxidation processes for the removal of pharmaceutical products from wastewater. In Maulin P. Shah, & Pooja Ghosh (Eds.), *Development in wastewater treatment research and processes* (pp. 287−308). Elsevier. Available from https://doi.org/10.1016/B978-0-443-19207-4.00003-3.

Das, P. P., Duarah, P., & Purkait, M. K. (2023). Fundamentals of food roasting process. In S. M. jafari (Ed.), *High-temperature processing of food products* (pp. 103−130). Elsevier. Available from https://doi.org/10.1016/B978-0-12-818618-3.00005-7.

Das, P. P., & Mondal, P. (2023). Membrane-assisted potable water reuses applications: Benefits and drawbacks. In M. Sillanpaa, A. Khadir, & K. Gurung (Eds.), *Resource recovery in drinking water treatment* (pp. 289−309). Elsevier. Available from https://doi.org/10.1016/B978-0-323-99344-9.00014-1.

Das, P. P., Mondal, P., Anweshan, A., & Purkait, M. K. (2021). Treatment of steel plant generated biological oxidation treated (BOT) wastewater by hybrid process. *Separation and Purification Technology*, *258*, 118013. Available from https://doi.org/10.1016/j.seppur.2020.118013.

Das, P. P., Mondal, P., & Purkait, M. K. (2022). *Recent advances in synthesis of iron nanoparticles via green route and their application in biofuel production* (pp. 79−104). Springer Science and Business Media LLC. Available from http://doi.org/10.1007/978-981-16-9356-4_4.

Das, P. P., Sharma, M., & Purkait, M. K. (2022). Recent progress on electrocoagulation process for wastewater treatment: A review. *Separation and Purification Technology*, *292*, 121058. Available from https://doi.org/10.1016/j.seppur.2022.121058.

Das, P. P., Sontakke, A. D., & Purkait, M. K. (2023). Electrocoagulation process for wastewater treatment: applications, challenges, and prospects. In Maulin P. Shah (Ed.), *Development in wastewater treatment research and processes* (pp. 23−48). Elsevier. Available from https://doi.org/10.1016/B978-0-323-95684-0.00015-4.

Dere Ozdemir, O., & Piskin, S. (2019). A novel synthesis method of zeolite X from coal fly ash: Alkaline fusion followed by ultrasonic-assisted synthesis method. *Waste and Biomass Valorization*, *10*(1), 143−154. Available from https://doi.org/10.1007/s12649-017-0050-7, http://www.springer.com/engineering/journal/12649.

Dhara, S., Das, P. P., Uppaluri, R., & Purkait, M. K. (2023). *Phosphorus recovery from municipal wastewater treatment plants. Development in wastewater treatment research and processes: Advances in industrial wastewater treatment technologies: Removal of contaminants and recovery of resources* (pp. 49−72). India: Elsevier. Available from https://www.sciencedirect.com/book/9780323956840.

Dhara, S., Shekhar Samanta, N., Das, P. P., Uppaluri, R. V. S., & Purkait, M. K. (2023). Ravenna grass-extracted alkaline lignin-based polysulfone mixed matrix membrane (MMM) for aqueous Cr(VI) removal. *ACS Applied Polymer Materials*, *5*(8), 6399−6411. Available from https://doi.org/10.1021/acsapm.3c00999, http://pubs.acs.org/journal/aapmcd.

Dhara, S., Das, P. P., Uppaluri, R., & Purkait, M. K. (2023). *Biological approach for energy self-sufficiency of municipal wastewater treatment plants* (pp. 235−260). Elsevier BV. Available from http://doi.org/10.1016/b978-0-323-99348-7.00006-0.

Dhara, S., Samanta, N. S., Uppaluri, R., & Purkait, M. K. (2023). High-purity alkaline lignin extraction from Saccharum ravannae and optimization of lignin recovery through response surface methodology. *International Journal of Biological Macromolecules*, *234*, 123594. Available from https://doi.org/10.1016/j.ijbiomac.2023.123594.

Duarah, P., Haldar, D., Patel, A. K., Dong, C. D., Singhania, R. R., & Purkait, M. K. (2022). A review on global perspectives of sustainable development in bioenergy generation. *Bioresource Technology*, *348*. Available from https://doi.org/10.1016/j.biortech.2022.126791, http://www.elsevier.com/locate/biortech.

Fotovat, F., Kazemian, H., & Kazemeini, M. (2009). Synthesis of Na-A and faujasitic zeolites from high silicon fly ash. *Materials Research Bulletin*, *44*(4), 913−917. Available from https://doi.org/10.1016/j.materresbull.2008.08.008.

Fukui, K., Nishimoto, T., Takiguchi, M., & Yoshida, H. (2006). Effects of NaOH concentration on zeolite synthesis from fly ash with a hydrothermal treatment method. *KONA Powder and Particle Journal*, *24*(March), 183−191. Available from https://doi.org/10.14356/kona.2006020, https://www.jstage.jst.go.jp/article/kona/24/0/24_2006020/_Pdf.

Gollakota, A. R. K., Volli, V., Munagapati, V. S., Wen, J. C., & Shu, C. M. (2020). Synthesis of novel ZSM-22 zeolite from Taiwanese coal fly ash for the selective separation of Rhodamine 6G. *Journal of Materials Research and Technology*, *9*(6), 15381−15393. Available from https://doi.org/10.1016/j.jmrt.2020.10.070, http://www.elsevier.com/journals/journal-of-materials-research-and-technology/2238-7854.

Flores, C. G., Schneider, H., Dornelles, J. S., Gomes, L. B., Marcilio, N. R., & Melo, P. J. (2021). Synthesis of potassium zeolite from rice husk ash as a silicon source. *Cleaner Engineering and Technology*, *4*, 100201. Available from https://doi.org/10.1016/j.clet.2021.100201.

Kim, J. K., & Lee, H. D. (2009). Effects of step change of heating source on synthesis of zeolite 4A from coal fly ash. *Journal of Industrial and Engineering Chemistry*, *15*(5), 736−742. Available from https://doi.org/10.1016/j.jiec.2009.09.055.

Leadbeater, N. E., & Torenius, H. M. (2002). A study of the ionic liquid mediated microwave heating of organic solvents. *Journal of Organic Chemistry*, *67*(9), 3145−3148. Available from https://doi.org/10.1021/jo016297g.

Li, Y., Peng, T., Man, W., Ju, L., Zheng, F., Zhang, M., & Guo, M. (2016). Hydrothermal synthesis of mixtures of NaA zeolite and sodalite from Ti-bearing electric arc furnace slag. *RSC*

*Advances*, *6*(10), 8358−8366. Available from https://doi.org/10.1039/c5ra26881h, http://pubs.rsc.org/en/journals/journalissues.

Li, Y., & Yang, W. (2008). Microwave synthesis of zeolite membranes: A review. *Journal of Membrane Science*, *316*(1−2), 3−17. Available from https://doi.org/10.1016/j.memsci.2007.08.054.

Lin, G., Zhuang, Q., Cui, Q., Wang, H., & Yao, H. (2015). Synthesis and adsorption property of zeolite FAU/LTA from lithium slag with utilization of mother liquid. *Chinese Journal of Chemical Engineering*, *23*(11), 1768−1773. Available from https://doi.org/10.1016/j.cjche.2015.10.001.

Liu, H. (2022). Conversion of harmful fly ash residue to zeolites: Innovative processes focusing on maximum activation, extraction, and utilization of aluminosilicate. *ACS Omega*, *7*(23), 20347−20356. Available from https://doi.org/10.1021/acsomega.2c02388, http://pubs.acs.org/journal/acsodf.

Liu, Y., Yan, C., Zhao, J., Zhang, Z., Wang, H., Zhou, S., & Wu, L. (2018). Synthesis of zeolite P1 from fly ash under solvent-free conditions for ammonium removal from water. *Journal of Cleaner Production*, *202*, 11−22. Available from https://doi.org/10.1016/j.jclepro.2018.08.128, https://www.journals.elsevier.com/journal-of-cleaner-production.

Madhu, J., Santhanam, A., Natarajan, M., & Velauthapillai, D. (2022). $CO_2$ adsorption performance of template free zeolite A and X synthesized from rice husk ash as silicon source. *RSC Advances*, *12*(36), 23221−23239. Available from https://doi.org/10.1039/d2ra04052b, http://pubs.rsc.org/en/journals/journal/ra.

Margeta, K., & Farkaš, A. (2020). *Introductory chapter: Zeol−tes - From discovery to new applications on the global market*. IntechOpen. Available from http://doi.org/10.5772/intechopen.92907.

McCusker, L. B., Liebau, F., & Englehardt, G. (2003). Nomenclature of structural and compositional characteristics of ordered microporous and mesoporous materials with inorganic hosts (IUPAC recommendations 2001): Physical chemistry division commision on colloid and surface chemistry including catalysis. *Microporous and Mesoporous Materials*, *58*(1), 3−13. Available from https://doi.org/10.1016/S1387-1811(02)00545-0.

McCusker, L. B., Olson, D. H., & Baerlocher, C. (2007). *Atlas of zeolite framework types atlas of zeolite framework types*. United States: Elsevier. Available from http://www.sciencedirect.com/science/book/9780444530646.

Misaelides, P. (2011). Application of natural zeolites in environmental remediation: A short review. *Microporous and Mesoporous Materials*, *144*(1−3), 15−18. Available from https://doi.org/10.1016/j.micromeso.2011.03.024, http://www.elsevier.com/inca/publications/store/6/0/0/7/6/0.

Mohsin Azim, M., & Mohsin, U. (2020). An efficient method for the ionothermal synthesis of aluminophospahte with the LTA framework type. *Microporous and Mesoporous Materials*, *295*, 109957. Available from https://doi.org/10.1016/j.micromeso.2019.109957.

Mondragon, F., Rincon, F., Sierra, L., Escobar, J., Ramirez, J., & Fernandez, J. (1990). New perspectives for coal ash utilization: synthesis of zeolitic materials. *Fuel*, *69*(2), 263−266. Available from https://doi.org/10.1016/0016-2361(90)90187-U.

Moreno, N., Querol, X., Ayora, C., Pereira, C. F., & Janssen-Jurkovicová, M. (2001). Utilization of zeolites synthesized from coal fly ash for the purification of acid mine waters. *Environmental Science and Technology*, *35*(17), 3526−3534. Available from https://doi.org/10.1021/es0002924.

Ng, T. Y. S., Chew, T. L., Yeong, Y. F., Jawad, Z. A., & Ho, C. D. (2019). Zeolite RHO synthesis accelerated by ultrasonic irradiation treatment. *Scientific Reports*, *9*(1). Available from https://doi.org/10.1038/s41598-019-51460-x, http://www.nature.com/srep/index.html.

Ojumu, T. V., Du Plessis, P. W., & Petrik, L. F. (2016). Synthesis of zeolite A from coal fly ash using ultrasonic treat−ent - A replacement for fusion step. *Ultrasonics Sonochemistry*, *31*, 342−349. Available from https://doi.org/10.1016/j.ultsonch.2016.01.016, http://www.elsevier.com/inca/publications/store/5/2/5/4/5/1.

Oluyinka, O. A., Patel, A. V., Shah, B. A., & Bagia, M. I. (2020). Microwave and fusion techniques for the synthesis of mesoporous zeolitic composite adsorbents from bagasse fly ash: sorption of p-nitroaniline and nitrobenzene. *Applied Water Science*, *10*(12). Available from https://doi.org/10.1007/s13201-020-01327-8, https://www.springer.com/journal/13201.

Omar, B. M., Bita, M., Louafi, I., & Djouadi, A. (2018). Esterification process catalyzed by ZSM-5 zeolite synthesized via modified hydrothermal method. *MethodsX*, *5*, 277−282. Available from https://doi.org/10.1016/j.mex.2018.03.004, http://www.journals.elsevier.com/methodsx/.

Oyebanji, J. A., Okekunle, P. O., & Fayomi, O. S. I. (2020). Synthesis and characterization of zeolite-Y using Ficus exasperata leaf: A preliminary study. *Case Studies in Chemical and Environmental Engineering*, *2*, 100063. Available from https://doi.org/10.1016/j.cscee.2020.100063.

Park, J., Kim, B. C., Park, S. S., & Park, H. C. (2001). Conventional versus ultrasonic synthesis of zeolite 4A from kaolin. *Journal of Materials Science Letters*, *20*(6), 531−533. Available from https://doi.org/10.1023/A:1010976416414.

Park, M., Choi, C. L., Lim, W. T., Kim, M. C., Choi, J., & Heo, N. H. (2000). Molten-salt method for the synthesis of zeolitic materials I. Zeolite formation in alkaline molten-salt system. *Microporous and Mesoporous Materials*, *37*(1−2), 81−89. Available from https://doi.org/10.1016/S1387-1811(99)00196-1.

Perreux, L., & Loupy, A. (2001). A tentative rationalization of microwave effects in organic synthesis according to the reaction medium, and mechanistic considerations. *Tetrahedron*, *57* (45), 9199−9223. Available from https://doi.org/10.1016/S0040-4020(01)00905-X.

Purnomo, C. W., Salim, C., & Hinode, H. (2012). Synthesis of pure Na-X and Na-A zeolite from bagasse fly ash. *Microporous and Mesoporous Materials*, *162*, 6−13. Available from https://doi.org/10.1016/j.micromeso.2012.06.007.

Qiu, S., Yu, J., Zhu, G., Terasaki, O., Nozue, Y., Pang, W., & Xu, R. (1998). Strategies for the synthesis of large zeolite single crystals. *Microporous and Mesoporous Materials*, *21*(4−6), 245−251. Available from https://doi.org/10.1016/S1387-1811(98)00048-1, http://www.elsevier.com/inca/publications/store/6/0/0/7/6/0.

Querol, X., Alastuey, A., López-Soler, A., Plana, F., Andrés, J. M., Juan, R., ... Ruiz, C. R. (1997). A fast method for recycling fly ash: Microwave-assisted zeolite synthesis. *Environmental Science and Technology*, *31*(9), 2527−2533. Available from https://doi.org/10.1021/es960937t.

Samanta, N. S., Anweshan, A., Mondal, P., Bora, U., & Purkait, M. K. (2023). Synthesis of precipitated calcium carbonate from LD-slag using $CO_2$. *Materials Today Communications*, *36*. Available from https://doi.org/10.1016/j.mtcomm.2023.106588, http://www.journals.elsevier.com/materials-today-communications/.

Samanta, N. S., Anweshan, A., & Purkait, M. K. (2023). *Techniques in removal of organics and emerging contaminants from wastewater for water reuse application. Development in wastewater treatment research and processes: Advances in industrial wastewater treatment technologies: Removal of contaminants and recovery of resources* (pp. 73−96). India: Elsevier. Available from https://www.sciencedirect.com/book/9780323956840.

Samanta, N. S., P. Das, P., Sharma, M., & Purkait, M. K. (2023). *Recycle of water treatment plant sludge and its utilization for wastewater treatment* (pp. 239−264). Elsevier BV. Available from http://doi.org/10.1016/b978-0-323-99344-9.00010-4.

Samanta, N. S., Banerjee, S., Mondal, P., Anweshan, A., Bora, U., & Purkait, M. K. (2021). Preparation and characterization of zeolite from waste Linz-Donawitz (LD) process slag of steel industry for removal of Fe3 + from drinking water. *Advanced Powder Technology*, *32*(9), 3372−3387. Available from https://doi.org/10.1016/j.apt.2021.07.023, http://www.elsevier.Com.

Samanta, N. S., Das, P. P., Mondal, P., Bora, U., & Purkait, M. K. (2022). Physico-chemical and adsorption study of hydrothermally treated zeolite A and FAU-type zeolite X prepared from LD (Linz−Donawitz) slag of the steel industry. *International Journal of Environmental Analytical Chemistry*. Available from https://doi.org/10.1080/03067319.2022.2079082, http://www.tandf.co.uk/journals/titles/03067319.Asp.

Samanta, N. S., Das, P. P., Mondal, P., Changmai, M., & Purkait, M. K. (2022). Critical review on the synthesis and advancement of industrial and biomass waste-based zeolites and their applications in gas adsorption and biomedical studies. *Journal of the Indian Chemical Society*, *99*(11). Available from https://doi.org/10.1016/j.jics.2022.100761, https://www.sciencedirect.com/journal/journal-of-the-indian-chemical-society.

Samanta, N. S., Mondal, P., & Purkait, M. K. (2023). *Nanofiltration technique for the treatment of industrial wastewater. Advanced application of nanotechnology to industrial wastewater* (pp. 165−190). India: Springer Nature. Available from https://link.springer.com/book/10.1007/978-981-99-3292-4.

Sharma, M., Das, P. P., Sood, T., Chakraborty, A., & Purkait, M. K. (2021). Ameliorated polyvinylidene fluoride based proton exchange membrane impregnated with graphene oxide, and cellulose acetate obtained from sugarcane bagasse for application in microbial fuel cell. *Journal of Environmental Chemical Engineering*, *9*, 106681. Available from https://doi.org/10.1016/j.jece.2021.106681.

Sharma, M., Das, P. P., Sood, T., Chakraborty, A., & Purkait, M. K. (2022a). Clean energy from salinity gradients using pressure retarded osmosis and reverse electrodialysis: A review. *Sustainable Energy Technologies and Assessments*, *49*, 101687. Available from https://doi.org/10.1016/j.seta.2021.101687.

Sharma, M., Das, P. P., Sood, T., Chakraborty, A., & Purkait, M. K. (2022b). Reduced graphene oxide incorporated polyvinylidene fluoride/cellulose acetate proton exchange membrane for energy extraction using microbial fuel cells. *Journal of Electroanalytical Chemistry*, *907*, 115890. Available from https://doi.org/10.1016/j.jelechem.2021.115890.

Sharma, M., Das, P. P., Sood, T., Chakraborty, A., & Purkait, M. K. (2023). Polyurethane foams as packing and insulating materials. In Ram K. Gupta (Ed.), *Polyurethanes: Preparation, properties, and applications* (pp. 83−99). American Chemical Society. Available from https://doi.org/10.1021/bk-2023-1454.ch004.

Sharma, M., Das, P. P., & Purkait, M. K. (2023). *Energy storage properties of nanomaterials. Advances in smart nanomaterials and their applications* (pp. 337−350). India: Elsevier. Available from https://www.sciencedirect.com/book/9780323995467, http://doi.org/10.1016/B978-0-323-99546-7.00005-7.

Sharma, M., Samanta, N. S., Chakraborty, A., & Purkait, M. K. (2023). *Simultaneous treatment of industrial wastewater and resource recovery using microbial fuel cell* (pp. 621−637). Elsevier BV. Available from http://doi.org/10.1016/b978-0-323-95327-6.00002-6.

Shekhar Samanta, N., Das, P. P., Dhara, S., & Purkait, M. K. (2023). An overview of precious metal recovery from steel industry slag: Recovery strategy and utilization. *Industrial and Engineering Chemistry Research*, *62*(23), 9006−9031. Available from https://doi.org/10.1021/acs.iecr.3c00604, http://pubs.acs.org/journal/iecred.

Sontakke, Ankush D., & Purkait, M. K. (2020). Fabrication of ultrasound-mediated tunable graphene oxide nanoscrolls. *Ultrasonics Sonochemistry*, *63*, 104976. Available from https://doi.org/10.1016/j.ultsonch.2020.104976.

Sontakke, A. D., Das, P. P., Mondal, P., & Purkait, M. K. (2021). Thin-film composite nanofil-tration hollow fiber membranes toward textile industry effluent treatment and environmental remediation applications: Review. *Emergent Materials*, *5*, 1409–1427. Available from https://doi.org/10.1007/s42247-021-00261-y.

Sontakke, A. D., Deepti., Samanta, N. S., & Purkait, M. K. (2023). *Smart nanomaterials in the medical industry*. *Advances in smart nanomaterials and their applications* (pp. 23–50). India: Elsevier. Available from https://www.sciencedirect.com/book/9780323995467.

Sugano, Y., Sahara, R., Murakami, T., Narushima, T., Iguchi, Y., & Ouchi, C. (2005). Hydrothermal synthesis of zeolite a using blast furnace slag. *ISIJ International*, *45*(6), 937–945. Available from https://doi.org/10.2355/isijinternational.45.937.

Sung, H. J., Ki, C. Y., Young, K. H., & Chang, J. S. (2007). Size control of silicone particles using sonochemical approaches. *Bulletin of the Korean Chemical Society*, *28*(12), 2401–2404. Available from https://doi.org/10.5012/bkcs.2007.28.12.2401, http://newjournal.kcsnet.or.kr/main/j_search/j_download.htm?code = B071243.

Taghizadeh, F., Ghaedi, M., Kamali, K., Sharifpour, E., Sahraie, R., & Purkait, M. K. (2013). Comparison of nickel and/or zinc selenide nanoparticle loaded on activated carbon as efficient adsorbents for kinetic and equilibrium study of removal of Arsenazo (III) dye. *Powder Technology*, *245*, 217–226. Available from https://doi.org/10.1016/j.powtec.2013.04.020.

Tanaka, H., Fujii, A., Fujimoto, S., & Tanaka, Y. (2008). Microwave-assisted two-step process for the synthesis of a single-phase Na-A zeolite from coal fly ash. *Advanced Powder Technology*, *19*(1), 83–94. Available from https://doi.org/10.1163/156855208X291783, http://www.elsevier.Com.

Tanaka, H., Sakai, Y., & Hino, R. (2002). Formation of Na-A and -X zeolites from waste solutions in conversion of coal fly ash to zeolites. *Materials Research Bulletin*, *37*(11), 1873–1884. Available from https://doi.org/10.1016/S0025-5408(02)00861-9.

Vaičiukynienė, D., Jakevičius, L., Kantautas, A., Vaitkevičius, V., Vaičiukynas, V., & Dvořák, K. (2021). Conversion of silica by-product into zeolites by thermo-sonochemical treatment. *Ultrasonics Sonochemistry*, *72*, 105426. Available from https://doi.org/10.1016/j.ultsonch.2020.105426.

Valtchev, V. P., & Bozhilov, K. N. (2004). Transmission electron microscopy study of the formation of FAU-type zeolite at room temperature. *Journal of Physical Chemistry B*, *108*(40), 15587–15598. Available from https://doi.org/10.1021/jp048341c.

Vassilev, S. V., & Vassileva, C. G. (2005). Methods for characterization of composition of fly ashes from coal-fired power stations: A critical overview. *Energy and Fuels*, *19*(3), 1084–1098. Available from https://doi.org/10.1021/ef049694d.

Vermeiren, W., & Gilson, J. P. (2009). Impact of zeolites on the petroleum and petrochemical industry. *Topics in Catalysis*, *52*(9), 1131–1161. Available from https://doi.org/10.1007/s11244-009-9271-8, http://springerlink.metapress.com/content/101754/.

Volli, V., & Purkait, M. K. (2015). Selective preparation of zeolite X and A from flyash and its use as catalyst for biodiesel production. *Journal of Hazardous Materials*, *297*, 101–111. Available from https://doi.org/10.1016/j.jhazmat.2015.04.066, http://www.elsevier.com/locate/jhazmat.

Wang, C. F., Li, J. S., Wang, L. J., & Sun, X. Y. (2008). Influence of NaOH concentrations on synthesis of pure-form zeolite A from fly ash using two-stage method. *Journal of Hazardous Materials*, *155*(1–2), 58–64. Available from https://doi.org/10.1016/j.jhazmat.2007.11.028.

Woolard, C. D., Strong, J., & Erasmus, C. R. (2002). Erratum: Evaluation of the use of modified coal ash as a potential sorbent for organic waste streams. *Applied Geochemistry*, *18*. Available from https://doi.org/10.1016/S0883-2927(03)00012-X.

Yan, Y., Li, J., Qi, M., Zhang, X., Yu, J., & Xu, R. (2009). Database of open-framework aluminophosphate syntheses: Introduction and application (I). *Science in China, Series B: Chemistry, 52*(11), 1734−1738. Available from https://doi.org/10.1007/s11426-009-0266-x.

Yang, L., Qian, X., Yuan, P., Bai, H., Miki, T., Men, F., . . . Nagasaka, T. (2019). Green synthesis of zeolite 4A using fly ash fused with synergism of NaOH and Na$_2$CO$_3$. *Journal of Cleaner Production, 212*, 250−260. Available from https://doi.org/10.1016/j.jclepro.2018.11.259, https://www.journals.elsevier.com/journal-of-cleaner-production.

Yoo, Y.S., Ban, H.J., Cheon, K.H., & Lee, J.L. (2009). 1 2009/01 Materials Science Forum 10.4028/http://www.scientific.net/MSF.620-622.225   16629752   225-228   Trans   Tech Publications Ltd South Korea. The effect of aging on synthesis of zeolite at high temperature. http://www.ttp.net/0255-5476.html 620.

Zhang, X., Tang, D., Zhang, M., & Yang, R. (2013). Synthesis of NaX zeolite: Influence of crystallization time, temperature and batch molar ratio SiO$_2$/Al$_2$O$_3$ on the particulate properties of zeolite crystals. *Powder Technology, 235*, 322−328. Available from https://doi.org/10.1016/j.powtec.2012.10.046.

Zhang, Z., Wang, H., & Provis, J. L. (2012). Quantitative study of the reactivity of fly ash in geopolymerization by FTIR. *Journal of Sustainable Cement-Based Materials, 1*(4), 154−166. Available from https://doi.org/10.1080/21650373.2012.752620.

# Chapter 2

# Solid waste as a potential source for zeolite synthesis

## 2.1 Introduction

The establishment of sustainable and environmentally friendly supply chains in industrial parks worldwide has become a crucial element in the global efforts toward achieving a circular economy (Bharti, Das, & Purkait, 2023; Das, Dhara, & Purkait, 2023, 2024; Das, Duarah, & Purkait, 2023; Das, Sharma, & Purkait, 2023; Das, Sontakke, & Purkait, 2023a, 2023b; Sharma, Das, Chakraborty, & Purkait, 2022; Sharma, Das, Sood, Chakraborty, & Purkait, 2022, 2021). With a focus on waste-to-resource initiatives, industrial parks are actively pursuing strategies that encourage the effective use of waste materials such as steel slags, coal fly ash, and municipal solid waste incineration generated fly ash. By developing green supply chains, these parks can reduce their environmental impact, while simultaneously contributing to the overall sustainability of the global economy. The adoption of such initiatives is pivotal in creating a more sustainable and equitable future (Li, Pan, Kim, Linn, & Chiang, 2015). As a green resource, steel slags are generated as a by-product during steel production. The four main types of modern steel include alloy, stainless, tool steels, and carbon with carbon steel being generated in either an electric arc furnace (EAF) or a basic oxygen furnace (BOF) and subsequently refined in a ladle furnace (LF) to obtain superior quality. In contrast, stainless steel can be generated in an LF, EAF, or argon oxygen decarburization (AOD) furnace (Huaiwei & Xin, 2011; Samanta et al., 2021). The production of stainless steels and carbon results in a substantial quantity of slag, amounting to approximately 15%−20% of the total steel manufacture, with the produced steel slags categorized as EAF slag, LF slag, and BOF slag (Caijun, 2004; Samanta, Anweshan, Mondal, Bora, & Purkait, 2023; Samanta, Das, Mondal, Bora, & Purkait, 2022; Samanta, Das, Mondal, Changmai, & Purkait, 2022).

Coal combustion fly ash, which is one of the leading industrial solid derivatives globally, is being widely used in the production of concrete and cement materials as a supplementary material because of its industrial benefits (Hemalatha, Mapa, George, & Sasmal, 2016). Coal fly ash (CFA), categorized as either class F or class C according to the American Society

Waste-based Zeolite. DOI: https://doi.org/10.1016/B978-0-443-22316-7.00002-4

for Testing and Materials (ASTM) system, has a spherical particle morphology that enhances the fluidity of cement-based products, filler effect that offers the creation of additional calcium silicate hydrates nuclei, and pozzolanic effect that results in the formation of secondary hydration products and enhanced durability and mechanical strength (Butler & Mearing, 1985). CFA has become a critical constituent in the production of high-performance concrete (Dhondy, Remennikov, & Shiekh, 2019; Monteiro, Miller, & Horvath, 2017). However, there is a shortage of CFA in many countries due to the ever-increasing demand for concrete infrastructure and the stringent regulations issued by environmental protection agencies to curb the discharges of $NO_x$ and $SO_2$ from this industry (Niladri Samanta, Das, Sharma, & Purkait, 2023). It is vital to address these issues constructively to ensure a sustainable and efficient supply of CFA for the making of concrete material.

Sugarcane bagasse ash (SCBA) is a valuable industrial by-product that is widely available in sugar-producing regions (Webber, White, Petrie, Shrefler, & Taylor, 2015). When sugar is produced, a considerable amount of bagasse ash can be burned to produce power and retained bagasse fly ash (BFA) after the process. Even though the ash content is <3% of the original bagasse mass, the large capacity of burned bagasse produces a significant quantity of BFA that can be put to good use (Webber et al., 2015). In Louisiana, the largest sugar-manufacturing area in the United States, up to 40,823 metric tons of SCBA are generated each year, and globally, approximately 5 million metric tons of BFA can be generated annually. Although this is not enough to replace CFA, SCBA can be utilized scientifically to partially compensate for the local scarcity of supplementary cementitious materials (SCMs) (Samanta, Das, Mondal, Changmai, et al., 2022). Therefore it is crucial to handle SCBA in an economically and environmentally feasible way. It is a resource that can be replenished if used properly and helps to mitigate environmental impacts. Thus it is essential to explore innovative ways to utilize SCBA effectively and sustainably (Rovani, Santos, Corio, & Fungaro, 2018).

Incineration is a highly efficient process for handling municipal solid waste (MSW) that can result in up to a 90% reduction in volume, a 60%−75% reduction in mass, and the destruction of harmful pathogens, while also potentially recovering energy. However, when MSW is incinerated, it can generate solid residues such as fly ash, which is considered a dangerous substance due to its high concentration of chlorinated organic materials and leachable detrimental heavy metals in some cases. Fly ash leaching tests reveal that the concentration of certain constituents, namely hazardous metals like Cd, Zn, and Cr surpasses regulatory limits and cannot be dumped in its current form (Shim, Rhee, & Lee, 2005). Hence, an initiation treatment is required before dumping (Karamanov, Pelino, & Hreglich, 2003). At the moment, landfilling is an increasingly common strategy for getting rid of fly ash, but it uses up a lot of space that cannot be used for other human activities. To avoid this, researchers have been working on developing reuse applications for fly ash as a preferred alternative to landfill disposal (Bethanis, Cheeseman, & Sollars, 2002; Park & Heo, 2002).

In this chapter, the characterization of various industrial solid wastes (namely steel slag and CFA) and MSW and their possibilities in the preparation of zeolite-like material have been explored.

## 2.2 Industrial waste

### 2.2.1 Steel industry slag

#### 2.2.1.1 *Characterization of steel slag*

##### 2.2.1.1.1 Physicochemical characterization of BOF slag

The steel-making industry produces solid waste like BOF slag that has variable chemical compositions, which can be improved by using different iron admixtures, ores, cooling processes, and steel-making methods. Table 2.1 shows the chemical constituents present in BOF slag. BOF slag mainly contains CaO, $SiO_2$, $Fe_2O_3$, $Al_2O_3$, and MgO, along with minor quantity of several oxides such as MnO, $P_2O_5$, $Na_2O$, and $SO_3$. The high concentration of MgO and CaO in this form of slag is important to minimize contaminations, whereas iron oxides come from the iron remnant that was not recovered when the molten iron is processed into steel (Yildirim and Prezzi, 2011). The higher alkaline oxide to acidic oxides ratio makes the slag more basic in nature. It comprises various mineral phases, for instance, CaO, dicalcium silicate ($C_2S$), MgO, tricalcium silicate ($C_3S$), and dicalcium ferrite ($C_2F$) (Wang, Yan, Yang, & Zhang, 2013).

**Abbreviations:-** indicates not found

##### 2.2.1.1.2 Physicochemical characterization of EAF slag

The EAF process has the potential to significantly reduce waste by utilizing steel scrap as feed material, thus making it an effective steel scrap reutilizing technique. The compositional analysis revealed the presence of a wider range of chemical constituents in electric arc furnace-generated solid waste rather than basic oxygen furnace-produced solid waste however, both EAF slag and BOF slag share various similar properties like physical appearance, mineral phases, and primary oxides. It was also noticed that stainless steel generates a high amount of EAF-S type slag, which contains high chromium concentration but low FeO content (Yildirim & Prezzi, 2011). Several mineral phases identified in EAF slag, such as olivine, wustite, $C_2S$, and $C_3S$, have been studied (Piatak, Parsons, & Seal, 2015; Santamaría, Rojí, Skaf, Marcos, & González, 2016). These studies have provided valuable insights into the mineralogy and chemical constituents of EAF-generated solid waste. Table 2.2 represents the inorganic constituents

**TABLE 2.1** Elemental analysis of BOF slag generated in various countries. Compositional analysis of steel industry generated BOF slag.

| Sources | $SiO_2$ | $Al_2O_3$ | CaO | MgO | $SO_3$ | MnO | $P_2O_5$ | f-CaO | Others | Treatment | References |
|---|---|---|---|---|---|---|---|---|---|---|---|
| India | 15.0 | 4.1 | 41.5 | 6.2 | 0.1 | – | – | 5.3 | 1.4 ($Na_2O$)/0.05 ($K_2O$) | Before weathering | Palankar, Ravi Shankar and Mithun (2016) |
| China | 14.8 | 5.5 | 46.7 | 6.3 | – | 2.8 | 1.7 | 7.5 | – | – | Pang et al. (2016) |
| China | 15.5 | 5.4 | 38.6 | 7.7 | 0.2 | 1.9 | 1.6 | – | – | – | Wang et al. (2013) |
| China | 11.0 | 1.4 | 41.4 | 8.6 | – | – | – | – | – | Hot stuffy technique for cooling | Liu, Liu, and Qi (2016) |
| China | 18.9 | 2.9 | 40.0 | 5.4 | 0.9 | 2.8 | 1.3 | – | – | – | Li, Zhao, Zhao, and He (2013) |
| UK | 11.5 | 2.3 | 37.4 | 9.3 | 0.3 | 3.7 | 1.3 | – | 0.37($TiO_2$)/0.03 ($Na_2O$)/0.01($K_2O$) | Weathered | Lizarazo-Marriaga, Claisse, and Ganjian (2011) |
| Iran | 10.4 | 2.0 | 56.4 | 1.7 | – | 2.5 | – | – | 3.1($TiO_2$)/0.2(S)/2.4 ($V_2O_5$)/0.3 ($Na_2O + K_2O$) | Magnetic separation | Monshi and Asgarani (1999) |

*Source:* From Jiang, Y., Ling, T. C., Shi, C., & Pan, S. Y. (2018). Characteristics of steel slags and their use in cement and concrete—A review. *Resources, Conservation and Recycling, 136*, 187–197. https://doi.org/10.1016/j.resconrec.2018.04.023.

**TABLE 2.2** Chemical constituents present in electric arc furnaces slag.

| Sources | $SiO_2$ | $Al_2O_3$ | CaO | MgO | MnO | $P_2O_5$ | ƒ-CaO | Others | Treatment | References |
|---|---|---|---|---|---|---|---|---|---|---|
| India | 23.3 | 6.1 | 30.8 | 12 | 1.5 | 0.6 | 0.4 | 0.9($TiO_2$) | Water quenched and cooled air | Muhmood, Vitta, and Venkateswaran (2009) |
| India | 29.0 | 5.9 | 38.8 | 21.4 | 1.4 | 0.5 | 0.1 | 0.7($TiO_2$) | Water quenched | Muhmood et al. (2009) |
| Malaysia | 26.4 | 4.84 | 43.4 | 16.9 | 1.86 | 2.66 | – | 0.15 ($Na_2O$) | Cooled air | Roslan, Ismail, Abdul-Majid, Ghoreishiamiri, and Muhammad (2016) |
| China | 24.9 | 4.89 | 54.0 | 7.26 | – | 0.05 | – | 0.15($SO_3$) | – | Li et al. (2013) |
| Egypt | 13.1 | 5.51 | 33 | 5.03 | 4.18 | 0.7 | – | 0.14 ($SO_3$)/0.6 ($TiO_2$) | – | Hekal, Abo-El-Enein, El-Korashy, Megahed, and El-Sayed (2013) |

*Source*: From Jiang, Y., Ling, T. C., Shi, C., & Pan, S. Y. (2018). Characteristics of steel slags and their use in cement and concrete—A review. *Resources, Conservation and Recycling, 136*, 187—197. https://doi.org/10.1016/j.resconrec.2018.04.023.

present in the steel industry-generated EAF slag which is being utilized for zeolite production as it contains an adequate amount of silica and alumina as zeolite precursor.

### 2.2.1.1.3 Physicochemical characterization of LF slag

The information presented in Table 2.3 highlights that LF slag is made up of oxides of MgO, CaO, $Al_2O_3$, and $SiO_2$. Compared to BOF and EAF-C slags, LF slag has a higher CaO content ranging from 44.5% to 58.4%, while iron-bearing components are much lower. LF slag also contains a minority of $TiO_2$ and $Cr_2O_3$, which can be attributed to the alloying of desired compounds (Kriskova et al., 2012). The calcium oxide/silica ratio of about 2 in LF slag makes $C_2S$ the primary mineral phase, which is present in the gamma polymorph ($\gamma$-$C_2S$) form. Other mineralogies including merwinite, bredigite, and periclase are also present in LF slag (Kriskova et al., 2012).

However, during the cooling process, LF slag undergoes self-pulverization, making it a white powder (Fig. 2.1). This transformation is caused by the transformation of $\beta$-$C_2S$ to $\gamma$-$C_2S$, leading to a volume rise of about 10% (Shi, 2002). Although this self-pulverization can cause handling and storing difficulties, stabilizers like phosphates and borates can be applied to avoid this technical problem (Tossavainen et al., 2007).

Overall, this information can be useful in understanding the composition and properties of LF slag and can help in developing effective strategies to mitigate any challenges that may arise during its handling and storage.

### 2.2.1.2 Scope of zeolite synthesis from steel slag

One viable route for the creation of sustainable materials is the synthesis of zeolites from steel slag that contains a high amount of silica and alumina (Tables 2.1−2.3) (Shekhar Samanta, Das, Dhara, & Purkait, 2023). This large quantity of steel slag presents a significant economic opportunity for industrial and construction sectors, as it can be utilized as a raw material for a wide range of applications, like road construction, cement production, and land reclamation. Moreover, steel slag has the added benefit of being an eco-friendly alternative to traditional materials, as it reduces the need for landfills and conserves natural resources. Researchers have the chance to simplify zeolite manufacturing and lessen reliance on conventional, resource-intensive precursor materials by using the natural silica and alumina content in steel slag. This strategy not only tackles waste management issues related to the manufacture of steel, but it also creates opportunities for economically and ecologically sound zeolite synthesis. The synthesized zeolite-like or zeolite-composite adsorbent can be further utilized for the decontamination of water (Dhara, Samanta, Uppaluri, & Purkait, 2023). Furthermore, the adaptability and potential influence of this novel strategy in the domains of materials science and sustainability is demonstrated by customizing zeolites from silica

**TABLE 2.3** Chemical elements present in ladle furnace (LF) slag.

| Sources | SiO$_2$ | Al$_2$O$_3$ | Fe | CaO | MgO | SO$_3$ | MnO | TiO$_2$ | Cr$_2$O$_3$ | References |
|---|---|---|---|---|---|---|---|---|---|---|
| Canada | 26.8 | 5.2 | 1.59 | 57.0 | 3.2 | 1.7 | 1.0 | 0.3 | – | Shi and Hu (2003) |
| Belgium | 28.3 | 1.2 | – | 51.5 | 11.3 | – | – | – | 3.9 | Kriskova et al. (2012) |
| Taiwan | 23.5 | 4.1 | 0.08 | 50.6 | 8.2 | – | – | 0.09 | 0.44 | Sheen, Le, and Sun (2015) |
| South Korea | 10.9 | 26.6 | 4.3 | 44.5 | 6.6 | – | 0.6 | – | – | Choi, Kim, Han, and Kim (2016) |

*Source*: From Jiang, Y., Ling, T. C., Shi, C., & Pan, S. Y. (2018). Characteristics of steel slags and their use in cement and concrete—A review. *Resources, Conservation and Recycling, 136*, 187–197. https://doi.org/10.1016/j.resconrec.2018.04.023.

**FIGURE 2.1** Image of steel industry-generated LF slag Image of steel industry-derived LF slag. *From Jiang, Y., Ling, T. C., Shi, C., & Pan, S. Y. (2018). Characteristics of steel slags and their use in cement and concrete—A review.* Resources, Conservation and Recycling, 136, *187–197. https://doi.org/10.1016/j.resconrec.2018.04.023.*

and alumina-rich steel slag for particular applications, such as environmental remediation or catalysis for biofuel production (Anweshan, Das, Dhara, & Purkait, 2023; Das, Mondal, & Purkait, 2022; Das, Deepti, & Purkait, 2023; Das, Dhara, & Purkait, 2023; Dhara, Das, Uppaluri, & Purkait, 2023; Dhara, Das, Uppaluri, & Purkait, 2023; Dhara, Shekhar Samanta, Das, Uppaluri, & Purkait, 2023; Duarah, Das, & Purkait, 2023; Mondal, Samanta, Kumar, & Purkait, 2020; Sharma, Das, Chakraborty, & Purkait, 2023). It is expected that as this field of study develops, it will open the door to new and economically viable zeolite products with a variety of industrial uses. For example, Wu and colleagues have prepared Na-Y (zeolite Y) type zeolite from steel slag via a two-step hydrothermal technique (Wu et al., 2022). In the zeolite preparation, they have also used red mud as a zeolite precursor. The synthetic zeolite shows a considerable surface area of 685 m$^2$/g and a pore volume of 0.07 cm$^3$/g. The as-synthesized zeolite was employed for the volatile organic compounds (VOCs) adsorption study and 60% of adsorption capacity with the addition of 5% zeolite Y was reported.

## 2.2.2 Power plant fly ash (coal fly ash)

### 2.2.2.1 Characterization of coal fly ash

#### 2.2.2.1.1 X-ray fluorescence analysis of CFA

Upon analyzing Fig. 2.2, it is evident that the fly ash comprised quartz and mullite mineral phases. The presence of amorphous phases, contributing over 50% of the total mass, can be observed between $2\theta = 20$ and $2\theta = 30$. It is worth noting that the mineralogy of the ash significantly influences the zeolite-like mineral preparation, as the various constituents in the ash dissolve with different degrees of ease (Fernández-Jiménez & Palomo, 2005).

#### 2.2.2.1.2 XPS analysis of CFA

The minor (Mg and K), major (aluminum and silicon), and trace elements (Zn, Sr, Cr, Cl, and Ti) in the CFA were quantified from X-ray photoelectron spectroscopy (XPS) analysis and the result is enlisted in Table 2.4. As is shown in Table 2.4 the oxygen was present in a very high quantity among others. In addition, it was also found that the zeolite precursor material

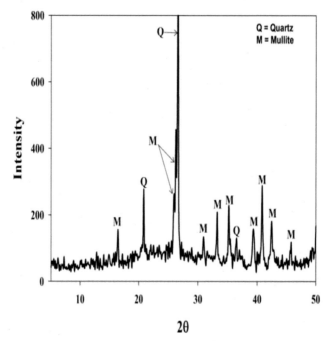

**FIGURE 2.2** XRD pattern of CFA XRD analysis of CFA. *XRD*, X-ray fluorescence. *From Du Plessis, P. W., Ojumu, T. V., & Petrik, L. F. (2013). Waste minimization protocols for the process of synthesizing zeolites from South African coal fly ash. Materials, 6(5), 1688–1703. https://doi.org/10.3390/ma6051688.*

**TABLE 2.4** XPS elemental analysis of coal fly ash.

| Elements of CFA | wt.% |
| --- | --- |
| C 1s | 12.01 |
| N 1s | 1.52 |
| O 1s | 49.53 |
| Na 1s | 1.14 |
| Mg 2s | 0.31 |
| Al 2p | 6.22 |
| Si 2p | 12.39 |
| P 2p | 0.11 |
| S 2p | 9.44 |
| Cl 2p | 0.21 |
| K 2p | 0.26 |
| Ca 2p | 3.86 |
| Ti 2p | 0.25 |
| Cr 2p | 0.6 |
| Mn 2p | 0.16 |
| Fe 2p | 1.58 |
| Zn 2p | 0.35 |
| Sr 3p | 0.08 |

*Source*: From Zhang, P., Liao, W., Kumar, A., Zhang, Q., & Ma, H. (2020). Characterization of sugarcane bagasse ash as a potential supplementary cementitious material: Comparison with coal combustion fly ash. *Journal of Cleaner Production, 277*, 123834. https://doi.org/10.1016/j.jclepro.2020.123834.

(mainly Al and Si) was present in a very low concentration. Fig. 2.3 shows the elemental analysis of CFA obtained from XPS analysis.

### 2.2.2.1.3 Field emission scanning electron microscopy analysis of CFA

The microscopic images of the power plant-generated CFA grains are illustrated in Fig. 2.4A and Fig. 2.4B, respectively. From the figure, the spherical shape of CFA particles was observed. It was also observed that the average particle size was found to be in the 20 nm range.

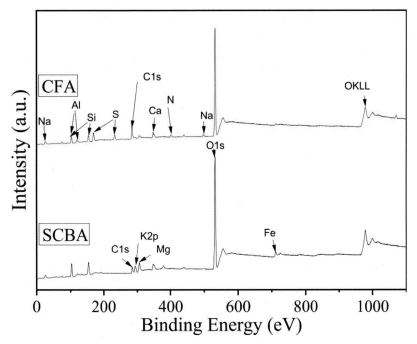

**FIGURE 2.3**  XPS analysis of CFA and bagasse fly ash XPS wide scan of power plant generated fly ash and sugarcane bagasse fly ash. *From Zhang, P., Liao, W., Kumar, A., Zhang, Q., & Ma, H. (2020). Characterization of sugarcane bagasse ash as a potential supplementary cementitious material: Comparison with coal combustion fly ash.* Journal of Cleaner Production, 277, 123834. https://doi.org/10.1016/j.jclepro.2020.123834.

### 2.2.2.1.4   Scope of zeolite formation from CFA

Zeolite-like mineral production utilizing thermal power plant-generated fly ash offers great potential for the creation of sustainable materials, especially when it includes considerable amounts of silica and alumina. The burning of coal produces enormous amounts of CFA, which is widely accessible worldwide. The CFA is directly used in zeolite synthesis since it contains an abundant amount of zeolite precursors, namely silica and alumina. Researchers can use hydrothermal and cutting-edge methods to make use of fly ash's natural composition and promote the crystallization of zeolitic structures. This not only lessens the environmental impact of coal ash disposal, but it also provides a greener option to current zeolite production techniques that frequently rely on energy-intensive extraction procedures. The produced zeolites or zeolite nanomaterials (Sontakke, Deepti, Samanta, & Purkait, 2023) may be employed for different purposes, for instance, gas adsorption, catalysis, and water filtration (Chakraborty, Das, & Mondal, 2023; Das & Mondal, 2023; Das, Sontakke, Samanta, & Purkait, 2023; Samanta, Anweshan, & Purkait, 2023; Samanta, Mondal, & Purkait, 2023; Sharma, Samanta,

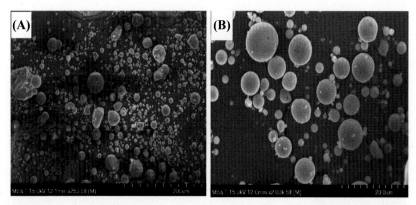

**FIGURE 2.4** Surface morphology of CFA at different magnification ranges FESEM images of coal fly ash. *FESEM*, Field emission scanning electron microscopy. *From Zhang, P., Liao, W., Kumar, A., Zhang, Q., & Ma, H. (2020). Characterization of sugarcane bagasse ash as a potential supplementary cementitious material: Comparison with coal combustion fly ash.* Journal of Cleaner Production, 277, *123834. https://doi.org/10.1016/j.jclepro.2020.123834.*

Chakraborty, & Purkait, 2023). For instance, Lankapati and their research team have synthesized zeolite-like useful minerals from CFA as silica and alumina feedstock via the traditional method. The synthesized zeolite material (FZS$^H$-5Al) is highly effective as a solid acid catalyst for synthesizing *n*-butyl levulinate from levulinic acid obtained from biomass. According to the findings, FZSH-5Al zeolite demonstrated impressive results under optimized reaction conditions, achieving an outstanding 98.5% conversion, 95.8% selectivity, and 94.8% yield. The study clearly proves that fly ash can be converted into a highly effective and safe catalyst in a cost-efficient manner.

## 2.3 Municipal solid waste

### 2.3.1 Physicochemical characteristics of municipal waste

#### 2.3.1.1 Elemental analysis of MSW

The compositional study of the MSW has been summarized in Table 2.5. The main components identified were Cl and Ca, which accounted for 15.82%, and 23.39, respectively. Interestingly, the high concentration of Ca is attributed to the calcium hydroxide solution used in the semidry scrubber. Other constituents found in the tested fly ash include Si, S, Fe, Na, Al, K, and Mg each accounting for about 1%−6%. Although detrimental metals Cu, Pb, Cr, and Zn were found in low amounts, around 0.16%−0.91%, Ni and Cd were found in even smaller quantities, around 0.021%−0.025%. Overall, this information is valuable for future studies and can help identify ways to optimize the use of baghouse fly ash (Y. Liu, Zheng, Li, & Xie, 2009).

**TABLE 2.5** Total elements present in municipal solid waste.

| Constituents | Concentration (mg/kg) | SD[a] | Constituents | Concentration (mg/kg) | SD[a] |
|---|---|---|---|---|---|
| Ca | 233,900 | 0.017 | Cu | 1600 | 0.006 |
| Na | 51,500 | 0.09 | Ce | 594 | 0.0027 |
| Cl | 158,200 | 0.018 | Sn | 1070 | 0.005 |
| K | 40,600 | 0.09 | La | 396 | 0.0032 |
| Sx | 27,200 | 0.05 | Mn | 435 | 0.0022 |
| Al | 12,600 | 0.04 | Cd | 252 | 0.001 |
| Si | 20,600 | 0.05 | Sr | 256 | 0.0013 |
| Fe | 10,900 | 0.04 | In | 162 | 0.0009 |
| Mg | 106,000 | 0.04 | Ni | 207 | 0.0012 |
| Zn | 9043 | 0.041 | Zr | 109 | 0.0005 |
| Br | 11,100 | 0.05 | Rb | 140 | 0.0007 |
| Px | 3080 | 0.015 | As | 70 | 0.0018 |
| Pb | 3277 | 0.015 | V | 30 | 0.0004 |
| Ti | 2390 | 0.012 | Pr | 45 | 0.0017 |
| Cr | 7040 | 0.04 | Nd | 89 | 0.0005 |
| F | 1610 | 0.047 | Bi | 22 | 0.0005 |
| Sb | 2430 | 0.012 | Mo | 35 | 0.0005 |

[a]Standard error.

Source: From Liu, Y., Zheng, L., Li, X. & Xie, S. (2009). SEM/EDS and XRD characterization of raw and washed MSWI fly ash sintered at different temperatures. *Journal of Hazardous Materials, 162*(1), 161–173. https://doi.org/10.1016/j.jhazmat.2008.05.029.

### 2.3.1.2    X-ray fluorescence and field emission scanning electron microscopy analysis of MSW

The XRF analysis revealed that the main components of the fly ash were $SO_3$ (8.97%), $Na_2O$ (5.99%), CaO (39.18%), $SiO_2$ (10.02%), and Cl (18.38%). Some components made up less than 5%, such as $Al_2O_3$ (3.81%), $K_2O$ (4.47%), $Fe_2O_3$ (2.11%), and MgO (3.75%), while the remaining was 3.32%. The high content of CaO was a result of the lime slurry injection to maintain the emission of toxic gases (HCl and $SO_2$) (Jiao et al., 2016). Similarly, the high concentration of chloride was generated mainly from the combustion of kitchen waste and plastic (Zhao, Hu, Tian, Chen, & Feng, 2020). The high amounts of sulfates and perchlorates make municipality solid waste ashes unsuitable for direct use in cement and concrete, which requires pretreatment before reutilization (Yakubu, Zhou, Ping, Shu, & Chen, 2018).

As illustrated in Fig. 2.5A, the most crystalline phases in municipality waste-generated fly ashes are $Ca(OH)_2$, NaCl, KCl, $CaCO_3$, $(Fe, Mg)_2Al_4Si_5O_{18}$, CaClOH, and $CaSO_4$. These results are consistent with those obtained from other research (Wongsa, Boonserm, Waisurasingha, Sata, & Chindaprasirt, 2017). Inorganic salts like NaCl, KCl, and $CaSO_4$ are basically found in burning fabric filter bags (Ma et al., 2017). During burning, NaCl and KCl readily volatilize and penetrate the fly ashes (Pöykiö, Mäkelä, Watkins, Nurmesniemi, & Dahl, 2016).

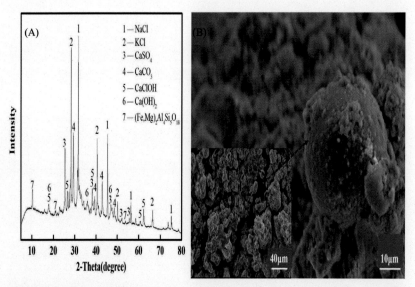

**FIGURE 2.5** XRD and FESEM analysis of municipal solid waste incineration fly ash: (A) XRD pattern and (B) FESEM microscopic analysis of the municipal solid waste. *FESEM*, Field emission scanning electron microscopy; *XRD*, X-ray fluorescence. *From Zhao, H., Tian, Y., Wang, R., Wang, R., Zeng, X., Yang, F., Wang, Z., Chen, M., & Shu, J. (2021). Seasonal variation of the mobility and toxicity of metals in Beijing's municipal solid waste incineration fly ash. Sustainability (Switzerland), 13(12). https://doi.org/10.3390/su13126532.*

Anhydrite Calcium sulfate, calcite, calcium hydroxy calcium chloride, and portlandite are formed when lime is sprayed to eliminate toxic gaseous compounds (Ma et al., 2017). The new crystalline phase (Fe, Mg)$_2$Al$_4$Si$_5$O$_{18}$ is created when Fe and Mg are combined with aluminosilicates at a temperature of 900°C for sintering (Liu et al., 2009). The absence of the crystalline phase in heavy metals (HMs) might be attributed to their low concentration in municipal waste-generated fly ash or to their amorphous appearance when embedded in silicates or aluminosilicates (Xiong, Zhu, Zhao, Jiang, & Zhang, 2014).

As shown in Fig. 2.5B, municipal solid waste incineration (MSWI) fly ashes have a unique and irregular shape with rough surfaces. Various tiny components are combined on the surface, making it highly porous. The municipal waste fly ashes contained several HMs, which can easily leach out from the solid surface, causing accumulation on the specific surface. However, with proper treatment, MSWI fly ashes can be reused in cement and concrete (Gong et al., 2017).

### 2.3.1.3 Thermogravimetric analysis of MSW

In Fig. 2.6, we can see that fly ashes produced by the incineration of MSW exhibit four distinct peaks, which are ascribed to the weight loss of the solid

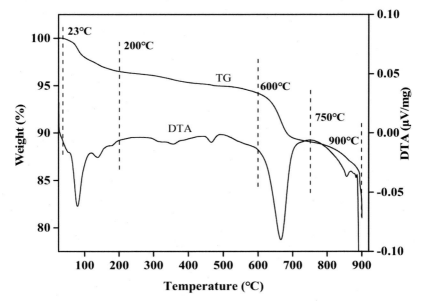

**FIGURE 2.6** TGA analysis of MSW fly ash TGA analysis of the collected municipal. *TGA,* Thermogravimetric analysis. *From Zhao, H., Tian, Y., Wang, R., Wang, R., Zeng, X., Yang, F., Wang, Z., Chen, M., & Shu, J. (2021). Seasonal variation of the mobility and toxicity of metals in Beijing's municipal solid waste incineration fly ash.* Sustainability (Switzerland), 13(12). *https://doi.org/10.3390/su13126532.*

sample. The first peak, occurring between 23°C and 200°C with a weight loss of 3.50%, is attributed to the evaporation of absorbed and residual water molecules in the ashes. This observation is consistent with previous studies (Rocca et al., 2013). The second peak, between 200°C and 600°C with a weight loss of 2.31%, can be ascribed to the emission of carbon dioxide and the breakdown of calcium hydroxide in the ashes (Zheng & Koziński, 2000). The third peak, occurring between 600°C and 750°C, with a weight loss of 5.13%, is due to the breakdown of carbonate such as $CaCO_3$ (Mangialardi, Piga, Schena, & Sirini, 1998). Finally, the fourth peak, occurring between 750°C and 900°C, with a weight loss of 6.86%, is primarily due to the escape of $SO_2$, release of $Cl^-$ ion, salt decomposition, and evaporation during $CaSO_4$ decomposition (Ma et al., 2017). These findings are important for understanding the behavior of MSWI fly ashes and can be used to inform future research in this area.

### 2.3.2 Municipal solid waste as a zeolite source

The production of zeolite presents an attractive and largely unexplored opportunity using MSW. Zeolites are crystalline aluminosilicate minerals with a highly porous structure used for several purposes, such as gas separation, wastewater treatment (Changmai et al., 2022; Das et al., 2021; Das, Anweshan, & Purkait, 2021; Das, Mondal, et al., 2021), and catalysis. The abundance of silicon and aluminum in MSW, two essential ingredients needed for the production of zeolites, makes it a possible source of zeolite. These components may be extracted from waste materials and purified using cutting-edge methods, diverting large amounts of garbage from landfills and incineration plants (Shekhar Samanta et al., 2023). Furthermore, the heterogeneous makeup of MSW provides a variety of feedstocks for the manufacture of zeolite, possibly producing a variety of zeolite varieties with different characteristics and uses. For example, high-quality zeolite type-P1 (Na-P1) was prepared via a microwave-facilitated hydrothermal process under optimal conditions by Zhou and colleagues (Zhou, Jiang, Qiu, Zhao, & Long, 2023). The cation exchange capacity (CEC) over the as-synthesized zeolite sample was found to be 2.83 meq/g. The prepared zeolite sample exhibited a high adsorption capacity of 84.55 and 84.65 mg/g for methylene blue and copper ($Cu^{2+}$), respectively. The pore volume and surface area of the Na-P1 zeolite were found to be 0.44 cm³/g and 61.42 m²/g, respectively. Overall, the microwave-irradiated hydrothermal treatment was found to be a cost-effective and promising technique for synthesizing zeolite-like useful material using MSWI fly ash as a raw material.

### 2.4 Conclusion

This chapter is dedicated to exploring the exciting potential of solid waste as a useful resource for creating zeolites. Zeolites are substances known for their porous structures and ion-exchange properties, offering a promising path toward waste management and the development of eco-friendly products. We begin by presenting an overview of the worldwide solid waste situation and highlighting the urgent need for new waste management techniques. We draw attention to specific waste streams, including steel slag, CFA, biomass ash, and municipal garbage that have unrealized potential for zeolite synthesis.

To investigate this approach, we thoroughly characterize the various solid wastes using various techniques such as XPS, scanning electron microscopy (SEM), and X-ray diffraction (XRD). Our findings indicate that waste precursors can be transformed into structures that resemble zeolites with useful features.

In conclusion, this chapter sheds light on the intriguing potential of solid waste as a feedstock for zeolite synthesis. Researchers and practitioners have the potential to contribute significantly to a more circular and ecologically conscientious approach to materials synthesis by using the inherent qualities of waste streams. This approach provides a creative solution to the dual problems of waste management and the improvement of sustainable products. Therefore this chapter offers a significant contribution to the development of sustainable materials and waste valorization, furthering our goal of a greener and cleaner future.

# References

Anweshan., Das, P. P., Dhara, S., & Purkait, M. K. (2023). *Nanosensors in food science and technology. Advances in smart nanomaterials and their applications* (pp. 247−272). India: Elsevier. Available from https://www.sciencedirect.com/book/9780323995467, https://doi.org/10.1016/B978-0-323-99546-7.00015-X.

Bethanis, S., Cheeseman, C. R., & Sollars, C. J. (2002). Properties and microstructure of sintered incinerator bottom ash. *Ceramics International*, *28*(8), 881−886. Available from https://doi.org/10.1016/S0272-8842(02)00068-8, 02728842.

Bharti, M., Das, P. P., & Purkait, M. K. (2023). A review on the treatment of water and wastewater by electrocoagulation process: Advances and emerging applications. *Journal of Environmental Chemical Engineering*, *11*, 111558. Available from https://doi.org/10.1016/j.jece.2023.111558.

Butler, W. Barry, & Mearing, M. A. (1985). Fly ash beneficiation and utilization in theory and in practice. *MRS Proceedings*, *65*. Available from https://doi.org/10.1557/proc-65-11, 0272-9172.

Caijun, S. (2004). Steel slag—Its production, processing, characteristics, and cementitious properties. *Journal of Materials Civil Engineering*, *16*, 899−1561.

Chakraborty, S., Das, P. P., & Mondal, P. (2023). *Recent advances in membrane technology for the recovery and reuse of valuable resources* (pp. 695−719). Elsevier BV. Available from https://doi.org/10.1016/b978-0-323-95327-6.00028-2.

Changmai, M., Das, P. P., Mondal, P., Pasawan, M., Sinha, A., Biswas, P., . . . Purkait, M. K. (2022). Hybrid electrocoagulation−microfiltration technique for treatment of nanofiltration rejected steel industry effluent. *International Journal of Environmental Analytical Chemistry*, *102*(1), 62−83. Available from http://www.tandf.co.uk/journals/titles/03067319.asp, https://doi.org/10.1080/03067319.2020.1715381.

Choi, S., Kim, J. M., Han, D., & Kim, J. H. (2016). Hydration properties of ladle furnace slag powder rapidly cooled by air. *Construction and Building Materials*, *113*, 682−690. Available from https://doi.org/10.1016/j.conbuildmat.2016.030.089, 09500618.

Das, P. P., Mondal, P., Anweshan., Sinha, A., Biswas, P., Sarkar, S., & Purkait, M. K. (2021). Treatment of steel plant generated biological oxidation treated (BOT) wastewater by hybrid process, *Separation and purification technology* (258). India: Elsevier B.V. Available from http://www.journals.elsevier.com/separation-and-purification-technology/, https://doi.org/10.1016/j.seppur.2020.118013.

Das, P. P., Anweshan, A., & Purkait, M. K. (2021). Treatment of cold rolling mill (CRM) effluent of steel industry, *Separation and purification technology* (274). India: Elsevier B.V.. Available from http://www.journals.elsevier.com/separation-and-purification-technology/, https://doi.org/10.1016/j.seppur.2021.119083.

Das, P. P., Sontakke, A. D., Samanta, N. S., & Purkait, M. K. (2023). Emerging contaminants in wastewater: Eco-toxicity and sustainability assessment. *Industrial wastewater reuse: Applications, prospects and challenges* (pp. 63−87). India: Springer Nature. Available from https://link.springer.com/book/10.1007/978-981-99-2489-9, https://doi.org/10.1007/978-981-99-2489-9_4.

Das, P. P., Dhara, S., & Purkait, M. K. (2023). Hybrid electrocoagulation and ozonation techniques for industrial wastewater treatment. *Sustainable industrial wastewater treatment and pollution control* (pp. 107−128). India: Springer Nature. Available from https://link.springer.com/book/10.1007/978-981-99-2560-5, https://doi.org/10.1007/978-981-99-2560-5_6.

Das, P. P., Dhara, S., & Purkait, M. K. (2023). The anaerobic ammonium oxidation process: Inhibition, challenges and opportunities. In Maulin P. Shah (Ed.), *Ammonia oxidizing bacteria: Applications in industrial wastewater treatment* (pp. 56−82). Royal Society of Chemistry. Available from https://doi.org/10.1039/BK9781837671960-00056.

Das, P. P., Dhara, S., & Purkait, M. K. (2024). Ozone-based oxidation processes for the removal of pharmaceutical products from wastewater. In Maulin P. Shah, & Pooja Ghosh (Eds.), *Development in wastewater treatment research and processes* (pp. 287−308). Elsevier. Available from https://doi.org/10.1016/B978-0-443-19207-4.00003-3.

Das, P. P., Duarah, P., & Purkait, M. K. (2023). Fundamentals of food roasting process. In S. M. Jafari (Ed.), *High-temperature processing of food products* (pp. 103−130). Elsevier. Available from https://doi.org/10.1016/B978-0-12-818618-3.00005-7.

Das, P. P., & Mondal, P. (2023). *Membrane-assisted potable water reuses applications: Benefits and drawbacks* (pp. 289−309). Elsevier BV. Available from https://doi.org/10.1016/b978-0-323-99344-9.00014-1.

Das, P. P., Anweshan., Mondal, P., Sinha, A., Biswas, P., Sarkar, S., & Purkait, M. K. (2021). Integrated ozonation assisted electrocoagulation process for the removal of cyanide from steel industry wastewater. *Chemosphere*, *263*128370. Available from https://doi.org/10.1016/j.chemosphere.2020.128370, 00456535.

Das, P. P., Mondal, P., & Purkait, M. K. (2022). *Recent advances in synthesis of iron nanoparticles via green route and their application in biofuel production* (pp. 79−104). Springer Science and Business Media LLC. Available from https://doi.org/10.1007/978-981-16-9356-4_4.

Das, P. P., Deepti., & Purkait, M. K. (2023). Industrial wastewater to biohydrogen production via potential bio-refinery route. In Shah Maulin P. (Ed.). *Biorefinery for water and wastewater treatment* (pp. 159−179). Springer Science and Business Media LLC. Available from http://doi.org/10.1007/978-3-031-20822-5_8.

Dhara, S., Shekhar Samanta, N., Das, P. P., Uppaluri, R. V. S., & Purkait, M. K. (2023). Ravenna grass-extracted alkaline lignin-based polysulfone mixed matrix membrane (MMM) for aqueous Cr(VI) removal. *ACS Applied Polymer Materials*, *5*(8), 6399−6411. Available from http://pubs.acs.org/journal/aapmcd, https://doi.org/10.1021/acsapm.3c00999.

Das, P. P., Sharma, M., & Purkait, M. K. (2023). Ammonia oxidizing bacteria in wastewater treatment. In Maulin P. Shah (Ed.), *Ammonia Oxidizing Bacteria: Applications in Industrial Wastewater Treatment* (pp. 83−102). Royal Society of Chemistry. Available from https://doi.org/10.1039/BK9781837671960-00083.

Das, P. P., Sontakke, A. D., & Purkait, M. K. (2023). Rice straw for biofuel production. In Maulin P. Shah (Ed.), *Green approach to alternative fuel for a sustainable future* (pp. 153−166). Elsevier. Available from https://doi.org/10.1016/B978-0-12-824318-3.00034-5.

Das, P. P., Sontakke, A. D., & Purkait, M. K. (2023). Electrocoagulation process for wastewater treatment: applications, challenges, and prospects. In Maulin P. Shah (Ed.), *Development in wastewater treatment research and processes* (pp. 23−48). Elsevier. Available from https://doi.org/10.1016/B978-0-323-95684-0.00015-4.

Dhara, S., Das, P. P., Uppaluri, R., & Purkait, M. K. (2023). Phosphorus recovery from municipal wastewater treatment plants. *Development in wastewater treatment research and processes: Advances in industrial wastewater treatment technologies: Removal of contaminants and recovery of resources* (pp. 49−72). India: Elsevier. Available from https://www.sciencedirect.com/book/9780323956840, https://doi.org/10.1016/B978-0-323-95684-0.00014-2.

Dhara, S., Samanta, N. S., Uppaluri, R., & Purkait, M. K. (2023). High-purity alkaline lignin extraction from *Saccharum ravannae* and optimization of lignin recovery through response surface methodology. *International Journal of Biological Macromolecules*, *234*123594. Available from https://doi.org/10.1016/j.ijbiomac.2023.123594, 01418130.

Dhara, S., Das, P. P., Uppaluri, R., & Purkait, M. K. (2023). *Biological approach for energy self-sufficiency of municipal wastewater treatment plants* (pp. 235−260). Elsevier BV. Available from https://doi.org/10.1016/b978-0-323-99348-7.00006-0.

Dhondy, T., Remennikov, A., & Shiekh, M. N. (2019). Benefits of using sea sand and seawater in concrete: a comprehensive review. *Australian Journal of Structural Engineering*, *20*(4), 280−289. Available from http://www.tandfonline.com/toc/TSEN20/current, https://doi.org/10.1080/13287982.2019.1659213.

Duarah, P., Das, P. P., & Purkait, M. K. (2023). Technological advancement in the synthesis and application of nanocatalysts. *Advanced application of nanotechnology to industrial wastewater* (pp. 191−214). India: Springer Nature. Available from https://link.springer.com/book/10.1007/978-981-99-3292-4, https://doi.org/10.1007/978-981-99-3292-4_10.

Fernández-Jiménez, A., & Palomo, A. (2005). Composition and microstructure of alkali activated fly ash binder: Effect of the activator. *Cement and Concrete Research*, *35*(10), 1984−1992. Available from https://doi.org/10.1016/j.cemconres.2005.030.003.

Gong, B., Deng, Y., Yang, Y., Tan, S. N., Liu, Q., & Yang, W. (2017). Solidification and biotoxicity assessment of thermally treated municipal solid waste incineration (MSWI) fly ash. *International Journal of Environmental Research and Public Health*, *14*(6). Available from http://www.mdpi.com/1660-4601/14/6/626/pdf, 10.3390/ijerph14060626.

Hekal, E. E., Abo-El-Enein, S. A., El-Korashy, S. A., Megahed, G. M., & El-Sayed, T. M. (2013). Hydration characteristics of Portland cement−Electric arc furnace slag blends. *HBRC Journal*, *9*(2), 118−124. Available from https://www.tandfonline.com/journals/thbr20, https://doi.org/10.1016/j.hbrcj.2013.05.006.

Hemalatha, T., Mapa, M., George, N., & Sasmal, S. (2016). Physico-chemical and mechanical characterization of high volume fly ash incorporated and engineered cement system towards developing greener cement. *Journal of Cleaner Production*, *125*, 268−281. Available from https://doi.org/10.1016/j.jclepro.2016.030.118, 09596526.

Huaiwei, Z., & Xin, H. (2011). An overview for the utilization of wastes from stainless steel industries. *Resources, Conservation and Recycling*, *55*(8), 745−754. Available from https://doi.org/10.1016/j.resconrec.2011.030.005, 18790658.

Jiao, F., Zhang, L., Dong, Z., Namioka, T., Yamada, N., & Ninomiya, Y. (2016). Study on the species of heavy metals in MSW incineration fly ash and their leaching behavior. *Fuel Processing Technology*, *152*, 108−115. Available from https://doi.org/10.1016/j.fuproc.2016.060.013, 03783820.

Karamanov, A., Pelino, M., & Hreglich, A. (2003). Sintered glass-ceramics from municipal solid waste-incinerator fly ashes-part I: The influence of the heating rate on the sinter- crystallisation. *Journal of the European Ceramic Society*, *23*(6), 827−832. Available from https://doi.org/10.1016/S0955-2219(02)00210-8.

Kriskova, L., Pontikes, Y., Cizer, O., Mertens, G., Veulemans, W., Geysen, D., ... Blanpain, B. (2012). Effect of mechanical activation on the hydraulic properties of stainless steel slags. *Cement and Concrete Research*, *42*(6), 778−788. Available from https://doi.org/10.1016/j.cemconres.2012.020.016.

Li, J., Pan, S. Y., Kim, H., Linn, J. H., & Chiang, P. C. (2015). Building green supply chains in eco-industrial parks towards a green economy: Barriers and strategies. *Journal of Environmental Management*, *162*, 158−170. Available from https://www.sciencedirect.com/journal/journal-of-environmental-management, https://doi.org/10.1016/j.jenvman.2015.07.030.

Li, Z., Zhao, S., Zhao, X., & He, T. (2013). Cementitious property modification of basic oxygen furnace steel slag. *Construction and Building Materials*, *48*, 575−579. Available from https://doi.org/10.1016/j.conbuildmat.2013.070.068, 09500618.

Liu, Q., Liu, J., & Qi, L. (2016). Effects of temperature and carbonation curing on the mechanical properties of steel slag-cement binding materials. *Construction and Building Materials*, *124*, 999−1006. Available from https://doi.org/10.1016/j.conbuildmat.2016.080.131, 09500618.

Liu, Y., Zheng, L., Li, X., & Xie, S. (2009). SEM/EDS and XRD characterization of raw and washed MSWI fly ash sintered at different temperatures. *Journal of Hazardous Materials*, *162*(1), 161−173. Available from https://doi.org/10.1016/j.jhazmat.2008.050.029, 03043894.

Lizarazo-Marriaga, J., Claisse, P., & Ganjian, E. (2011). Effect of steel slag and portland cement in the rate of hydration and strength of blast furnace slag pastes. *Journal of Materials in Civil Engineering*, *23*(2), 153−160. Available from https://doi.org/10.1061/(ASCE)MT.1943-5533.0000149, 08991561.

Ma, W., Fang, Y., Chen, D., Chen, G., Xu, Y., Sheng, H., & Zhou, Z. (2017). Volatilization and leaching behavior of heavy metals in MSW incineration fly ash in a DC arc plasma furnace. *Fuel*, *210*, 145−153. Available from http://www.journals.elsevier.com/fuel/, https://doi.org/10.1016/j.fuel.2017.07.091.

Mangialardi, T., Piga, L., Schena, G., & Sirini, P. (1998). Characteristics of MSW incinerator ash for use in concrete, Mary Ann Liebert Inc., Italy*Environmental Engineering Science*, *15*(4), 291−297. Available from http://www.liebertonline.com/ees, https://doi.org/10.1089/ees.1998.15.291.

Mondal, P., Samanta, N. S., Kumar, A., & Purkait, M. K. (2020). Recovery of $H_2SO_4$ from wastewater in the presence of NaCl and $KHCO_3$ through pH responsive polysulfone membrane: Optimization approach. *Polymer Testing*, *86*. Available from https://www.journals.elsevier.com/polymer-testing, https://doi.org/10.1016/j.polymertesting.2020.106463.

Monshi, A., & Asgarani, M. K. (1999). Producing portland cement from iron and steel slags and limestone. *Cement and Concrete Research*, *29*(9), 1373−1377. Available from https://doi.org/10.1016/S0008-8846(99)00028-9, 00088846.

Monteiro, P. J. M., Miller, S. A., & Horvath, A. (2017). Towards sustainable concrete. *Nature Materials*, *16*(7), 698−699. Available from http://www.nature.com/nmat/, http://doi.org/10.1038/nmat4930.

Muhmood, L., Vitta, S., & Venkateswaran, D. (2009). Cementitious and pozzolanic behavior of electric arc furnace steel slags. *Cement and Concrete Research*, *39*(2), 102−109. Available from https://doi.org/10.1016/j.cemconres.2008.110.002, 00088846.

Palankar, N., Ravi Shankar, A. U., & Mithun, B. M. (2016). Durability studies on eco-friendly concrete mixes incorporating steel slag as coarse aggregates. *Journal of Cleaner Production*, *129*, 437−448. Available from https://doi.org/10.1016/j.jclepro.2016.040.033, 09596526.

Pang, B., Zhou, Z., Hou, P., Du, P., Zhang, L., & Xu, H. (2016). Autogenous and engineered healing mechanisms of carbonated steel slag aggregate in concrete. *Construction and Building Materials*, *107*, 191−202. Available from https://doi.org/10.1016/j.conbuildmat.2015.120.191.

Park, Y. J., & Heo, J. (2002). Vitrification of fly ash from municipal solid waste incinerator. *Journal of Hazardous Materials*, *91*(1-3), 83−93. Available from https://doi.org/10.1016/S0304-3894(01)00362-4, 03043894.

Piatak, N. M., Parsons, M. B., & Seal, R. R. (2015). Characteristics and environmental aspects of slag: A review. *Applied Geochemistry*, *57*, 236−266. Available from http://www.journals.elsevier.com/applied-geochemistry, https://doi.org/10.1016/j.apgeochem.2014.040.009.

Pöykiö, R., Mäkelä, M., Watkins, G., Nurmesniemi, H., & Dahl, O. (2016). Heavy metals leaching in bottom ash and fly ash fractions from industrial-scale BFB-boiler for environmental risks assessment. *Transactions of Nonferrous Metals Society of China (English Edition)*, *26*(1), 256−264. Available from https://doi.org/10.1016/S1003-6326(16)64112-2, 10036326.

Rocca, S., Zomeren, A. v, Costa, G., Dijkstra, J. J., Comans, R. N. J., & Lombardi, F. (2013). Mechanisms contributing to the thermal analysis of waste incineration bottom ash and quantification of different carbon species. *Waste Management*, *33*(2), 373−381. Available from https://doi.org/10.1016/j.wasman.2012.110.004, 18792456.

Roslan, N. H., Ismail, M., Abdul-Majid, Z., Ghoreishiamiri, S., & Muhammad, B. (2016). Performance of steel slag and steel sludge in concrete. *Construction and Building Materials*, *104*, 16−24. Available from https://doi.org/10.1016/j.conbuildmat.2015.120.008, 09500618.

Rovani, S., Santos, J. J., Corio, P., & Fungaro, D. A. (2018). Highly pure silica nanoparticles with high adsorption capacity obtained from sugarcane waste ash. *ACS Omega*, *3*(3), 2618−2627. Available from http://pubs.acs.org/journal/acsodf, https://doi.org/10.1021/acsomega.8b00092.

Samanta, N. S., Anweshan., & Purkait, M. K. (2023). Techniques in removal of organics and emerging contaminants from wastewater for water reuse application. *Development in wastewater treatment research and processes: Advances in industrial wastewater treatment technologies: Removal of contaminants and recovery of resources* (pp. 73−96). India: Elsevier. Available from https://www.sciencedirect.com/book/9780323956840, https://doi.org/10.1016/B978-0-323-95684-0.00009-9.

Samanta, N. S., Banerjee, S., Mondal, P., Anweshan., Bora, U., & Purkait, M. K. (2021). Preparation and characterization of zeolite from waste Linz-Donawitz (LD) process slag of steel industry for removal of Fe3 + from drinking water. *Advanced Powder Technology*, *32*(9), 3372−3387. Available from http://www.elsevier.com, https://doi.org/10.1016/j.apt.2021.07.023.

Samanta, N. S., Das, P. P., Mondal, P., Changmai, M., & Purkait, M. K. (2022). Critical review on the synthesis and advancement of industrial and biomass waste-based zeolites and their applications in gas adsorption and biomedical studies. *Journal of the Indian Chemical Society*, *99*(11). Available from https://www.sciencedirect.com/journal/journal-of-the-indian-chemical-society, https://doi.org/10.1016/j.jics.2022.100761.

Samanta, N. S., Das, P. P., Mondal, P., Bora, U., & Purkait, M. K. (2022). Physico-chemical and adsorption study of hydrothermally treated zeolite A and FAU-type zeolite X prepared from

LD (Linz−Donawitz) slag of the steel industry. *International Journal of Environmental Analytical Chemistry*. Available from http://www.tandf.co.uk/journals/titles/03067319.asp, https://doi.org/10.1080/03067319.2022.2079082.

Samanta, N. S., Mondal, P., & Purkait, M. K. (2023). Nanofiltration technique for the treatment of industrial wastewater. *Advanced application of nanotechnology to industrial wastewater* (pp. 165−190). India: Springer Nature. Available from https://link.springer.com/book/10.1007/978-981-99-3292-4, https://doi.org/10.1007/978-981-99-3292-4_9.

Samanta, N. S., Anweshan., Mondal, P., Bora, U., & Purkait, M. K. (2023). Synthesis of precipitated calcium carbonate from LD-slag using CO2. *Materials Today Communications, 36*. Available from http://www.journals.elsevier.com/materials-today-communications/, https://doi.org/10.1016/j.mtcomm.2023.106588.

Samanta, N. S., Das, P. P., Sharma, M., & Purkait, M. K. (2023). *Recycle of water treatment plant sludge and its utilization for wastewater treatment* (pp. 239−264). Elsevier BV. Available from https://doi.org/10.1016/b978-0-323-99344-9.00010-4.

Santamaría, A., Rojí, E., Skaf, M., Marcos, I., & González, J. J. (2016). The use of steelmaking slags and fly ash in structural mortars. *Construction and Building Materials, 106*, 364−373. Available from https://doi.org/10.1016/j.conbuildmat.2015.120.121, 09500618.

Sharma, M., Das, P. P., Sood, T., Chakraborty, A., & Purkait, M. K. (2021). Ameliorated polyvinylidene fluoride based proton exchange membrane impregnated with graphene oxide, and cellulose acetate obtained from sugarcane bagasse for application in microbial fuel cell. *Journal of Environmental Chemical Engineering, 9*, 106681. Available from https://doi.org/10.1016/j.jece.2021.106681.

Sharma, M., Das, P. P., Sood, T., Chakraborty, A., & Purkait, M. K. (2022). Reduced graphene oxide incorporated polyvinylidene fluoride/cellulose acetate proton exchange membrane for energy extraction using microbial fuel cells. *Journal of Electroanalytical Chemistry, 907*, 115890. Available from https://doi.org/10.1016/j.jelechem.2021.115890.

Sharma, M., Samanta, N. S., Chakraborty, A., & Purkait, M. K. (2023). *Simultaneous treatment of industrial wastewater and resource recovery using microbial fuel cell* (pp. 621−637). Elsevier BV. Available from https://doi.org/10.1016/b978-0-323-95327-6.00002-6.

Sharma, M., Das, P. P., Chakraborty, A., & Purkait, M. K. (2022). Clean energy from salinity gradients using pressure retarded osmosis and reverse electrodialysis: A review. *Sustainable Energy Technologies and Assessments, 49*, 101687. Available from https://doi.org/10.1016/j.seta.2021.101687.

Sharma, M., Das, P. P., Chakraborty, A., & Purkait, M. K. (2023). *Extraction of clean energy from industrial wastewater using bioelectrochemical process* (pp. 601−620). Elsevier BV. Available from https://doi.org/10.1016/b978-0-323-95327-6.00003-8.

Sheen, Y. N., Le, D. H., & Sun, T. H. (2015). Innovative usages of stainless steel slags in developing self-compacting concrete. *Construction and Building Materials, 101*, 268−276. Available from https://doi.org/10.1016/j.conbuildmat.2015.100.079, 09500618.

Shekhar Samanta, N., Das, P. P., Dhara, S., & Purkait, M. K. (2023). An overview of precious metal recovery from steel industry slag: Recovery strategy and utilization. *Industrial and Engineering Chemistry Research, 62*(23), 9006−9031. Available from http://pubs.acs.org/journal/iecred, https://doi.org/10.1021/acs.iecr.3c00604.

Shi, C. (2002). Characteristics and cementitious properties of ladle slag fines from steel production. *Cement and Concrete Research, 32*(3), 459−462. Available from https://doi.org/10.1016/S0008-8846(01)00707-4, 00088846.

Shi, C., & Hu, S. (2003). Cementitious properties of ladle slag fines under autoclave curing conditions. *Cement and Concrete Research, 33*(11), 1851−1856. Available from https://doi.org/10.1016/S0008-8846(03)00211-4.

Shim, Y. S., Rhee, S. W., & Lee, W. K. (2005). Comparison of leaching characteristics of heavy metals from bottom and fly ashes in Korea and Japan. *Waste Management*, 25(5), 473−480. Available from https://doi.org/10.1016/j.wasman.2005.030.002.

Sontakke, A. D., Deepti., Samanta, N. S., & Purkait, M. K. (2023). Smart nanomaterials in the medical industry. *Advances in smart nanomaterials and their applications* (pp. 23−50). India: Elsevier. Available from https://www.sciencedirect.com/book/9780323995467, https://doi.org/10.1016/B978-0-323-99546-7.00025-2.

Tossavainen, M., Engstrom, F., Yang, Q., Menad, N., Lidstrom Larsson, M., & Bjorkman, B. (2007). Characteristics of steel slag under different cooling conditions. *Waste Management*, 27 (10), 1335−1344. Available from https://doi.org/10.1016/j.wasman.2006.080.002, 0956053X.

Wang, Q., Yan, P., Yang, J., & Zhang, B. (2013). Influence of steel slag on mechanical properties and durability of concrete. *Construction and Building Materials*, 47, 1414−1420. Available from https://doi.org/10.1016/j.conbuildmat.2013.060.044, 09500618.

Webber, C. L., III, White, P. M., Petrie, E. C., Shrefler, J. W., & Taylor, M. J. (2015). Sugarcane bagasse ash as a seedling growth media component. *Journal of Agricultural Science*, 8(1), 1. Available from https://doi.org/10.5539/jas.v8n1p1.

Wongsa, A., Boonserm, K., Waisurasingha, C., Sata, V., & Chindaprasirt, P. (2017). Use of municipal solid waste incinerator (MSWI) bottom ash in high calcium fly ash geopolymer matrix. *Journal of Cleaner Production*, 148, 49−59. Available from https://doi.org/10.1016/j.jclepro.2017.010.147, 09596526.

Wu, R., Xiao, Y., Zhang, P., Lin, J., Cheng, G., Chen, Z., & Yu, R. (2022). Asphalt VOCs reduction of zeolite synthesized from solid wastes of red mud and steel slag. *Journal of Cleaner Production*, 345131078. Available from https://doi.org/10.1016/j.jclepro.2022.131078, 09596526.

Xiong, Y., Zhu, F., Zhao, L., Jiang, H., & Zhang, Z. (2014). Heavy metal speciation in various types of fly ash from municipal solid waste incinerator. *China Journal of Material Cycles and Waste Management*, 16, 608−615. Available from http://link.springer.de/link/service/journals/10163/index.htm, https://doi.org/10.1007/s10163-014-0274-6.

Yakubu, Y., Zhou, J., Ping, D., Shu, Z., & Chen, Y. (2018). Effects of pH dynamics on solidification/stabilization of municipal solid waste incineration fly ash. *Journal of Environmental Management*, 207, 243−248. Available from https://www.sciencedirect.com/journal/journal-of-environmental-management, https://doi.org/10.1016/j.jenvman.2017.11.042.

Yildirim, I. Z., & Prezzi, M. (2011). Chemical, mineralogical, and morphological properties of steel slag. *Advances in Civil Engineering*, 2011. Available from https://doi.org/10.1155/2011/463638, 16878094.

Zhao, K., Hu, Y., Tian, Y., Chen, D., & Feng, Y. (2020). Chlorine removal from MSWI fly ash by thermal treatment: Effects of iron/aluminum additives. *Chinese Academy of Sciences, China Journal of Environmental Sciences (China)*, 88, 112−121. Available from https://doi.org/10.1016/j.jes.2019.080.006, http://www.journals.elsevier.com/journal-of-environmental-sciences/.

Zheng, G., & Koziński, J. A. (2000). Thermal events occurring during the combustion of biomass residue. *Fuel*, 79(2), 181−192. Available from https://doi.org/10.1016/S0016-2361(99)00130-1, 00162361.

Zhou, Q., Jiang, X., Qiu, Q., Zhao, Y., & Long, L. (2023). Synthesis of high-quality Na P1 zeolite from municipal solid waste incineration fly ash by microwave-assisted hydrothermal method and its adsorption capacity. *Science of the Total Environment*, 855158741. Available from https://doi.org/10.1016/j.scitotenv.2022.158741, 00489697.

# Chapter 3

# Preparation of different types of zeolite from steel slag

Steel slag (SS) is a by-product of the steelmaking process obtained when molten steel is separated from impurities in a steelmaking furnace. Steel is produced in integrated steel plants by using two different kinds of furnaces: a basic oxygen furnace (BOF) and an electric arc furnace (EAF). Based on the process employed, there are three types of SS defined below:

1. In a BOF, the formation of slag occurs as a result of injecting lime (CaO) and dolomite lime (CaO·MgO) along with blast furnace (BF) metal scrap into the furnace under high-pressure oxygen. This oxygen initiates the conversion of impurities into metal oxides, which then further react with lime (CaO) and dolomite lime (CaO·MgO) to produce the BOF slag (Carvalho, Vernilli, Almeida, Demarco, & Silva, 2017).
2. EAF slag, generated in facilities utilizing an EAF for stainless steel production, can be categorized into two distinct types: oxidation slag and reduction slag (Faleschini, Brunelli, Zanini, Dabalà, & Pellegrino, 2016).
3. Linz-Donawitz (LD) slag is generated when refined liquid steel undergoes further refining through a ladle. In this process, the ladle is poured with refined liquid steel containing some additional impurities (Shi, 2002).

A general composition of SS in terms of oxide components is presented in Table 3.1

The utilization of the silicate content of SS for zeolite synthesis presents another avenue of exploration (Leuchtenmueller, Schatzmann, & Steinlechner, 2020; Samanta, Das, Mondal, Changmai, & Purkait, 2022). Zeolites are crystalline solids composed of oxides and alumina silicate minerals, with the chemical formula $M_m[Al_mSi_nO_{2(m+n)}] \cdot xH_2O$, where M represents an alkali cation. These ions form a three-dimensional framework with intermolecular cavities, granting them catalytic and ion-exchange properties. Zeolites find extensive applications across various fields, including catalysis, water treatment, and agriculture (Bharti, Das, & Purkait, 2023; Chakraborty, Gautam, Das, & Hazarika, 2019; Changmai & Das, 2022; Das, Anweshan, & Mondal, 2021; Das, Anweshan, & Purkait, 2021; Das & Mondal, 2021; Das, Sharma, & Purkait, 2022; Sharma & Das, 2021, 2022a,

Waste-based Zeolite. DOI: https://doi.org/10.1016/B978-0-443-22316-7.00003-6

**TABLE 3.1** Inorganic metal oxides present in steel slag.

| Metal oxides | Slag type | | |
|---|---|---|---|
| | LD slag | BOF slag | EAF slag |
| CaO | 44−45 | 39−45 | 25−35 |
| FeO | 23 − 24 | 16−27 | 7−25 |
| MgO | 7 − 8 | 6−9.5 | 2−9 |
| MnO | 0.7−0.8 | 2.7−4 | 3.7−6 |
| $SiO_2$ | 13−14 | 11.3−12 | 10−20 |
| $P_2O_5$ | 1.5−3 | 1−1.5 | 0.5−0.8 |
| $TiO_2$ | 0.4−0.8 | 0.4−0.45 | 0.8−0.9 |
| $Na_2O$ | − | 0.23−0.25 | 0.1 − 0.4 |
| $Cr_2O_3$ | 0.27 | 1.6−2 | 0.3−1.1 |
| $K_2O$ | − | 0.20 | 0.4 |
| $Al_2O_3$ | 2−4 | 1−2 | 3−10 |
| $V_2O_5$ | 0.028 | 1.2−2.9 | 2−3 |
| $Fe_2O_3$ | − | − | 10−30 |

*Source*: From Das, P., Upadhyay, S., Dubey, S., & Singh, K. K. (2021). Waste to wealth: Recovery of value-added products from steel slag. *Journal of Environmental Chemical Engineering, 9*(4). https://doi.org/10.1016/j.jece.2021.105640.

2022b; Sontakke & Das, 2021). They can be produced from materials with high silica content such as SS, BF slag, and fly ash (Dhara, Samanta, Uppaluri, & Purkait, 2023; Dhara, Samanta, Das, Uppaluri, & Purkait, 2023; Samanta, Das, Sharma, & Purkait, 2023; Samanta, Anweshan, & Purkait, 2023). Various types of zeolites can be synthesized using SS, including hydroxyapatite-zeolite (HAP-ZE) and zeolite-hydroxyapatite-activated palm ash (Z-HAP-AA). For instance, HAP-ZE can be synthesized by the hydrothermal synthesis of SS, phosphoric acid, and sodium hydroxide (Kuwahara, Ohmichi, Kamegawa, Mori, & Yamashita, 2009). An example of a flowsheet for producing zeolite from SS is shown in Fig. 3.1.

A blend of EAF slag and palm oil ash was used as source materials to create mesoporous Z-HAP-AA. Z-HAP-AA serves as an effective adsorbent for eliminating toxic contaminants like tetracyclines (Huang, Wang, Shi, Huang, & Zhang, 2013). Additionally, its mesoporosity provides a substantial surface area for the uptake of ions. To produce Z-HAP-AA, a mixture of oil palm ash and sodium hydroxide in a 1:3 ratio was mixed with EAF slag. Subsequently, the mixture underwent treatment with phosphoric acid and

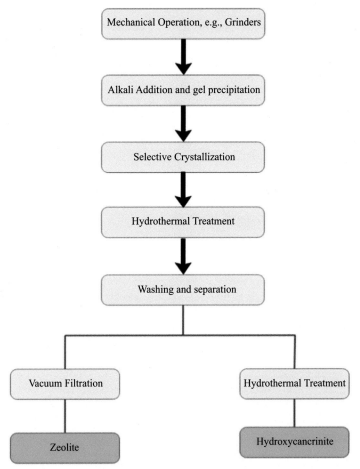

**FIGURE 3.1** Flowchart showing an outline of zeolite production from steel slag A flowsheet for the production of zeolites from steel slag. *From Liu, W., Aldahri, T., Ren, S., Xu, C. C., Rohani, S., Liang, B., & Li, C. (2020). Solvent-free synthesis of hydroxycancrinite zeolite microspheres during the carbonation process of blast furnace slag.* Journal of Alloys and Compounds, 847. *https://doi.org/10.1016/j.jallcom.2020.156456.*

sodium aluminate to yield mesoporous zeolite (Zhang, Wei, & Yu, 2005). Furthermore, Faujasite zeolite (Y-zeolite, X-zeolite), synthesized through an alkaline hydrothermal process using SS, can be transformed into a support for a nanoporous titania photocatalyst (Kuwahara, Ohmichi, Mori, Katayama, & Yamashita, 2008). Producing valuable zeolites from SS by harnessing its silica—alumina components offers an economically feasible approach to waste valorization, fostering broad-scale utilization. Although challenges such as lower surface area and catalytic activity compared with commercially prepared zeolites exist, the use of SS for zeolite synthesis may prove economically viable, especially when considering cost-effective silica sources and environmental benefits (Das, Deepti, & Purkait, 2023a; Das, Dhara, & Purkait, 2023b, 2023c, 2024; Das, Mondal, & Purkait, 2022; Das & Mondal, 2023; Das, Samanta, Dhara, & Purkait, 2023d; Das, Sharma, & Purkait, 2023e; Das, Sontakke, & Purkait, 2023a, 2023b; Das, Sontakke, Samanta, & Purkait, 2023). Prior studies extracting silica gel have relied on high concentrations and volumes of acid, resulting in increased raw material and construction expenses. Additionally, the obtained silica gel matrix may require further refinement. This presents an opportunity for process development to prioritize cost-effective, high-purity purification methods.

## 3.1   Zeolite synthesis from various slag sources

### 3.1.1   Blast furnace slag

Liu and their research team have synthesized sole gismondine-type zeolite (zeolite P) during the $CO_2$ mineralization process by using blast furnace slag (BFS) as feedstock (Liu, Aldahri, Xu, Li, & Rohani, 2021). The composition of the BFS is shown in Table 3.2. Response surface methodology was employed to examine the impact of hydrothermal temperature, the molarity of the alkaline solution, and hydrothermal duration on zeolite crystallinity. The findings indicated that hydrothermal temperature played the most significant role in zeolite P formation. Elevated temperatures enabled the production of well-crystallized zeolite products with a shorter hydrothermal duration, even with lower alkaline concentrations. Nevertheless, excessively high temperatures and alkaline concentrations encouraged the formation of sodalite, leading to a decrease in the relative zeolite P content within the final product. A zeolite P product with a uniform shape of octahedron was obtained under mild conditions as shown in Fig. 3.2 (NaOH concentration of 1.5 mol/L, 120°C and 2 − 4 hours). The process of generating zeolite from BFS can be categorized into four distinct stages: the dissolution of aluminate and silicate ions from the raw material, the creation of an amorphous intermediate aluminosilicate gel precursor, the restructuring of the aluminosilicate precursor's structure to produce zeolite nuclei, and the subsequent growth of

**TABLE 3.2** Chemical compositions of blast furnace slag used in this study (wt.%).

| CaO* | Al$_2$O$_3$* | MgO* | SiO$_2$* | TiO$_2$ | Fe$_2$O$_3$ | MnO | S | Others |
|------|------|------|------|------|------|------|------|------|
| 41.84 | 13.98 | 6.74 | 32.08 | 1.33 | 0.87 | 0.19 | 1.04 | 1.91 |

*Source*: From Liu, W., Aldahri, T., Xu, C., Li, C., & Rohani, S. (2021). Synthesis of sole gismondine-type zeolite from blast furnace slag during $CO_2$ mineralization process. *Journal of Environmental Chemical Engineering, 9*(1), 104652. https://doi.org/10.1016/j.jece.2020.104652.

**FIGURE 3.2** Zeolite P morphological analysis SEM images of zeolite P obtained at optimized conditions from response surface methodology. *From Liu, W., Aldahri, T., Xu, C., Li, C., & Rohani, S. (2021). Synthesis of sole gismondine-type zeolite from blast furnace slag during $CO_2$ mineralization process. Journal of Environmental Chemical Engineering, 9(1), 104652. https:// doi.org/10.1016/j.jece.2020.104652.*

crystals. According to the findings of this investigation, the primary components of BFS, including magnesium and calcium for $CO_2$ storage, as well as aluminum and silicon for zeolite synthesis, were efficiently harnessed. This approach reduces the overall expenses associated with $CO_2$ mineralization.

In a similar manner, Kuwahara et al. (2009) employed BFS to create a composite material known as HAP-ZE, using specific chemical reagents, $H_3PO_4$ and NaOH. The material was characterized through X-ray diffraction (XRD), Fourier-transform infrared (FTIR) spectroscopy, scanning electron microscopy (SEM), energy dispersive X-ray (EDX), and elemental mapping techniques, revealing that during the initial stages of the aging process, calcium, a prominent component in steel slag, rapidly engaged in a reaction with phosphate, involving magnesium, resulting in the formation of magnesium-substituted HAP with $(Ca^+ Mg)/P = 1.67$. After 2 days of aging, well-crystallized HAP and faujasite-type zeolite with $SiO_2/Al_2O_3 = 2.4$ were separately formed via a nonsimultaneous crystallization process, while the yield of the zeolite phase was rather low because of the inherent $SiO_2/Al_2O_3$ ratio in steel slag. Extended aging resulted in a phase transition from FAU-zeolite to Pl-zeolite. The trace elements present in steel slag, such as iron (Fe) and manganese (Mn), were determined to have a negligible impact on the synthesis of HAP-ZE. Additionally, practical experiments substantiated that HAP-ZE, synthesized under optimal conditions, exhibited exceptional adsorption capabilities for volatile organic compounds (VOCs), fatty acids, and proteins. This suggests its practical suitability as an adsorbent, even though the zeolite phase content is relatively low.

The SEM images of HAP-ZE (after 48 hours of aging) with EDX results are shown in Fig. 3.3. The SEM image of the biphasic region reveals the coexistence of two distinct crystal types at the microscale. Considering their crystal sizes and structural characteristics, it is reasonable to attribute the observed octahedral crystals, with an average particle size of 6 mm, to the FAU-type zeolite. The round-shaped crystals, exhibiting varying particle sizes, are identified as HAP. Most of these HAP crystals have grown into large grains, while some are evenly dispersed on the zeolite crystals. This morphology is believed to result from a sequential crystallization process, wherein HAP crystals formed first, followed by the subsequent growth of zeolite crystals around them. Further support for this interpretation comes from the EDX analysis, which shows that the octahedral particles are rich in Si, Al, and Na, while the roundish particles are abundant in Ca and P.

Conversely, Sugano et al. (2005) delved into the alkali hydrothermal synthesis of zeolite A using BF slag. Their initial experiment involved the use of a synthetic slag composed of $SiO_2$, $Al_2O_3$, and CaO powders. They established that the most favorable slag compositions for synthesizing zeolite A were characterized by a molar ratio of Si to Al (Si/Al) at 1 and a reduction in CaO content to 15% by mass. The optimal hydrothermal treatment conditions encompassed a temperature range of 328−358 K, a 1 M NaOH solution, and a ratio of the volume of NaOH solution to the total mass of slag (Vsol/Wslag) at 15 mL/g. It was also noted that synthetic slag with a higher CaO content, such as 40%, resulted in the formation of tobermorite and hydrogarnet phases.

Zeolite A was successfully produced from BF slag, as illustrated in Fig. 3.4, through a process that involved optimizing both the compositions of the raw material powders and the hydrothermal treatment conditions. The optimization of raw material powder compositions, such as Si or Al content, was achieved by carefully adding the appropriate amount of $SiO_2$ powder or $NaAlO_2$ powder, serving as a source of Si and Al. This study introduced a ball milling-type reaction vessel containing numerous small SiC balls, which was found to be highly effective in accelerating the synthetic reaction rate. As a result, it significantly reduced the time required for the complete synthesis of zeolite A. To further investigate the properties of the zeolite A powder synthesized using BF slag, the temperature-dependent heat capacity was measured after vapor absorption at ambient temperature. The measurements revealed endothermic behavior with a peak occurring at approximately 473 K. Fig. 3.4 shows the hydrothermally synthesized A-type zeolite (Na-A) using steel industry-processed BFS as precursor material. The study suggests

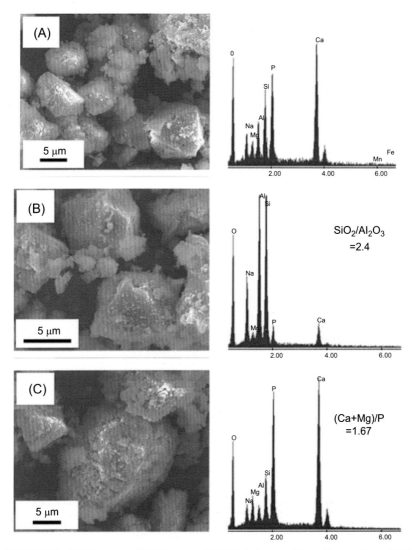

**FIGURE 3.3**  SEM images and EDX spectra of (A) biphasic area, (B) zeolite crystals, and (C) HAP crystals in HAP-ZE. *From Kuwahara, Y., Ohmichi, T., Kamegawa, T., Mori, K., & Yamashita, H. (2009). A novel synthetic route to hydroxyapatite-zeolite composite material from steel slag: Investigation of synthesis mechanism and evaluation of physicochemical properties.* Journal of Materials Chemistry, 19(39), 7263–7272. https://doi.org/10.1039/b911177h.

that silica and alumina-rich steel industry-by-product may be utilized as a potential feedstock of zeolite-like mineral preparation (Samanta, Anweshan, Mondal, Bora, & Purkait, 2023; Samanta, Das, Mondal, Bora, & Purkait, 2022; Shekhar Samanta, Das, Dhara, & Purkait, 2023).

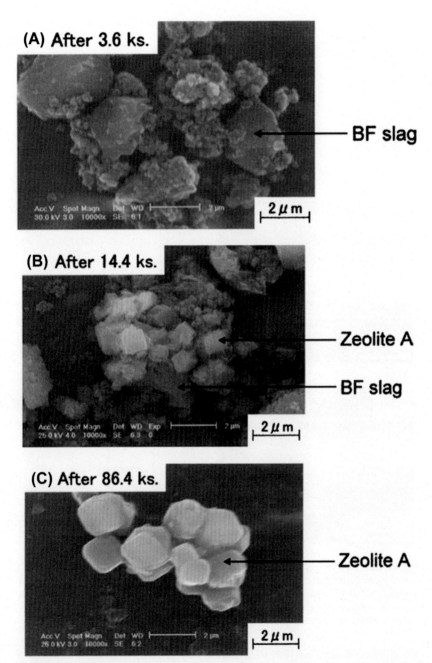

**FIGURE 3.4**   FESEM analysis of BFS-derived zeolite type-A. Scanning electron micrograms of the products obtained from direct hydrothermal treatment of blast furnace slag: (A) SEM analysis of BF slag after 3.6 ks; (B) SEM analysis of zeolite A and BF slag after 14.4 ks; and (C) SEM analysis of zeolite A after 86.4 ks. *From Sugano, Y., Sahara, R., Murakami, T., Narushima, T., Iguchi, Y., & Ouchi, C. (2005). Hydrothermal synthesis of zeolite a using blast furnace slag. ISIJ International, 45(6), 937−945. https://doi.org/10.2355/isijinternational.45.937.*

**TABLE 3.3** Composition of blast furnace slag utilized for zeolite synthesis.

| Components | CaO | SiO₂ | Al₂O₃ | MgO | SO₃ | Fe₂O₃ | TiO₂ | K₂O | MnO |
|---|---|---|---|---|---|---|---|---|---|
| Concentration (wt.%) | 42.9 | 34.5 | 13.7 | 6.1 | 1.1 | 0.6 | 0.4 | 0.3 | 0.3 |

*Source:* From Wajima, T. (2014). Synthesis of zeolite from blast furnace slag using alkali fusion with addition of EDTA. *Advanced Materials Research,* 1044−1045, 124−127. https://doi.org/10.4028/www.scientific.net/AMR.1044-1045.124.

Furthermore, in a subsequent study, Wajima (2014) attempted to create a zeolite-like substance from iron-steel industrial solid waste material (BFS). This involved high-temperature fusion with the inclusion of NaOH as an alkaline source. The investigation assessed the impact of ethylenediamine tetraacetic acid (EDTA) during the zeolitization process. Without the addition of EDTA, a combination of hydroxy sodalite and calcite formed, whereas the introduction of EDTA resulted in the synthesis of zeolite A and zeolite X. The composition of BF slag utilized in the study is shown in Table 3.3.

### 3.1.2 Electric arc furnace slag

Ti-bearing electric arc furnace slag (Ti-bearing EAF slag) is a significant solid waste product produced during the direct reduction iron-making process. This slag contains various elements, including Ti, Mg, Al, and Si. Current efforts primarily focus on extracting titanium using methods such as acid leaching, selective enrichment, and alkali fusion. Unfortunately, little attention has been given to the comprehensive recycling of other valuable elements, potentially leading to the generation of secondary waste. To fully utilize the valuable elements present in Ti-bearing EAF slag, Li et al. (2016) have proposed a novel process that involves alkali fusion, water leaching, and acidolysis. During the alkali fusion stage, the elements within Ti-bearing EAF slag are transformed into their respective sodium salts. Subsequently, the water-leaching process yields an alkaline solution containing sodium aluminate and sodium silicate, serving as an ideal precursor for the production of Na-A zeolite.

Using Ti-bearing EAF slag, Li and colleagues successfully synthesized both Na-A zeolite ($6Na_2O \cdot 6Al_2O_3 0.12SiO_2$) and sodalite (SOD, $4Na_2O \cdot 3Al_2O_3 0.6SiO_2$). They systematically investigated the effects of several factors, including the $SiO_2/Al_2O_3$ molar ratio ($n(SiO_2)/n(Al_2O_3)$), $H_2O/Na_2O$ molar ratio ($n(H_2O)/n(Na_2O)$), hydrothermal temperature, and duration, on the crystal phase and microstructure of the prepared zeolites. The results demonstrated that under conditions with a fixed $n(SiO_2)/n(Al_2O_3)$ ratio of 2.0:1 and a $n(H_2O)/n(Na_2O)$ ratio of 100:1, Na-A zeolite with excellent crystallinity and a cubic morphology was achieved at 140°C

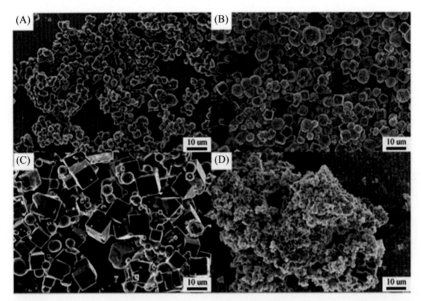

**FIGURE 3.5** Morphological analysis of Na-A-type zeolite synthesized from EAF slag under different working conditions SEM images of samples prepared with different $n(H_2O)/n(Na_2O)$. (A) 40: 1, (B) 70: 1, (C) 100: 1, (D) 130: 1 ($n(SiO_2)/n(Al_2O_3) = 2.0$: 1, $T = 140°C$, $t = 3$ h). *From Li, Y., Peng, T., Man, W., Ju, L., Zheng, F., Zhang, M., & Guo, M. (2016). Hydrothermal synthesis of mixtures of NaA zeolite and sodalite from Ti-bearing electric arc furnace slag. RSC Advances, 6(10), 8358–8366. https://doi.org/10.1039/c5ra26881h.*

for 3 hours. Decreasing the $n(H_2O)/n(Na_2O)$ ratio, increasing the hydrothermal temperature, and prolonging the hydrothermal time favored the formation of spherical SOD zeolite. Notably, the samples obtained during the process exhibited a coexistence of phase transformation between Na-A and SOD zeolite. Figs. 3.5–3.7 show the morphological variations of synthesized zeolites at different ratios of $n(H_2O)/n(Na_2O)$, hydrothermal temperature, and hydrothermal time, respectively.

The results indicated that Na-A zeolite with good crystallinity and cubic morphology could be obtained at 140°C for 3 hours with $n(SiO_2)/n(Al_2O_3)$ and $n(H_2O)/n(Na_2O)$ fixed at 2.0:1 and 100:1, respectively. Decreasing $n(H_2O)/n(Na_2O)$, raising the hydrothermal temperature, and prolonging the hydrothermal time was beneficial for the formation of spherical SOD zeolite.

The investigation into the removal capacity of $Cu^{2+}$ using the synthesized zeolite, as depicted in Fig. 3.8, was carried out at 120°C for 180 minutes, resulting in an impressive removal capacity of approximately 1.346 mmol/g. This performance significantly surpassed the results obtained at 140°C or 160°C, which were only 0.722 mmol/g and 0.673 mmol/g, respectively. The removal of $Cu^{2+}$ from wastewater was attributed to distinct mechanisms involving ion-exchange and adsorption processes. In the ion-

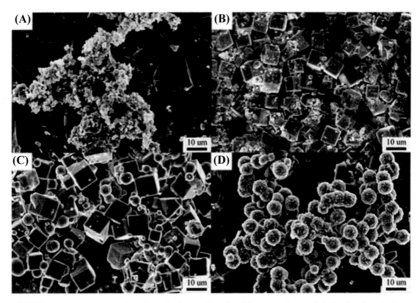

**FIGURE 3.6** FESEM results of synthetic zeolites SEM images of samples obtained at different hydrothermal temperatures. (A) 100°C, (B) 120°C, (C) 140°C, (D) 160°C ($n(SiO_2)/n(Al_2O_3)$ = 2.0:1, $n(H_2O)/n(Na_2O)$ = 100: 1, $t$ = 3 h). *From Li, Y., Peng, T., Man, W., Ju, L., Zheng, F., Zhang, M., & Guo, M. (2016). Hydrothermal synthesis of mixtures of NaA zeolite and sodalite from Ti-bearing electric arc furnace slag.* RSC Advances, 6(10), 8358−8366. *https://doi.org/10.1039/c5ra26881h.*

exchange process, $Cu^{2+}$ traverses the zeolite's pores and lattice channels, engaging in ion exchange with $Na^+$ present in the zeolite. The diffusion occurs more rapidly through the pores but becomes slower when ions move through the narrower channels. Upon closer examination, it was determined that the pores and channels in SOD zeolites were considerably smaller compared to those in NaA zeolites. As a result, the movement of $Cu^{2+}$ for ion exchange with $Na^+$ ions was hindered, leading to the superior removal performance of Na-A zeolite for $Cu^{2+}$ compared to SOD zeolite.

### 3.1.3    Linz-Donawitz slag

Recently, Samanta and colleagues conducted an investigation into the synthesis of zeolite using LD slag from the steel industry as the primary source of aluminum (Al) and silicon (Si) (Samanta et al., 2021). This synthesis process involved alkali fusion followed by a hydrothermal method, resulting in the production of zeolite A as the final product. To enhance activation, the slag and alkali (NaOH) were maintained at a ratio of 1:1.2, which led to the formation of cubical-shaped zeolite A. The study focused on evaluating the removal performance of iron ($Fe^{3+}$), with varying factors such as adsorbent dosage,

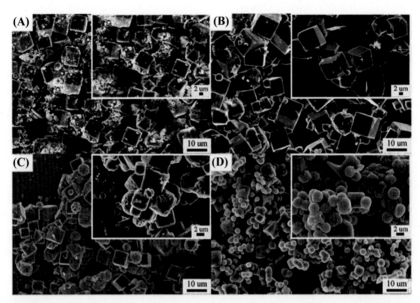

**FIGURE 3.7** FESEM images of as-synthesized zeolite samples prepared at different times SEM images of samples synthesized at different hydrothermal times. (A) 3 h, (B) 6 h, (C) 9 h, (D) 12 h ($n(SiO_2)/n(Al_2O_3) = 2.0:1$, $n(H_2O)/n(Na_2O) = 100: 1$, $T = 120°C$). *From Li, Y., Peng, T., Man, W., Ju, L., Zheng, F., Zhang, M., & Guo, M. (2016). Hydrothermal synthesis of mixtures of NaA zeolite and sodalite from Ti-bearing electric arc furnace slag. RSC Advances, 6 (10), 8358−8366. https://doi.org/10.1039/c5ra26881h.*

temperature, and pH. The maximum adsorption capacity of 27.55 mg/g was observed at an adsorbent concentration of 1.4 g/L and an initial $Fe^{3+}$ concentration of 10 ppm. The structural morphology of the synthesized product was analyzed using field-emission scanning electron microscopy (FESEM) and field-emission transmission electron microscopy (FETEM). FESEM and FETEM images of the prepared sample have been shown in Fig. 3.9A and B, respectively, from which it was clear that the geometric structure of the obtained zeolite was cubical (inset Fig. 3.9A). The crystalline cubical structure of the resultant sample can be identified as zeolite A.

With increasing adsorbate concentration, the sorption capacity decreased due to the zeolite surface reaching maximum saturation. The study also performed a zeolite stability test under varying pH levels and treatment conditions. In an acidic environment, the zeolite's cage structure underwent contraction after 3 days of treatment due to deformation and the formation of Si−O−Al and Si−OH bonds, respectively. However, no changes were observed on the zeolite surface under neutral and alkaline conditions. The results indicated that the adsorption behavior of the prepared zeolite sample might slow down the synthesis process due to instability under acidic conditions, but not under neutral or alkaline conditions. As mentioned earlier, the presence of metal

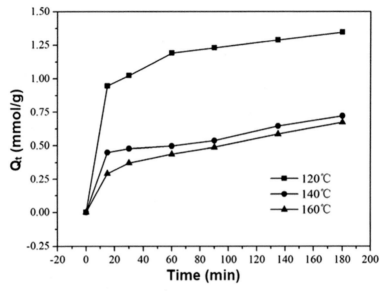

**FIGURE 3.8** Removal capacity of Cu $^{2+}$ with contact time over the prepared zeolite sample Relationship between contact time and removal capacity of $Cu^{2+}$ by zeolite synthesized at different temperatures. *From Li, Y., Peng, T., Man, W., Ju, L., Zheng, F., Zhang, M., & Guo, M. (2016). Hydrothermal synthesis of mixtures of NaA zeolite and sodalite from Ti-bearing electric arc furnace slag. RSC Advances, 6(10), 8358–8366. https://doi.org/10.1039/c5ra26881h.*

oxides may impede the zeolite nucleation process. Therefore calcination and acid leaching of the slag were performed to remove volatile and undesirable components, thereby enhancing zeolite purity.

In a separate investigation, Samanta and colleagues synthesized sodium-rich zeolite A and zeolite X (FAU-type) using LD slag through a fusion-assisted hydrothermal treatment process (Samanta et al., 2022). The physico-chemical and thermal stability of the prepared samples were assessed through various characterization techniques, including FTIR spectroscopy, XRD, and thermogravimetric analysis (TGA) at different pH levels and treatment durations. Furthermore, the longevity of the crystalline phase and the corresponding zeolite structure was evaluated using XRD, FTIR, TGA, and FESEM analysis. The morphological analysis of zeolite A and zeolite X are shown in Fig. 3.10. Zeolite A depicts a unique cubical structure and is thermally more stable as compared to zeolite type-X.

Furthermore, zeolite A exhibited the highest efficiency in removing dye, with a removal rate of 98.13%, compared to 94.47% for zeolite X. Equilibrium sorption capacities were also measured, with values of 25.30 mg/g for zeolite A and 23.57 mg/g for zeolite X. The study suggests that both synthesized adsorbents are not only effective but also economically sustainable for the adsorption of cationic methylene blue. Moreover, the

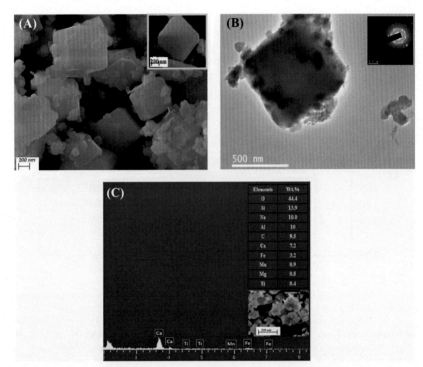

**FIGURE 3.9** (A) FESEM (B) TEM, and (C) EDX analysis of the prepared zeolite A. *From Samanta, N. S., Banerjee, S., Mondal, P., Anweshan, Bora, U., & Purkait, M. K. (2021). Preparation and characterization of zeolite from waste Linz-Donawitz (LD) process slag of steel industry for removal of $Fe^{3+}$ from drinking water. Advanced Powder Technology, 32(9), 3372–3387. https://doi.org/10.1016/j.apt.2021.07.023.*

adsorption of methylene blue followed a multistage diffusion process, in line with a pseudo-second-order kinetic model (with $R^2$ values of 0.999 for zeolite A and 0.996 for zeolite X). The Langmuir isotherm model was found to best fit the equilibrium data, revealing monolayer adsorption capacities of 20 mg/g for zeolite A and 25.40 mg/g for zeolite X, respectively.

## 3.2 Major limitations and challenges

The synthesis of zeolites from steel slag, a by-product of the steel manufacturing process, is a promising approach for recycling and reusing this waste material. However, there are several limitations and challenges associated with this process, which are mentioned below:

1. Composition variability: Steel slag composition can vary significantly depending on the steel production process and the source of the slag. This variability can make it challenging to consistently produce high-

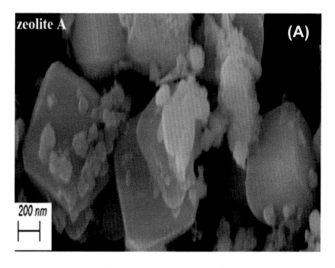

**FIGURE 3.10**  FESEM images of (A) zeolite A and (B) zeolite X. *From Samanta, N. S., Das, P. P., Mondal, P., Bora, U., & Purkait, M. K. (2022). Physico-chemical and adsorption study of hydrothermally treated zeolite A and FAU-type zeolite X prepared from LD (Linz-Donawitz) slag of the steel industry.* International Journal of Environmental Analytical Chemistry. *https://doi. org/10.1080/03067319.2022.2079082.*

quality zeolites, as the feedstock composition may not always be suitable for zeolite synthesis.

2. Impurities: Steel slag can contain various impurities, such as heavy metals and other contaminants. These impurities can be incorporated into the synthesized zeolites, making them unsuitable for certain applications, especially in catalysis or adsorption where purity is crucial.

3. Energy intensive: The synthesis of zeolites from steel slag typically involves high-temperature treatments, which can be energy intensive. This can make the process less environmentally friendly and economically viable, especially if energy costs are high (Chakraborty, Das, & Mondal, 2023; Dhara, Das, Uppaluri, & Purkait, 2023a, 2023b).
4. Kinetic constraints: Zeolite synthesis from steel slag may face kinetic limitations, as the transformation of steel slag components into zeolites can be slow and incomplete. Specialized techniques may be required to enhance reaction kinetics.
5. Scaling up: Although laboratory-scale synthesis of zeolites from steel slag may be feasible, scaling up the process for industrial applications can be challenging. Maintaining product quality and consistency at a larger scale while managing the variability of feedstock can be complex.
6. Characterization and testing: Accurate characterization of the synthesized zeolites is critical to ensure their suitability for specific applications. However, characterizing zeolites derived from steel slag can be challenging due to potential impurities and variations in composition.
7. Utilization of by-products: The by-products generated during zeolite synthesis from steel slag, such as the remaining slag or unconverted materials, may need to be managed or disposed of properly. Finding environmentally responsible ways to handle these by-products is essential.
8. Market demand: The commercial viability of zeolite synthesis from steel slag depends on the demand for zeolite products in various industries. If there is insufficient demand, it may not be economically viable to produce zeolites from steel slag.
9. Regulatory compliance: Meeting regulatory standards for the use of zeolites derived from steel slag is essential, especially when the material is intended for use in applications like water treatment or soil remediation. Compliance with environmental regulations can be a challenge.
10. Economic viability: The overall economic feasibility of the process may depend on factors such as the cost of raw materials, energy, and the value of the end products. Determining the cost-effectiveness of zeolite synthesis from steel slag is crucial.

In summary, while zeolite synthesis from steel slag offers the potential for recycling and reusing a waste material, it comes with various challenges related to composition variability, impurities, energy consumption, kinetics, scaling, and economic factors. Addressing these challenges will be essential for realizing the full potential of this approach.

## 3.3 Future prospective and conclusion

The future perspective of zeolite synthesis from steel slag holds promise, as it can contribute to sustainable resource utilization and address environmental concerns. Here are some potential developments and trends in this field:

1. Improved processes and technologies: Researchers are likely to focus on developing more efficient and environmentally friendly synthesis methods for zeolites from steel slag. This could involve innovative catalysts, reaction conditions, or alternative energy sources to reduce the energy intensity of the process.

2. Optimized feedstock management: To overcome the variability in steel slag composition, methods for sorting and selecting suitable slag sources for zeolite synthesis may be developed. This could help ensure consistent product quality.

3. Impurity control: Techniques for reducing and controlling impurities in synthesized zeolites will continue to evolve. This is essential to make the zeolites suitable for applications where purity is crucial, such as in catalysts and adsorbents.

4. Tailored zeolite products: Zeolites can be synthesized to have specific properties for various applications. Future research may focus on tailoring zeolite properties according to the requirements of industries like petrochemicals, water treatment, and environmental remediation.

5. Scalability: Overcoming the challenges of scaling up zeolite synthesis from steel slag will be important. Developing cost-effective and reliable processes for large-scale production is a key area for improvement.

6. Waste reduction: Minimizing the production of waste by-products during zeolite synthesis and finding sustainable uses for any remaining slag or unconverted materials will be a priority.

7. Advanced characterization techniques: Improved methods for characterizing the synthesized zeolites, especially in terms of purity, crystallinity, and pore structure, will continue to be developed.

8. Regulatory compliance: As the use of zeolites in various applications grows, regulatory standards and guidelines for zeolites derived from steel slag may become more well defined. Compliance with these standards will be important for market acceptance.

9. Market expansion: The demand for zeolites is likely to grow in various industries, including water treatment, petrochemicals, and environmental remediation. Zeolites synthesized from steel slag may become a cost-effective and sustainable choice for these applications.

10. Collaboration and investment: Collaboration between researchers, industry, and government bodies will play a significant role in advancing the field. Investment in research and development, as well as pilot projects, can help drive progress.

11. Circular economy practices: The synthesis of zeolites from steel slag aligns with the principles of the circular economy, where waste materials are transformed into valuable products. Future developments may further integrate this approach into industrial and environmental strategies.

In conclusion, the future of zeolite synthesis from steel slag is likely to involve advancements in processes, materials, and applications, contributing to a more sustainable and resource-efficient industrial landscape. As environmental concerns and the need for sustainable practices continue to grow, the utilization of waste materials like steel slag for valuable zeolite production is expected to gain momentum.

# References

Bharti, M., Das, P. P., & Purkait, M. K. (2023). A review on the treatment of water and wastewater by electrocoagulation process: Advances and emerging applications. *Journal of Environmental Chemical Engineering*, *11*, 111558. Available from https://doi.org/10.1016/j.jece.2023.111558.

Carvalho, S. Z., Vernilli, F., Almeida, B., Demarco, M., & Silva, S. N. (2017). The recycling effect of BOF slag in the portland cement properties. *Resources, Conservation and Recycling*, *127*, 216–220. Available from https://doi.org/10.1016/j.resconrec.2017.08.021, http://www.elsevier.com/locate/resconrec.

Chakraborty, S., Das, P. P., & Mondal, P. (2023). Recent advances in membrane technology for the recovery and reuse of valuable resources. In M. Sillanpaa, A. Khadir, & K. Gurung (Eds.), *Resource recovery in industrial waste waters* (pp. 695–719). Elsevier. Available from https://doi.org/10.1016/B978-0-323-95327-6.00028-2.

Chakraborty, S., Gautam, S. P., Das, P. P., & Hazarika, Manuj K. (2019). Instant controlled pressure drop (DIC) treatment for improving process performance and milled rice quality. *Journal of The Institution of Engineers (India): Series A*, *100*, 683–695. Available from https://doi.org/10.1007/s40030-019-00403-w.

Changmai, M., Das, P. P., Mondal, P., Pasawan, M., Sinha, A., Biswas, P., Sarkar, S., & Purkait, M. K. (2022). Hybrid electrocoagulation–microfiltration technique for treatment of nanofiltration rejected steel industry effluent. *International Journal of Environmental Analytical Chemistry*, *102*, 62–83. Available from https://doi.org/10.1080/03067319.2020.1715381.

Das, P. P., Anweshan, A., Mondal, P., & Purkait, M. K. (2021). Integrated ozonation assisted electrocoagulation process for the removal of cyanide from steel industry wastewater. *Chemosphere*, *263*, 128370. Available from https://doi.org/10.1016/j.chemosphere.2020.128370.

Das, P. P., Anweshan., & Purkait, M. K. (2021). Treatment of cold rolling mill (CRM) effluent of steel industry. *Separation and Purification Technology*, *274*, 119083. Available from https://doi.org/10.1016/j.seppur.2021.119083.

Das, P. P., Deepti., & Purkait, M. K. (2023). Industrial wastewater to biohydrogen production via potential bio-refinery route. In Maulin P. Shah (Ed.), *Biorefinery for water and wastewater treatment* (pp. 159–179). Springer. Available from https://doi.org/10.1007/978-3-031-20822-5_8.

Das, P. P., Dhara, S., & Purkait, M. K. (2023). Hybrid electrocoagulation and ozonation techniques for industrial wastewater treatment. In Maulin P. Shah (Ed.), *Sustainable industrial wastewater treatment and pollution control* (pp. 107–128). Springer. Available from https://doi.org/10.1007/978-981-99-2560-5_6.

Das, P. P., Dhara, S., & Purkait, M. K. (2023). The anaerobic ammonium oxidation process: Inhibition, challenges and opportunities. In Maulin P. Shah (Ed.), *Ammonia Oxidizing Bacteria: Applications in Industrial Wastewater Treatment* (pp. 56–82). Royal Society of Chemistry. Available from https://doi.org/10.1039/BK9781837671960-00056.

Das, P. P., Dhara, S., & Purkait, M. K. (2024). Ozone-based oxidation processes for the removal of pharmaceutical products from wastewater. In Maulin P. Shah, & Pooja Ghosh (Eds.), *Development in wastewater treatment research and processes* (pp. 287−308). Elsevier. Available from https://doi.org/10.1016/B978-0-443-19207-4.00003-3.

Das, P. P., Mondal, P., Anweshan, A., & Purkait, M. K. (2021). Treatment of steel plant generated biological oxidation treated (BOT) wastewater by hybrid process. *Separation and Purification Technology, 258*, 118013. Available from https://doi.org/10.1016/j.seppur.2020.118013.

Das, P. P., & Mondal, P. (2023). Membrane-assisted potable water reuses applications: Benefits and drawbacks. In M. Sillanpaa, A. Khadir, & K. Gurung (Eds.), *Resource recovery in drinking water treatment* (pp. 289−309). Elsevier. Available from https://doi.org/10.1016/B978-0-323-99344-9.00014-1.

Das, P. P., Mondal, P., & Purkait, M. K. (2022). Recent advances in synthesis of iron nanoparticles via green route and their application in biofuel production. In M. Srivastava, M. A. Malik, & P. K. Mishra (Eds.), *Green nano solution for bioenergy production enhancement* (pp. 79−104). Springer. Available from https://doi.org/10.1007/978-981-16-9356-4_4.

Das, P. P., Samanta, N. S., Dhara, S., & Purkait, M. K. (2023). Biofuel production from algal biomass. In Maulin P. Shah (Ed.), *Green approach to alternative fuel for a sustainable future* (pp. 167−179). Elsevier. Available from https://doi.org/10.1016/B978-0-12-824318-3.00009-6.

Das, P. P., Sharma, M., & Purkait, M. K. (2022). Recent progress on electrocoagulation process for wastewater treatment: A review. *Separation and Purification Technology, 292*, 121058. Available from https://doi.org/10.1016/j.seppur.2022.121058.

Das, P. P., Sharma, M., & Purkait, M. K. (2023). Ammonia oxidizing bacteria in wastewater treatment. In Maulin P. Shah (Ed.), *Ammonia Oxidizing Bacteria: Applications in Industrial Wastewater Treatment* (pp. 83−102). Royal Society of Chemistry. Available from https://doi.org/10.1039/BK9781837671960-00083.

Das, P. P., Sontakke, A. D., & Purkait, M. K. (2023). Electrocoagulation process for wastewater treatment: applications, challenges, and prospects. In Maulin P. Shah (Ed.), *Development in wastewater treatment research and processes* (pp. 23−48). Elsevier. Available from https://doi.org/10.1016/B978-0-323-95684-0.00015-4.

Das, P. P., Sontakke, A. D., & Purkait, M. K. (2023). Rice straw for biofuel production. In Maulin P. Shah (Ed.), *Green approach to alternative fuel for a sustainable future* (pp. 153−166). Elsevier. Available from https://doi.org/10.1016/B978-0-12-824318-3.00034-5.

Das, P. P., Sontakke, A. D., Samanta, N. S., & Purkait, M. K. (2023). Emerging contaminants in wastewater: Eco-toxicity and sustainability assessment. In Maulin P. Shah (Ed.), *Industrial wastewater reuse* (pp. 63−87). Springer. Available from https://doi.org/10.1007/978-981-99-2489-9_4.

Dhara, S., Das, P. P., Uppaluri, R., & Purkait, M. K. (2023). Biological approach for energy self-sufficiency of municipal wastewater treatment plants. In M. Sillanpaa, A. Khadir, & K. Gurung (Eds.), *Resource recovery in municipal waste waters* (pp. 235−260). Elsevier. Available from https://doi.org/10.1016/B978-0-323-99348-7.00006-0.

Dhara, S., Das, P. P., Uppaluri, R., & Purkait, M. K. (2023). Phosphorus recovery from municipal wastewater treatment plants. In Maulin P. Shah (Ed.), *Development in wastewater treatment research and processes* (pp. 49−72). Elsevier. Available from https://doi.org/10.1016/B978-0-323-95684-0.00014-2.

Dhara, S., Samanta, N. S., Das, P. P., Uppaluri, R. V. S., & Purkait, M. K. (2023). Ravenna grass-extracted alkaline lignin-based polysulfone mixed matrix membrane (MMM) for aqueous Cr(VI) removal. *ACS Applied Polymer Materials*, 5(8), 6399−6411. Available from https://doi.org/10.1021/acsapm.3c00999, http://pubs.acs.org/journal/aapmcd.

Dhara, S., Samanta, N. S., Uppaluri, R., & Purkait, M. K. (2023). High-purity alkaline lignin extraction from Saccharum ravannae and optimization of lignin recovery through response surface methodology. *International Journal of Biological Macromolecules*, 234123594. Available from https://doi.org/10.1016/j.ijbiomac.2023.123594.

Faleschini, F., Brunelli, K., Zanini, M. A., Dabalà, M., & Pellegrino, C. (2016). Electric arc furnace slag as coarse recycled aggregate for concrete production. *Journal of Sustainable Metallurgy*, 2(1), 44−50. Available from https://doi.org/10.1007/s40831-015-0029-1. Available from: https://link.springer.com/journal/40831.

Huang, L., Wang, M., Shi, C., Huang, J., & Zhang, B. (2013). Adsorption of tetracycline and ciprofloxacin on activated carbon prepared from lignin with $H_3PO_4$ activation. *Desalination and Water Treatment*, 52(13−15), 2678−2687. Available from https://doi.org/10.1080/19443994.2013.833873.

Kuwahara, Y., Ohmichi, T., Kamegawa, T., Mori, K., & Yamashita, H. (2009). A novel synthetic route to hydroxyapatite-zeolite composite material from steel slag: Investigation of synthesis mechanism and evaluation of physicochemical properties. *Journal of Materials Chemistry*, 19(39), 7263−7272. Available from https://doi.org/10.1039/b911177h.

Kuwahara, Y., Ohmichi, T., Mori, K., Katayama, I., & Yamashita, H. (2008). Synthesis of zeolite from steel slag and its application as a support of nano-sized $TiO_2$ photocatalyst. *Journal of Materials Science*, 43(7), 2407−2410. Available from https://doi.org/10.1007/s10853-007-2073-0.

Leuchtenmueller, M., Schatzmann, W., & Steinlechner, S. (2020). A kinetic study to recover valuables from hazardous ISF slag. *Journal of Environmental Chemical Engineering*, 8(4) 103976. Available from https://doi.org/10.1016/j.jece.2020.103976.

Li, Y., Peng, T., Man, W., Ju, L., Zheng, F., Zhang, M., & Guo, M. (2016). Hydrothermal synthesis of mixtures of NaA zeolite and sodalite from Ti-bearing electric arc furnace slag. *RSC Advances*, 6(10), 8358−8366. Available from https://doi.org/10.1039/c5ra26881h. Available from: http://pubs.rsc.org/en/journals/journalissues.

Liu, W., Aldahri, T., Xu, C., Li, C., & Rohani, S. (2021). Synthesis of sole gismondine-type zeolite from blast furnace slag during $CO_2$ mineralization process. *Journal of Environmental Chemical Engineering*, 9(1)104652. Available from https://doi.org/10.1016/j.jece.2020.104652.

Samanta, N. S., Anweshan., Mondal, P., Bora, U., & Purkait, M. K. (2023). Synthesis of precipitated calcium carbonate from LD-slag using $CO_2$. *Materials Today Communications*, 36. Available from https://doi.org/10.1016/j.mtcomm.2023.106588. Available from, http://www.journals.elsevier.com/materials-today-communications/.

Samanta, N. S., Anweshan., & Purkait, M. K. (2023). *Techniques in removal of organics and emerging contaminants from wastewater for water reuse application. Development in wastewater treatment research and processes: Advances in industrial wastewater treatment technologies: Removal of contaminants and recovery of resources* (pp. 73−96). India: Elsevier Available from. Available from https://www.sciencedirect.com/book/9780323956840, 10.1016/B978-0-323-95684-0.00009-9.

Samanta, N. S., Banerjee, S., Mondal, P., Anweshan., Bora, U., & Purkait, M. K. (2021). Preparation and characterization of zeolite from waste Linz-Donawitz (LD) process slag of steel industry for removal of $Fe^{3+}$ from drinking water. *Advanced Powder Technology*, 32 (9), 3372−3387. Available from https://doi.org/10.1016/j.apt.2021.07.023. Available from: http://www.elsevier.com.

Samanta, N. S., Das, P. P., Sharma, M., & Purkait, M. K. (2023). *Recycle of water treatment plant sludge and its utilization for wastewater treatment. Resource recovery in drinking water treatment* (pp. 239−264). Elsevier BV. Available from 10.1016/b978-0-323-99344-9.00010-4.

Samanta, N. S., Das, P. P., Mondal, P., Bora, U., & Purkait, M. K. (2022). Physico-chemical and adsorption study of hydrothermally treated zeolite A and FAU-type zeolite X prepared from LD (Linz−Donawitz) slag of the steel industry. *International Journal of Environmental Analytical Chemistry*, *10290397*. Available from https://doi.org/10.1080/03067319.2022.2079082. Available from, http://www.tandf.co.uk/journals/titles/03067319.asp.

Samanta, N. S., Das, P. P., Mondal, P., Changmai, M., & Purkait, M. K. (2022). Critical review on the synthesis and advancement of industrial and biomass waste-based zeolites and their applications in gas adsorption and biomedical studies. *Journal of the Indian Chemical Society*, *99*(11). Available from https://doi.org/10.1016/j.jics.2022.100761, ISSN: 0019-4522. Available from, https://www.sciencedirect.com/journal/journal-of-the-indian-chemical-society.

Sharma, M., Das, P. P., Sood, T., Chakraborty, A., & Purkait, M. K. (2021). Ameliorated polyvinylidene fluoride based proton exchange membrane impregnated with graphene oxide, and cellulose acetate obtained from sugarcane bagasse for application in microbial fuel cell. *Journal of Environmental Chemical Engineering*, *9*, 106681. Available from https://doi.org/10.1016/j.jece.2021.106681.

Sharma, M., Das, P. P., Sood, T., Chakraborty, A., & Purkait, M. K. (2022a). Clean energy from salinity gradients using pressure retarded osmosis and reverse electrodialysis: A review. *Sustainable Energy Technologies and Assessments*, *49*, 101687. Available from https://doi.org/10.1016/j.seta.2021.101687.

Sharma, M., Das, P. P., Sood, T., Chakraborty, A., & Purkait, M. K. (2022b). Reduced graphene oxide incorporated polyvinylidene fluoride/cellulose acetate proton exchange membrane for energy extraction using microbial fuel cells. *Journal of Electroanalytical Chemistry*, *907*, 115890. Available from https://doi.org/10.1016/j.jelechem.2021.115890.

Shekhar Samanta, N., Das, P. P., Dhara, S., & Purkait, M. K. (2023). An overview of precious metal recovery from steel industry slag: Recovery strategy and utilization. *Industrial and Engineering Chemistry Research*, *62*(23), 9006−9031. Available from https://doi.org/10.1021/acs.iecr.3c00604, http://pubs.acs.org/journal/iecred.

Shi, C. (2002). Characteristics and cementitious properties of ladle slag fines from steel production. *Cement and Concrete Research*, *32*(3), 459−462. Available from https://doi.org/10.1016/S0008-8846(01)00707-4.

Sontakke, A. D., Das, P. P., Mondal, P., & Purkait, M. K. (2021). Thin-film composite nanofiltration hollow fiber membranes toward textile industry effluent treatment and environmental remediation applications: Review. *Emergent Materials*, *5*, 1409−1427. Available from https://doi.org/10.1007/s42247-021-00261-y.

Sugano, Y., Sahara, R., Murakami, T., Narushima, T., Iguchi, Y., & Ouchi, C. (2005). Hydrothermal synthesis of zeolite a using blast furnace slag. *ISIJ International*, *45*(6), 937−945. Available from https://doi.org/10.2355/isijinternational.45.937.

Wajima, T. (2014). Synthesis of zeolite from blast furnace slag using alkali fusion with addition of EDTA. *Advanced Materials Research*, *1044−1045*, 124−127. Available from https://doi.org/10.4028/www.scientific.net/AMR.1044-1045.124.

Zhang, X., Wei, G., & Yu, F. (2005). Thermal radiative properties of xonotlite insulation material. *Journal of Thermal Science*, *14*(3), 281−284. Available from https://doi.org/10.1007/s11630-005-0015-1.

# Chapter 4

# Synthesis of zeolite from coal fly ash

## 4.1 Introduction

Over the last three decades, the global energy mix has predominantly consisted of nuclear, hydro, wind, solar, biofuels, waste, natural gas combustion, coal combustion, and oil combustion. Although there is growing attention directed toward cleaner energy sources such as nuclear, hydro, wind, solar, biofuels, and waste due to environmental concerns, fossil fuels, namely oil, natural gas, and coal, still make up approximately 80% of the global energy supply. This is primarily because of their high-energy content and cost-effectiveness, suggesting that the consumption of fossil fuels will remain substantial for the foreseeable future. Among fossil fuels, coal, as the second most utilized source, has exhibited a relatively higher growth rate, particularly since 2002. Despite a slight dip in 2015, it is anticipated that coal consumption will inevitably increase in accordance with current trends. Fig. 4.1 illustrates global and regional projections for coal consumption, indicating a steady upward trajectory in coal consumption over the next 3 years.

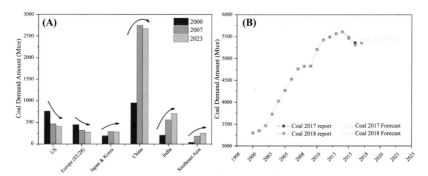

FIGURE 4.1 **Forecasts for coal demand amount: (A) in both 2017 and 2018 based on the reports; (B) in different countries or regions** . A scenario of how coal demand rises over years in different countries. *From Ju, T., Meng, Y., Han, S., Lin, L., & Jiang, J. (2021). On the state of the art of crystalline structure reconstruction of coal fly ash: A focus on zeolites.* Chemosphere, *283, 131010. https://doi.org/10.1016/j.chemosphere.2021.131010.*

Waste-based Zeolite. DOI: https://doi.org/10.1016/B978-0-443-22316-7.00004-8
**85**

Significant quantities of coal fly ash (CFA) result from the process of burning coal. It is approximated that, in coal-fired power facilities, for every 2 tons of coal consumed, 1 ton of CFA is generated (Ltd, 2019). Managing and disposing of this substantial volume of CFA is a challenging task, as it occupies a considerable amount of land space. Furthermore, due to the minuscule size of the particles, CFA can easily disperse into the atmosphere, water bodies, and potentially into human lungs, leading to severe air and water pollution concerns (Mondal, Samanta, Kumar, & Purkait, 2020; Purkait, Sinha, Mondal, & Singh, 2018a, 2018b; Sun et al., 2019) and posing risks to human health (Das, Anweshan, & Mondal, 2021; Das, Anweshan, & Purkait, 2021; Das, Mondal, & Anweshan, 2021; Das, Sharma, & Purkait, 2022).

To provide a clear understanding of the concepts related to coal ash, including coal ash (CA), CFA, coal bottom ash (CBA), and fly ash (FA), let us break them down:

- CA serves as a collective term for the entire residue remaining after coal combustion. Within CA, the lighter component is known as CFA, typically captured by a dust collector from exhaust gases. Conversely, the heavier portion is referred to as CBA, which is collected from the bottom of the combustion chamber (Samanta, Das, Mondal, Bora, & Purkait, 2022; Samanta, Das, Mondal, Changmai, & Purkait, 2022).
- FA, on the other hand, encompasses fly ash generated from various combustion systems, including biomass, municipal solid waste, and coal. It is worth noting that the composition and properties of FA may differ from CA. However, when it comes to coal combustion, FA and CFA are essentially the same (Das, Dhara, & Purkait, 2023, 2024).
- In specific situations, such as fluidized bed combustion, CFA is designated as fluidized bed combustion ash (FBC ash). This implies that, in a narrower sense, CFA pertains to fly ash produced from pulverized coal furnaces, which is the most common type and is also known as pulverized coal ash (PC ash) or ordinary fly ash (OF ash in practical terms).

Furthermore, European Coal Combustion Products Association (ECOBA) defines CFA as fly ash obtained through electrostatic or mechanical precipitation of dust-like particles from the flue gas of coal or lignite-fired furnaces operating at temperatures between 1100°C and 1400°C, which is often referred to as pulverized fuel ash (PFA).

There are primarily four types of furnaces utilized for coal combustion: grate combustion furnace, fluidized bed furnace, pulverized fuel furnace, and cyclone furnace (Chakraborty, Gautam, Das, & Hazarika, 2019; Changmai & Das, 2022). The choice of furnace type is typically determined based on the characteristics of the coal being used and other relevant factors. In the coal-burning industry, pulverized fuel furnaces are the preferred choice, especially in cases where combustion temperatures are high, often exceeding 1400°C.

In these furnaces, coal is first ground into powder with particles smaller than 100 μm in a coal mill before being introduced into the combustion chamber. Subsequently, a mixture of coal powder and air is injected into the chamber to ensure more thorough combustion (Kleinhans, Wieland, Frandsen, & Spliethoff, 2018). Pulverized fuel furnaces are favored due to their ability to capture approximately 80% of CA in collectors, resulting in the formation of CFA (Palmer, 2015). As a result, they find extensive practical use. Once collected, the CFA can be transported for storage or further applications. One of the most significant strategies for reducing CFA waste is to incorporate it into construction materials, leveraging its pozzolanic properties, as depicted in Fig. 4.2 (Messina, Ferone, Colangelo, Roviello, & Cioffi, 2018).

CFA finds applications in various fields, such as serving as sorbents (Jin et al., 2018; Samanta, Anweshan, Mondal, Bora, & Purkait, 2023; Shekhar Samanta, Das, Dhara, & Purkait, 2023), acting as a catalyst foundation (Park, Dattatraya Saratale, Cho, & Bae, 2020), and contributing to soil remediation efforts (Song, Lin, & Takahashi, 2020). A recent discovery highlights the potential of injecting CFA back into furnaces, achieving a dual purpose of reducing heavy metal emissions, and minimizing CFA production (Samanta et al., 2021; Wang, Zhang, Wang, Xu, & Pan, 2020). As the analysis of CFA advances and CFA storage increases, there is a growing demand for more intricate utilization, particularly involving the reconstruction of its crystalline structure. The concept of transforming coal fly ash into zeolite was initially mentioned by Anner, Nydegger, Schmocker, Lambert, and Miescher (1972), while Höller and Wirsching first described the formation of zeolite from CFA in 1985 (Holler & Wirsching, 1985). Over the past few decades, numerous researchers have reported optimal reaction conditions and improved methods for this transformation (Collins, Rozhkovskaya, Outram, & Millar, 2020). During these years, a variety of zeolite types, including A, Na-X, Na-Y, Na-P, HS, ZSM-5, W, ZK-5, and Analcimes (Supelano et al., 2020), have been successfully synthesized from CFA.

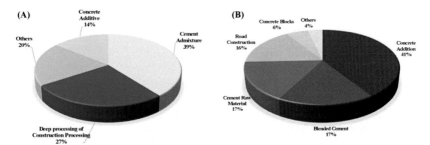

**FIGURE 4.2  Current main applications of CFA: (A) in China; (B) in Europe** . Various main applications of Coal Fly Ash in China and Europe. *From Ju, T., Meng, Y., Han, S., Lin, L., & Jiang, J. (2021). On the state of the art of crystalline structure reconstruction of coal fly ash: A focus on zeolites.* Chemosphere, 283, *131010. https://doi.org/10.1016/j.chemosphere.2021.131010.*

## 4.2 Classification of coal fly ash

Coal originates from prehistoric plant matter, progressing through stages from plant material to peat, then to lignite, followed by bituminous coal, and finally, reaching anthracite, also known as "blind coal." This evolutionary sequence is termed the degree of coalification. Each stage of coal exhibits distinct combustion characteristics, leading to variations in the properties of CFA. The diverse nature of CFA attributes itself to discrepancies in coal types and the conditions under which they are burned, resulting in a spectrum of characteristics, as illustrated in Table 4.1.

American Society of Testing Materials (ASTM) has established a standard for CFA based on the characteristics of the coal used, with the primary objective of providing a basis for its application in concrete production. According to this standard, CFA is categorized into two classes: Class F, which possesses pozzolanic properties and is primarily derived from the combustion of anthracite (blind coal) or bituminous coal, and Class C, which lacks both pozzolanic and cementing properties and is obtained from the combustion of lignite or subbituminous coal. Notably, some Class C CFAs contain over 10% CaO content (ASTM C618, 2015). Several scholars have put forth similar classification methods, such as the McCarthy method (McCarthy, Zheng, Dhir, & Tella, 2018) and the Majko method. Due to the natural variations in coal composition across the globe, unique types of fly ash have emerged. For instance, China has defined a distinctive category

**TABLE 4.1 Compositions of CFA from different kinds of coal.**

| Composition (%) | Bituminous | Subbituminous | Lignite | Anthracite |
|---|---|---|---|---|
| $SiO_2$ | 20−60 | 40−60 | 15−45 | 28−57 |
| $Al_2O_3$ | 5−35 | 20−30 | 10−25 | 18−36 |
| $Fe_2O_3$ | 10−40 | 4−10 | 4−15 | 3−16 |
| CaO | 1−12 | 5−30 | 15−40 | 1−27 |
| MgO | 0−5 | 1−6 | 3−10 | 1−4 |
| $SO_3$ | 0−4 | 0−2 | 0−10 | 0−9 |
| $Na_2O$ | 0−4 | 0−2 | 0−6 | 0−1 |
| $K_2O$ | 0−3 | 0−4 | 0−4 | 0−4 |
| LOI (loss on ignition) | 0−15 | 0−3 | 0−5 | 1−8 |

Source: From Ju, T., Meng, Y., Han, S., Lin, L., & Jiang, J. (2021). On the state of the art of crystalline structure reconstruction of coal fly ash: A focus on zeolites. Chemosphere, 283, 131010. https://doi.org/10.1016/j.chemosphere.2021.131010.

known as high-aluminum fly ash, which is remarkably rich in alumina, with levels reaching up to 50%. This classification framework provides valuable insights into the composition of CFA.

## 4.3 Surface morphology

The formation of CFA can be segmented into three distinct phases: the primary constituents evolve from combustion remnants to porous carbon, then transition to porous glass, and ultimately solidify into glass beads. Consequently, CFA particles comprised combustion residues, porous carbon, porous glass, and glass beads, which manifest in various forms such as spherical particles, irregularly shaped agglomerates, and porous structures. This macroscopic appearance of CFA is depicted in Fig. 4.3, where it takes the form of fine powders.

As depicted in the scanning electron microscopy (SEM) images presented in Fig. 4.4, it is evident that glass beads in CFA predominantly take the form of cenospheres, and they play a crucial role in shaping the composition of CFA. Cenospheres are essentially hollow spheres that form during the combustion process, characterized by wall thickness-to-diameter ratios ranging between 2.5% and 10.5% (Ngu, Wu, & Zhang, 2007).

Research into cenospheres has revealed a close relationship between their formation and factors such as the Si/Al ratio and cooling rate. Specifically, a rapid cooling rate during coal combustion has been found to lead to the fragmentation of cenospheres, resulting in smaller CFA particles (Wu & Li, 2019). Furthermore, high $SiO_2/Al_2O_3$ ratios, coupled with low levels of $TiO_2$ and $Fe_2O_3$, also contribute to the production of smaller cenospheres (Li, Gao, & Wu, 2013). Smaller cenospheres tend to exhibit a predominantly single-ring structure, whereas larger cenospheres typically display a network structure. The formation of cenospheres involves a process of structural reconstruction during the creation of CFA. The distribution and structure of pores also play a significant role in this context. The initial CFA samples reveal a microporous structure, as illustrated in Fig. 4.5, characterized by an average pore diameter of 2 nm or less. This structure presents limitations for the application of CFA as porous materials for sorption and catalysis. Upon calcination, the pore structure undergoes a collapse, transforming into a

**FIGURE 4.3 Apparent raw and calcined CFA** . Physical observation of raw CFA and calcined CFA. *From Ju, T., Meng, Y., Han, S., Lin, L., & Jiang, J. (2021). On the state of the art of crystalline structure reconstruction of coal fly ash: A focus on zeolites.* Chemosphere, 283, *131010. https://doi.org/10.1016/j.chemosphere.2021.131010.*

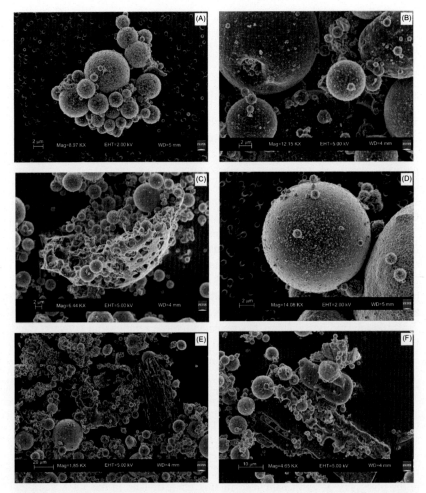

**FIGURE 4.4** **SEM images of coal fly ash.** Morphological structure of raw coal fly ash. *From Belviso, C. (2018). State-of-the-art applications of fly ash from coal and biomass: A focus on zeolite synthesis processes and issues.* Progress in Energy and Combustion Science, 65, *109−135 (Belviso, 2018a). https://doi.org/10.1016/j.pecs.2017.10.004.*

nonporous configuration. However, the application of acid or alkali treatment may lead to an expansion of the pore structure. A typical CFA composition includes mullite, quartz, corundum, and amorphous silica, and these components remain unchanged after the calcination process.

## 4.4 The crystalline structure of coal fly ash

The visible morphology of CFA is intricately linked to its crystalline structure, which varies in accordance with the mineral phase composition of the

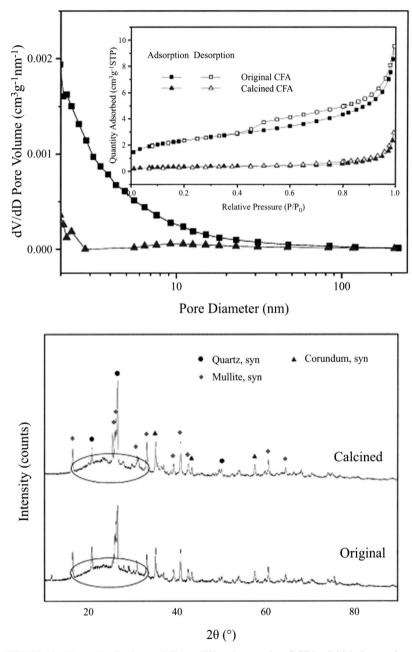

**FIGURE 4.5 Pore distribution and X-ray diffraction results of CFA**. Initial characterization of CFA in terms of pore size distribution and mineral phases present. *From Ju, T., Jiang, J., Meng, Y., Yan, F., Xu, Y., Gao, Y., & Aihemaiti, A. (2020). An investigation of the effect of ultrasonic waves on the efficiency of silicon extraction from coal fly ash.* Ultrasonics Sonochemistry, *60, 104765. https://doi.org/10.1016/j.ultsonch.2019.104765.*

CFA. Spherical particles comprise aluminosilicates with magnetite crystals situated on their surfaces, while irregular agglomerates primarily consist of aluminosilicates. The porous structures observed within CFA are typically composed of carbon particles (Valeev, Kunilova, Alpatov, Varnavskaya, & Ju, 2019).

Zeolites and other porous materials like M41S belong to the category of crystalline aluminosilicates with a framework structure featuring tetrahedral units $[SiO_4]^{4-}$ and $[AlO_4]^{5-}$ interconnected through the sharing of oxygen atoms (Visa, 2016). Hence, the transformation of CFA into porous materials is not only reasonable but also potentially feasible. Various types of CFA share common crystalline phases, including mullite, quartz, corundum, and amorphous components, albeit in varying proportions (Wang, Dai, Zou, French, & Graham, 2019). As these crystals are primarily composed of silicon and aluminum, alkali solutions are commonly used to disrupt their structure. Mullite and quartz, due to their strong internal bonds, are less soluble, while amorphous content is more readily dissolved.

CFA contains a variety of elements, effectively rendering it an urban mine. The elemental characteristics of CFA are of significance in the context of its recycling as an urban resource. From an elemental perspective, Liu et al. have explored the behavior of rare earth elements during the combustion process (Liu et al., 2020). Numerous studies have been conducted to recover major and trace elements, as well as alumina and rare earth elements present in CFA (Hailu, McCrindle, Seopela, & Combrinck, 2019; Pan et al., 2020). Subsequent operations following elemental extraction have also been undertaken. Yan et al. prepared nano-$Al_2O_3$ and nanosilica using extracted aluminum and silicon from CFA (Yan et al., 2017), while Panek et al. synthesized MCM-41 from CFA (Panek et al., 2017).

Crystalline structural alterations occur during the process of elemental extraction. Research has investigated the formation of new crystalline phases when extracting silicon from CFA (Ju et al., 2020). However, it is important to note that these changes in crystalline structure are not the root cause of element extraction, but rather a concurrent part of the process. This study, in particular, concentrates on the processes and products directly resulting from the reconstruction of crystalline structure.

## 4.5 Zeolite synthesis from coal fly ash utilizing various techniques

### 4.5.1 Traditional hydrothermal method

The direct hydrothermal approach to transform CFA into zeolite involves several key steps. First, CFA is dissolved in an alkaline solution, typically using substances like NaOH or KOH, to extract aluminate and silicate components (Das, Sontakke, & Purkait, 2023; Dhara, Das, Uppaluri, & Purkait, 2023).

Subsequently, a heat treatment is applied to generate zeolite crystals from the resulting mixture. Adjustments to the Si/Al ratio can be made by introducing aluminate or silicate compounds to yield the desired type of zeolite. The specific type of zeolite produced, such as zeolite-X, zeolite-A, zeolite-P, analcime, chabazite, or hydrosodalite (Murayama, Takahashi, Shuku, Lee, & Shibata, 2008; Park et al., 2020; Sakthivel, Reid, Goldstein, Hench, & Seal, 2013), can vary depending on various reaction conditions including temperature, pressure, reaction duration, Si/Al ratio, alkaline solution used, pH levels, and the presence of seeding agents.

In Wirsching's study, volcanic glass was employed in the hydrothermal treatment within an alkaline solution at different temperatures (180°C, 200°C, and 250°C) to produce chabazite, phillipsite, and analcime (Wirsching, 1975). This experiment demonstrated that zeolites could be generated from raw materials rich in alumina and silica. Following this, Holler and Wirsching (1985) utilized a similar approach to treat CFA for zeolite production, marking the beginning of a continued scientific exploration into CFA conversion to zeolite (Holler & Wirsching, 1985).

Another significant study by Hollman, Steenbruggen, and Janssen-Jurkovičová (1999) involved the direct hydrothermal synthesis of zeolite Na-P1. This process involved mixing 500 g of fly ash with a 2-M NaOH solution in a volume of 1.25 $dm^3$, followed by 96 hours of heat treatment at 90°C. Samples were periodically drawn from the reaction mixture for analysis. During the conversion process, the silicon concentration peaked at 10,000 mg/$dm^3$ before stabilizing at 6800 mg/$dm^3$. X-ray diffraction (XRD) results indicated that this maximum Si concentration aligned with the commencement of crystallization (Hollman et al., 1999). Aluminum concentration, on the other hand, reached its peak at the onset of the reaction and diminished nearly to zero within the first hour of crystallization. This phenomenon is attributed to an initial induction period during which fly ash particles dissolved, releasing their Al and Si contents. Zeolite nuclei formed on the surface of these particles, and crystallization began when these nuclei reached a specific minimum size.

The process was extended to a two-step approach, where the fly ash was first mixed with a 2-M sodium hydroxide solution and incubated at 90°C for 6 hours. After filtration, the solid residue was dried, and the filtrate was adjusted to achieve a Si/Al molar ratio of 0.8−2. The adjusted filtrate was then incubated at 90°C for 48 hours, after which the resulting mixture was filtered, and the residues were dried. The new filtrate was added to the fly ash residue from the initial 6-hour incubation, followed by 24-hour incubation at 90°C, filtration, and drying. Alternatively, a one-step direct hydrothermal synthesis was carried out by treating the same reaction mixture for 96 hours at 90°C. Numerous studies have explored the conversion of CFA into zeolite using conventional heating methods. Fig. 4.6 shows the various transformation of raw CFA after it converts to zeolite Na-P1, and zeolite Na-X through one-step and two-step conversion.

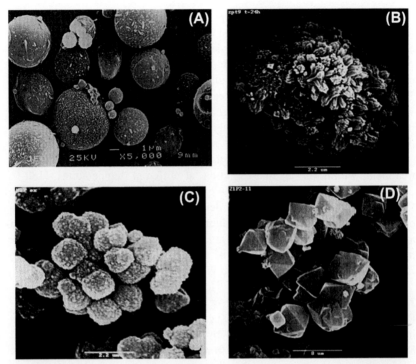

**FIGURE 4.6** SEM images of (A) raw CFA, (B) zeolite Na-P1 obtained by a one-step conversion process, (C) zeolite Na-P1 obtained by a two-step conversion process, and (D) zeolite Na-X obtained by a two-step conversion process . Morphological structures of raw CFA and zeolite Na-P1 obtained after one-step and two-step conversion process along with zeolite Na-X structural variation obtained after two-step conversion process. *From Hollman, G. G., Steenbruggen, G., & Janssen-Jurkovičová, M. (1999). Two-step process for the synthesis of zeolites from coal fly ash. Fuel, 78(10), 1225–1230. https://doi.org/10.1016/S0016-2361(99) 00030-7.*

## 4.5.2 Microwave assistance technique

Microwave technology harnesses the characteristics of high-frequency alternating electric fields, offering significant advantages in energy conversion and transmission (Dhara, Das, Uppaluri, & Purkait, 2023). When subjected to external electric fields, polar molecules undergo directional polarization, while nonpolar molecules experience displacement polarization. Consequently, microwave radiation uniformly distributes electromagnetic energy, converting it into heat and facilitating the restructuring of CFA, as observed in the study by Yanguang et al. (2015). Much like ultrasound, microwave heating enhances dissolution and crystallization processes while inhibiting nucleation. This unique behavior leads to the formation of irregular sodalite with large crystalline grains when microwave heating is applied independently. Conversely, when a conventional one-step hydrothermal heating process is employed, a mixture of zeolite 4A and sodalite is obtained. For the highest purity zeolite 4A, a combination of microwave and hydrothermal heating is the preferred method, as demonstrated by Yanguang et al. (2015).

In contemporary applications, microwave technology has been expanded as a viable hydrothermal process for synthesizing fine oxide powders, as indicated by Sōmiya and Roy (2000). The use of microwave-assisted hydrothermal processes has proven effective in zeolite synthesis from silica-alumina gel precursors, as exemplified by Jansen, Arafat, Barakat, and Bekkum (1992). Additionally, it has been reported that microwave heating can expedite the zeolitization of CFA, reducing reaction times, as discussed in Querol et al.'s work from 1997 (Querol et al., 1997).

Microwave energy is directly absorbed by the water solvent, allowing for rapid heating in comparison to traditional heating methods (Bharti, Das, & Purkait, 2023; Dhara, Samanta, & Das, 2023). The electromagnetic waves induce rotation of water dipoles, potentially activating water molecules by breaking hydrogen bonds (Das, Deepti, & Purkait, 2023; Das, Dhara, & Purkait, 2023). This capability makes short-duration zeolitization a preferable choice for efficient industrial production, as prolonged reactions consume more energy. A two-step approach, beginning with premicrowave treatment followed by conventional heating, has been found to be effective in enhancing the formation of zeolite from CFA, as demonstrated in Inada et al.'s study (Inada, Tsujimoto, Eguchi, Enomoto, & Hojo, 2005). In their report, Inada et al. explored the impact of microwave irradiation on zeolite formation, with a particular focus on changes in zeolite yield throughout the reaction. The experiment involved mixing fly ash with a 2-M NaOH solution and subjecting it to heating using either an oil bath or a microwave for a duration of 2 hours. Interestingly, while zeolite Na-P1 formed as a result of the conventional oil bath treatment, no zeolitic product was obtained when employing microwave heating alone. However, when microwave irradiation was applied during the hydrothermal treatment, it notably accelerated zeolitization during the initial stages of the process. This acceleration was attributed to the enhanced dissolution of $SiO_2$ and $Al_2O_3$ from CFA due to microwave stimulation (Sharma & Das, 2023; Sontakke & Das, 2021).

On the contrary, microwave irradiation during the middle to later stages of the reaction had a retarding effect on zeolite crystallization. Therefore it was concluded that microwave heating is an effective method for expediting zeolite production from CFA within a shorter timeframe, provided that the irradiation schedule is carefully controlled to emphasize the early stages of the process. The composition of the raw CFA utilized is provided in Table 4.2.

Fig. 4.7 shows the SEM images of the surface of CFA particles. Initially, the fly ash particles exhibited a smooth surface due to the presence of the

**TABLE 4.2** Chemical and mineralogical compositions of coal fly ash (wt.%).

| Chemical composition | | | | | Mineralogical composition | | | |
|---|---|---|---|---|---|---|---|---|
| | | | | | Crystal phase | | Glass phase | |
| $SiO_2$ | $Al_2O_3$ | CaO | $Fe_2O_3$ | Others | Quartz | Mullite | $SiO_2$ | $Al_2O_3$ |
| 38.3 | 34.8 | 11.0 | 8.1 | 7.8 | 2.2 | 27.2 | 28.4 | 15.3 |

*Source:* From Inada, M., Tsujimoto, H., Eguchi, Y., Enomoto, N., & Hojo, J. (2005). Microwave-assisted zeolite synthesis from coal fly ash in hydrothermal process. *Fuel, 84*(12–13), 1482–1486. https://doi.org/10.1016/j.fuel.2005.02.002.

glass phase encapsulating them. However, following the traditional heating process, granular substances, presumably zeolite Na-P1, were noted adhering to the particle surfaces. This indicates that the glass phase dissolved in the alkaline solution and then precipitated as zeolite onto the particles. Similarly, during microwave heating, particulate deposits were observed on the fly ash particle surfaces, even though XRD analysis did not detect any zeolite products.

Continuous microwave irradiation delays the formation of zeolite in its crystalline structure. The resulting product is believed to be an amorphous aluminosilicate gel, indicating that microwave heating hinders the crystallization of zeolite from the intermediate gel. Partial microwave irradiation yields various effects during the conventional heating process. Early-stage microwave heating, for instance, promotes zeolite formation, attributed to the accelerated dissolution of $SiO_2$ and $Al_2O_3$ from CFA. This is facilitated by the rapid heating rate and possibly the robust interaction of active water molecules with the glass phase. Conversely, microwave heating during the middle stage significantly retards zeolite formation due to the inhibition of nucleation within the intermediate aluminosilicate gel. Therefore the combination of early-stage microwave heating followed by conventional heating proves effective in enhancing the zeolitization of CFA.

In a similar vein, CFA was transformed into zeolite (CFAZP) under experimental conditions conducted at atmospheric pressure. This process involved an initial six-hour conventional hydrothermal heating phase at a relatively low temperature ($90 \pm 3°C$), followed by a subsequent 30-minute microwave irradiation step, as described in Aldahri et al.'s work (Aldahri, Behin, Kazemian, & Rohani, 2017). The produced materials were subjected to characterization using various techniques, including XRD, thermogravimetric analysis (TGA) and differential thermal analysis (DTA), SEM, particle size distribution (PSD), Brunauer–Emmett–Teller (BET) analysis, and cation-exchange capacity (CEC) measurements.

The impact of microwave irradiation on the crystal growth of nucleated CFAZP was examined at different solid/liquid ratios, which represent the

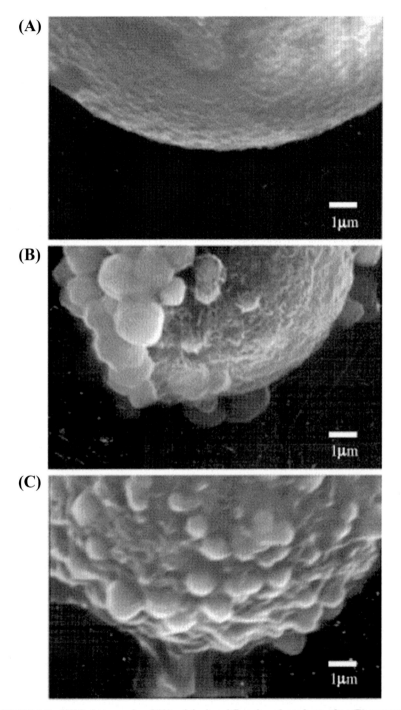

**FIGURE 4.7** **SEM photographs of (A) original coal fly ash and products after (B) conventional hydrothermal treatment and (C) microwave heating** . Morphological structure of raw CFA and CFA after hydrothermal treatment (conventional) and microwave treatment. *From Inada, M., Tsujimoto, H., Eguchi, Y., Enomoto, N., & Hojo, J. (2005). Microwave-assisted zeolite synthesis from coal fly ash in hydrothermal process.* Fuel, 84(12–13), 1482–1486. https:// doi.org/10.1016/j.fuel.2005.02.002.

suspended CFA mass relative to the volume of NaOH solution (given in grams per milliliter, g/mL). To study this, a three-variable, three-level central composite statistical experimental design was employed to explore how independent variables influenced the response function. This function was defined as the ratio of the characteristic peak intensity at 2θ: 28 degrees of a sample in comparison to the same peak of a sample subjected to a 24-hour conventional heating process. Under optimum experimental conditions, an impressive relative peak intensity of CFAZP, reaching as high as 97%, was attained. These conditions included a NaOH concentration of 1 M, a 6-hour period of conventional heating followed by 30 minutes of microwave irradiation, and a solid/liquid ratio of 0.40 g/mL. It was noted that, with a constant microwave energy input, higher solid/liquid ratios led to an increased relative peak intensity of the product.

Fig. 4.8A illustrates SEM micrographs depicting both the original CFA and the zeolitized CFA, prepared under optimal conditions (i.e., M: 1; solid/liquid ratio: 0.40 g/mL; MW irradiation: 30 minutes), which yielded the highest relative peak intensity of 97%. These images reveal that the CFA utilized in this study originally had a quasispherical shape with a smooth surface. Notable transformations are observed on the surface of the CFA during the zeolitization process.

In Fig. 4.8B, we observe the surface of the CFA following a 6-hour hydrothermal treatment, showing the initial formation of zeolite particles on the CFA's surface, partially altering the morphology of the fly ash spheres. Subsequently, in Fig. 4.8C, the micrographs depict Na-P zeolite particles completely covering the entire surface of the CFA, which was achieved by applying microwave (MW) irradiation after the conventional heating process. The incorporation of MW heating significantly accelerates the crystal growth rate of zeolite CFAZP. The micrographs of CFAZP reveal a uniform distribution of zeolite particles and the presence of small plate-like crystallites, which exhibit a pseudospherical form. Zeolite CFAZP can be identified on the surface of the pseudospherical CFA forms as small plates, forming polycrystals with an average size of approximately 1–1.5 μm.

### 4.5.3 Ultrasound energy technique

Reconstructing CFA requires a substantial amount of energy to activate structural changes and facilitate the transformation process. Ultrasound emerges as a compelling option due to its high-energy density and additional features, such as mechanical stirring and cavitation, which reduce reaction temperatures and shorten reaction times. Ultrasound finds application in three main contexts:

- **Enhancing dissolution:** The first application involves using ultrasound as a supplementary pretreatment, directly accelerating the dissolution

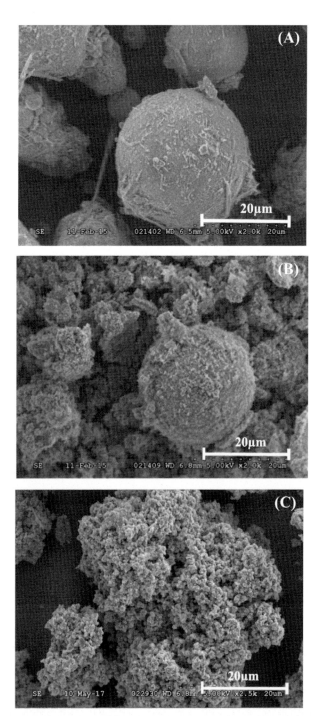

**FIGURE 4.8  SEM images of the CFA and microwave-assisted synthesized zeolites samples after 30 min irradiation: (A) raw CFA, (B) CFAZP synthesized hydrothermally for 6 h, and (C) CFAZP synthesized hydrothermally for 6 h followed by 30 min microwave irradiation.** Various structural variations of raw CFA after only hydrothermal treatment and hydrothermal-assisted microwave irradiation treatment. *From Aldahri, T., Behin, J., Kazemian, H., & Rohani, S. (2017). Effect of microwave irradiation on crystal growth of zeolitized coal fly ash with different solid/liquid ratios.* Advanced Powder Technology, 28(11), 2865–2874. *https://doi.org/10.1016/j.apt.2017.08.013.*

phase. Studies have shown that ultrasound can effectively replace the prefusion step, reducing pretreatment time from 1.5 hours to just 10 minutes, as demonstrated by Ojumu, Du Plessis, and Petrik (2016).

- **Boosting crystallization:** The second application employs ultrasound to enhance the crystallization stage. For instance, Aldahri et al.found that zeolite Na-P formed when ultrasound was applied after the hydrothermal process. When comparing preultrasound and postultrasound, it becomes evident that ultrasound can enhance both dissolution and crystallization. Preultrasound accelerates dissolution but does not match the crystallization rate, whereas postultrasound promotes crystallization and speeds up zeolite formation while somewhat inhibiting nucleation.
- **Holistic ultrasound assistance:** The third application involves leveraging ultrasound's advantages across the entire hydrothermal process, injecting energy into dissolution, condensation, nucleation, and crystallization stages. Furthermore, ultrasound has been employed for monitoring the CFA-derived zeolite formation process, as exemplified by Hums et al.

Wang et al. (2019) utilized CFA sourced from a Western Australian coal power plant with a Si/Al ratio of 1.9. Their experiments involved subjecting the CFA to an alkaline solution with varying NaOH concentrations ranging from 1 to 5 M, along with a solid-to-liquid ratio of 0.5 g/cm$^3$. The slurry mixture was then subjected to sonication at room temperature for different durations, ranging from 30 minutes to 2 hours, in an ultrasonic bath with a power output of 300 W. XRD analysis of both the sonicated and the unsonicated alkaline CFA slurry did not reveal any phase changes. However, it was observed that the sonicated alkaline-treated CFA exhibited a significantly larger surface area compared to raw CFA, measuring at 35.4 m$^2$/g as opposed to 5.6 m$^2$/g. Moreover, the ultrasonically treated CFA demonstrated superior performance in adsorbing methylene blue compared to untreated CFA, with values of $1.5 \times 10^{-5}$ mol/g and $8.0 \times 10^{-6}$ mol/g, respectively.

Additionally, Belviso (2018a) reported that ultrasound offers a more expeditious approach to enhance the transformation from CFA into zeolite and from faujasite and A-type zeolite into stable sodalite. This suggests that ultrasound expedites the transformation into a more stable state, representing a distinct mechanism from the conventional hydrothermal process. Figs. 4.9 and 4.10 show the variation in morphology of various types of zeolite when treated differently with ultrasound and hydrothermally after 2 hours, respectively.

In a prior investigation carried out by Woolard, Strong, and Erasmus (2002), CFA was acquired from a South African power plant, with a Si/Al ratio of 1.37. The treatment duration was extended to 21 hours, in contrast to the 6-hour duration, and a more concentrated alkaline solution of 7 M was employed, compared to the 5 M solution. Importantly, no ultrasound energy was utilized in this study. The outcome revealed a transformation of CFA

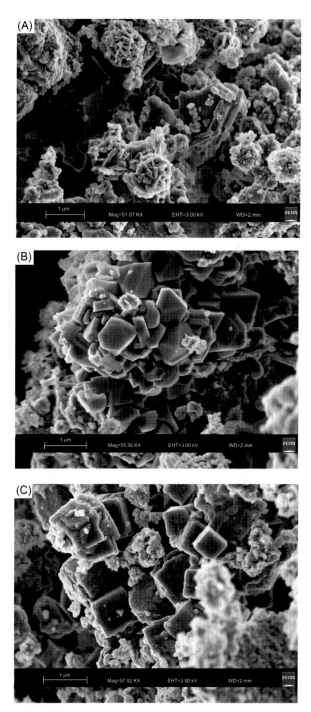

**FIGURE 4.9** **SEM images of: (A) sodalite, (B) faujasite, and (C) A-type zeolite formed after 2 h of ultrasonic treatment** . Morphological structures of sodalite, faujasite and zeolite-A after ultrasonic treatment for 2 h. *From Belviso, C. (2018). Ultrasonic vs hydrothermal method: Different approaches to convert fly ash into zeolite. How they affect the stability of synthetic products over time?* Ultrasonics Sonochemistry, 43, 9–14 (Belviso, 2018b). *https://doi.org/ 10.1016/j.ultsonch.2017.12.050.*

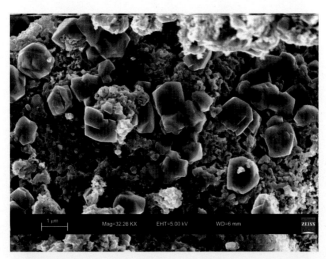

**FIGURE 4.10    SEM image of zeolites (faujasite and A-type zeolite) formed after 2 h of hydrothermal process** . Morphological structures of zeolites after 2 h of hydrothermal process. *From Belviso, C. (2018). Ultrasonic vs hydrothermal method: Different approaches to convert fly ash into zeolite. How they affect the stability of synthetic products over time?* Ultrasonics Sonochemistry, 43, 9–14 (Belviso, 2018b). https://doi.org/10.1016/j.ultsonch.2017.12.050.

into hydroxysodalite. Furthermore, the surface area of the untreated CFA increased from 0.9 to 7.4 m$^2$/g after the introduction of ultrasound treatment. Notably, both sets of results demonstrated comparable adsorption capacities while significantly reducing the treatment duration through the incorporation of ultrasonic energy.

### 4.5.4    Fusion followed by hydrothermal treatment

Shigemoto et al. introduced a preliminary fusion step before the hydrothermal treatment to create zeolites from CFA. In this approach, the initial material was fused with solid-phase NaOH at high temperatures. Subsequently, this solid fused mixture was combined with an aqueous solution and allowed to age. The alkali solution in the fusion and aging steps effectively dissolved all of the mineral phases, including mullite and quartz, originally present in the CFA. This dissolution occurred due to the disruption of the mineral structures, as reported by Wajim et al. The dissolution of these previously insoluble mineral phases introduced a higher concentration of aluminosilicates into the solution, ultimately increasing the conversion rate of the CFA. Furthermore, it was observed that the BET surface area of the crystallized product, when the fusion step was incorporated, increased by a factor of 10 compared to crystallization without fusion, as reported by Sakthivel et al. (2013). Additionally, zeolites produced through the fusion method were found to exhibit a higher CEC and improved crystallinity. It was also noted that the fusion method favored the production of zeolite-X or zeolite-A, particularly in cases of aluminum-rich CFA, as opposed to the nonfusion method, which resulted in a mixture of Na-P1, Na-X, and hydroxysodalite

Bukhari et al. conducted experiments involving the use of microwaves in the hydrothermal treatment process following the fusion step. In their fusion synthesis experiments for producing zeolite Na-A from CFA, they fused 2.18 g of sodium hydroxide granules with 1.82 g of CFA at 550°C for 2 hours. Following this fusion step, 17 cm$^3$ of deionized water was added to the mixture, which was then mechanically stirred for 2 hours at 20°C. Subsequently, 3 cm$^3$ of a sodium aluminate solution with a concentration of 0.155 g/dm$^3$ was introduced to the reaction mixture, and this mixture was subjected to single-mode microwave irradiation (CEM Corp., Discover, USA) for varying durations and power settings under reflux conditions. The resulting product, denoted as CFAZ, was found to be zeolite Na-A with a CEC value of 2.42 meq/g. Furthermore, CFAZ exhibited the ability to immobilize heavy metal cations, including Ba, Cu, Cr, Mn, Ni, Pb, and V, which were originally present in the CFA. The study also revealed that the power and duration of microwave irradiation increased the crystallinity of the CFAZ product. The BET surface area of the final product significantly surpassed that of the raw CFA, measuring at 42.44 m$^2$/g in comparison to the CFA's surface area of 15.47 m$^2$/g. Notably, CFAZ proved effective in immobilizing heavy metal ions present in the original CFA.

Park et al. reported their use of kaolin, an aluminosilicate source, in the production of zeolite Na-A with enhanced crystallinity through the utilization of ultrasonic (US) energy, as opposed to conventional heating methods. The kaolin used had a Si/Al ratio of 1.15, and the molar batch composition of the reaction mixture was $Al_2O_3:1.94SiO_2:4Na_2O:100H_2O$. Ultrasonic treatment in an ultrasound bath operating at a frequency of 47 kHz and a power of 130 W resulted in a relative crystallinity of 68%, in contrast to 0% under the same conditions of 60°C and 4 hours when subjected to conventional treatment. Complete crystallinity, reaching 100%, was achieved through ultrasonic treatment at 70°C and 4 hours, compared to 82% under conventional heating conditions. This underscores the superior outcomes in terms of crystallinity attained through ultrasonic treatment compared to traditional heating methods.

Musyoka et al. successfully transformed CFA into zeolite-A with the assistance of US energy. They employed CFA obtained from a South African power plant, characterized by a Si/Al ratio of 1.65. The initial step involved fusing this CFA with NaOH at a ratio of 1−1.2 at a temperature of 550°C for 1.5 hours. Subsequently, they conducted hydrothermal treatment for varying durations, ranging from 1 to 2 hours, at a temperature of 100°C. Importantly, the introduction of ultrasonic energy during the aging phase for 40 minutes before the hydrothermal synthesis played a pivotal role in reducing the synthesis time required to attain single-phase zeolite-A. It effectively shortened the time from 2 hours to just 1 hour by enhancing the crystallization process.

Belviso et al. (2016) also embarked on similar research, converting CFA acquired from Italian power plants, with varying Si/Al ratios of 1.65, 1.86, and 1.95, into zeolite. Following the same fusion procedure with NaOH at a ratio of 1–1.2 at 550°C for 1 hour, they subjected the fused CFA slurry to 1 hour of sonication treatment in an ultrasonic bath with an output of 240 W. The ultrasonically treated fused CFA slurry was then allowed to incubate for 4 days at different temperatures, ranging from 25°C to 60°C. Notably, the US-treated samples demonstrated the ability to produce zeolite-X at a lower temperature, specifically at 25°C and above, in contrast to the untreated samples, which exhibited no crystallization of zeolite until temperatures exceeded 40°C. The study also reported a phase transition after the US treatment, with the observed phase being hydroxysodalite. However, hydroxysodalite was not observed after a 4-day incubation period for all samples. Interestingly, different CFAs obtained from distinct power plants yielded varying levels of crystallinity under identical reaction conditions, highlighting the significance of the Si/Al ratio of CFA in zeolite production. Furthermore, Belviso et al. ( 2016) conducted experiments using seawater and reached the same conclusion that US treatment reduces the required temperature for crystallization.

### 4.5.5 Reagents addition method

To achieve products with more precise structures, additional reagents are employed, and recent articles highlight four primary functions in this regard:

1. **Providing seed crystals:** One essential function is the provision of seed crystals to facilitate faster nucleus formation. For instance, aluminosilicate gels (with a composition of $15Na_2O:1Al_2O_3:15SiO_2:300H_2O$) were added to the solution as seed crystals, promoting the formation of zeolite Y with a maximum crystallinity of 72% and some zeolite-P as the main impurity, as demonstrated by Wang et al. (2020). Similarly, Na-P1 zeolite was directly introduced as seeds to generate numerous Na-P1 crystals, as illustrated by Cardoso, Paprocki, Ferret, Azevedo, and Pires (2015).

2. **Guiding crystalline structure growth:** Templates play a pivotal role in forming ordered structures. By utilizing macromolecules, the desired structure based on the templates is developed, followed by calcination or ion-exchange processes to eliminate the templates and obtain the target products, as described by Belviso, Cavalcante, Lettino, Ragone, and Belviso (2016).

3. **Adjusting elemental ratios:** Maintaining the appropriate Si/Al ratio is crucial for the growth of specific crystalline structures. However, the

composition of the original CFA varies due to various factors. To address this, materials containing Al or Si are introduced into the system to adjust the Si/Al ratio. Examples include $NaAlO_2$ (Ojumu et al., 2016), Tetraethyl orthosilicate (TEOS), and $Al/Al_2O_3$ powder (Feng et al., 2019). In cases where a zeolite like ZSM-5 requires a high Si/Al ratio, TEOS is commonly added. After the crystalline structures have developed, ion exchange and calcination are employed to remove the template, as demonstrated by Feng et al. (2019).

4. **Expanding product functions:** As reconstruction products derived from CFA find applications across various fields, there is a growing demand for additional functions to enhance performance. For instance, active materials such as $H_2(OESPz)$ (2,3,7,8,12, 13,17,18-octakis(ethylthio)-5,10,15,20−21H,23H-porfirazine) and $Cu(OESPz)$ (2,3,7,8,12,13,17,18-octakis(copper(II))) have been loaded into CFA-derived zeolite for catalytic purposes (Belviso et al., 2016). Additionally, alumina loading on zeolite through the wet-impregnation method using aluminum chloride hexahydrate has been explored to enhance sorption characteristics, as studied by Koshy and Singh (2016).

Wang et al. (2020), in their study, proposed the utilization of two distinct CFA types, categorized as high-calcium (HCFA) and low-calcium (LCFA), as additives in silico-aluminophosphate (SAP) geopolymers with the aim of enhancing early performance. To gain insight into the chemical interactions of CFA within a phosphate-rich environment, the researchers investigated CFA/orthophosphoric acid suspensions employing XRD, SEM, and NMR analyses. The findings unveiled that HCFA had the capacity to act as seeds for active calcium sources. This prompted acid-base and hydrolysis reactions between the alkaline calcium species found in HCFA and the acidic phosphate species present in the activator. This interaction resulted in the formation of calcium phosphate compounds, such as brushite and monetite, which expedited the setting process (with initial and final setting times of 18 and 26 minutes, respectively) and fostered early strength development (1-day and 3-day strengths increased to 15 and 29 MPa, respectively) for the HCFA-SAP geopolymer when compared to a CFA-free SAP geopolymer.

Conversely, the inclusion of LCFA was observed to enhance workability but had a detrimental effect on the setting and strength properties of the LCFA-SAP geopolymers. Fig. 4.11 illustrates the morphology of the raw CFA particles, distinguishing between LCFA particles (A) and HCFA particles (B).

Meanwhile, Fig. 4.12 presents microscopic images of both CFA types immersed in a phosphate solution for 28 days. In Fig. 4.12A, it is evident that phosphate species aggregated on the LCFA, which were originally isolated spherical particles. This suggests the formation of binding phases through geopolymerization between LCFA and phosphate. Furthermore, the phosphate treatment of HCFAs resulted in the generation of interparticle binding phases, as illustrated in Fig. 4.12B.

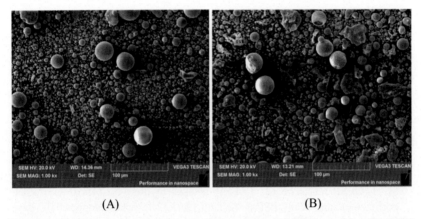

(A)                                    (B)

**FIGURE 4.11 Microscopic images of raw CFA particles: (A) LCFA particles; (B) HCFA particles** . Raw CFA particles morphological structure. *From Wang, Y. S., Alrefaei, Y., & Dai, J. G. (2020). Influence of coal fly ash on the early performance enhancement and formation mechanisms of silico-aluminophosphate geopolymer.* Cement and Concrete Research, 127. *https://doi.org/10.1016/j.cemconres.2019.105932.*

## 4.6 Critical assessment and future perspective

CFA is not only a waste product but also poses environmental hazards, making its disposal a costly burden. With the growing global demand for electricity and the associated increase in CFA production, it is crucial to find ways to turn this financial liability into a valuable resource. One promising avenue is the conversion of CFA into zeolites, which have a wide range of applications, including sorption, catalysis, ion exchange, and the creation of eco-friendly products. This transformation not only leads to the development of environmentally friendly technologies but also makes coal power a cleaner energy source. This shift toward cleaner coal power production can enable regions rich in coal deposits to harness their natural resources for cost-effective and greener energy production.

Numerous studies have explored the utilization of zeolites derived from CFA for various applications. However, CFA typically contains only 40%−60% aluminosilicate components, resulting in zeolites with residual impurities that limit their applications. Nonetheless, these CFA-based zeolites have proven effective for gas adsorption, such as $SO_2$, and can be further functionalized for tasks like oil spill remediation. Moreover, the production of CFA-based zeolites with high CEC is of significant interest due to their ability to remove heavy metal ions from wastewater. Combining microwave energy with conventional heating shows promise in producing CFA-based zeolites with high CEC values, and further exploration of these innovative energy sources is needed to create versatile products for various applications.

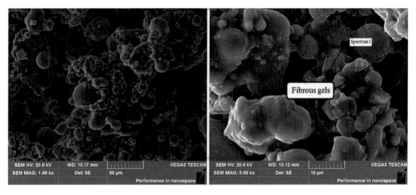

(A) LCFA sediment

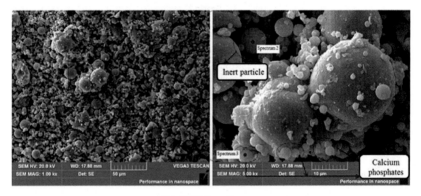

(B) HCFA sediment

**FIGURE 4.12** **Microscopic images of (A) LCFA and (B) HCFA after phosphate treatment for 28 days** . Effect of phosphate solution dipping of LCFA and HCFA for 28 days. *From Wang, Y. S., Alrefaei, Y., & Dai, J. G. (2020). Influence of coal fly ash on the early performance enhancement and formation mechanisms of silico-aluminophosphate geopolymer.* Cement and Concrete Research, 127. *https://doi.org/10.1016/j.cemconres.2019.105932.*

The mechanism of converting CFA into zeolites through hydrothermal reactions is intricate and not yet fully understood. Numerous experiments have explored the impact of different reaction conditions on zeolite crystallization from CFA, including aging of the mixture, temperature, alkalinity, cations in the reaction mixture, and reaction time. These studies have shown that adjusting these parameters can yield various zeolitic materials. However, the fundamental mechanism of zeolitization remains unclear, and it appears to differ for various zeolite types. Therefore further research is needed to delve into the individual mechanisms of different zeolite types and their frameworks.

Building on previous research that has successfully enhanced reactions using microwave and ultrasonic energies, the aim is to conduct additional experiments and research to refine these techniques. Microwave and ultrasound energies have been used to accelerate chemical reactions and improve product quality efficiently, including the production of zeolites from both pure chemicals and CFA. Microwave energy, when combined with conventional heating, has been shown to yield zeolites with enhanced CEC and BET surface areas, while reducing reaction times and producing smaller crystal sizes. Microwave irradiation can also increase the dissolution of amorphous CFA, leading to improved conversion. Ultrasonic energy has similarly reduced reaction times and enhanced the performance of CFA-based zeolites compared to conventional methods. Even in the absence of phase change, ultrasound treatment can boost CFA's adsorption capacity for organic dyes. Further studies are warranted to refine the use of these novel methods for CFA-based zeolite production. Combining both methods for CFA zeolitization is proposed for exploration. Additionally, while most experiments have been conducted on a laboratory scale, scaling up to pilot and industrial levels is essential to realize the potential of utilizing novel energies for zeolite production from abundant CFA worldwide. This transition to larger-scale production may come with challenges that need to be overcome, but the benefits for the environment and the future of greener technology are significant if these novel methods can be perfected and made feasible for industrial use.

# References

Aldahri, T., Behin, J., Kazemian, H., & Rohani, S. (2017). Effect of microwave irradiation on crystal growth of zeolitized coal fly ash with different solid/liquid ratios. *Advanced Powder Technology*, *28*(11), 2865–2874. Available from https://doi.org/10.1016/j.apt.2017.080.013, http://www.elsevier.com.

Anner, R. M., Nydegger, U. E., Schmocker, K., Lambert, P. H., & Miescher, P. A. (1972). Experimental and clinical study of the NBT test. *Schweizerische Medizinische Wochenschrift*, *102*, 1606–1607.

ASTM C618. (2015). Standard specification for coal fly ash and raw or calcined natural pozzolan for use in concrete. American Society for Testing and Materials.

Belviso, C. (2018a). State-of-the-art applications of fly ash from coal and biomass: A focus on zeolite synthesis processes and issues. *Progress in Energy and Combustion Science*, *65*, 109–135. Available from https://doi.org/10.1016/j.pecs.2017.10.004.

Belviso, C. (2018b). Ultrasonic vs hydrothermal method: Different approaches to convert fly ash into zeolite. How they affect the stability of synthetic products over time? *Ultrasonics Sonochemistry*, *43*, 9–14. Available from https://doi.org/10.1016/j.ultsonch.2017.12.050.

Belviso, S., Cavalcante, F., Lettino, A., Ragone, P., & Belviso, C. (2016). Fly ash as raw material for the synthesis of zeolite-encapsulated porphyrazine and metallo porphyrazine tetrapyrrolic macrocycles, Italy*Microporous and Mesoporous Materials*, *236*, 228–234. Available from http://www.elsevier.com/inca/publications/store/6/0/0/7/6/0, https://doi.org/10.1016/j.micromeso.2016.08.044.

Bharti, M., Das, P. P., & Purkait, M. K. (2023). A review on the treatment of water and wastewater by electrocoagulation process: Advances and emerging applications. *Journal of Environmental Chemical Engineering*, *11*, 111558. Available from https://doi.org/10.1016/j.jece.2023.111558.

Cardoso, A. M., Paprocki, A., Ferret, L. S., Azevedo, C. M. N., & Pires, M. (2015). Synthesis of zeolite Na-P1 under mild conditions using Brazilian coal fly ash and its application in wastewater treatment. *Fuel*, *139*, 59−67. Available from http://www.journals.elsevier.com/fuel/, https://doi.org/10.1016/j.fuel.2014.08.016.

Chakraborty, S., Gautam, S. P., Das, P. P., & Hazarika, M. K. (2019). Instant Controlled Pressure Drop (DIC) Treatment for Improving Process Performance and Milled Rice Quality. *Journal of The Institution of Engineers (India): Series A*, *100*, 683−695. Available from https://doi.org/10.1007/s40030-019-00403-w.

Changmai, M., Das, P. P., et al. (2022). Hybrid electrocoagulation−microfiltration technique for treatment of nanofiltration rejected steel industry effluent. *International Journal of Environmental Analytical Chemistry*, *102*, 62−83. Available from https://doi.org/10.1080/03067319.2020.1715381.

Collins, F., Rozhkovskaya, A., Outram, J. G., & Millar, G. J. (2020). A critical review of waste resources, synthesis, and applications for Zeolite LTA. *Microporous and Mesoporous Materials*, *291*. Available from http://www.elsevier.com/inca/publications/store/6/0/0/7/6/0, 10.1016/j.micromeso.2019.109667.

Das, P. P., Anweshan, A., Mondal, P., et al. (2021). Integrated ozonation assisted electrocoagulation process for the removal of cyanide from steel industry wastewater. *Chemosphere*, *263*, 128370. Available from https://doi.org/10.1016/j.chemosphere.2020.128370.

Das, P. P., Anweshan, A., & Purkait, M. K. (2021). Treatment of cold rolling mill (CRM) effluent of steel industry. *Separation and Purification Technology*, *274*, 119083. Available from https://doi.org/10.1016/j.seppur.2021.119083.

Das, P. P., Deepti., & Purkait, M. K. (2023). Industrial Wastewater to Biohydrogen Production via Potential Bio-refinery Route. In Maulin P. Shah (Ed.), *Biorefinery for Water and Wastewater Treatment* (pp. 159−179). Springer. Available from https://doi.org/10.1007/978-3-031-20822-5_8.

Das, P. P., Dhara, S., & Purkait, M. K. (2023). Hybrid Electrocoagulation and Ozonation Techniques for Industrial Wastewater Treatment. In Maulin P. Shah (Ed.), *Sustainable Industrial Wastewater Treatment and Pollution Control* (pp. 107−128). Singapore: Springer. Available from https://doi.org/10.1007/978-981-99-2560-5_6.

Das, P. P., Dhara, S., & Purkait, M. K. (2023). The Anaerobic Ammonium Oxidation Process: Inhibition, Challenges and Opportunities. In Maulin P. Shah (Ed.), *Ammonia Oxidizing Bacteria: Applications in Industrial Wastewater Treatment* (pp. 56−82). Royal Society of Chemistry. Available from https://doi.org/10.1039/BK9781837671960-00056.

Das, P. P., Dhara, S., & Purkait, M. K. (2024). Ozone-based oxidation processes for the removal of pharmaceutical products from wastewater. In Maulin P. Shah, & Pooja Ghosh (Eds.), *Development in Wastewater Treatment Research and Processes* (pp. 287−308). Elsevier. Available from https://doi.org/10.1016/B978-0-443-19207-4.00003-3.

Das, P. P., Mondal, P., Anweshan, A., et al. (2021). Treatment of steel plant generated biological oxidation treated (BOT) wastewater by hybrid process. *Separation and Purification Technology*, *258*, 118013. Available from https://doi.org/10.1016/j.seppur.2020.118013.

Das, P. P., Sharma, M., & Purkait, M. K. (2022). Recent progress on electrocoagulation process for wastewater treatment: A review. *Separation and Purification Technology*, *292*, 121058. Available from https://doi.org/10.1016/j.seppur.2022.121058.

Das, P. P., Sontakke, A. D., & Purkait, M. K. (2023). Electrocoagulation process for wastewater treatment: applications, challenges, and prospects. In Maulin P. Shah (Ed.), *Development in Wastewater Treatment Research and Processes* (pp. 23−48). Elsevier. Available from https://doi.org/10.1016/B978-0-323-95684-0.00015-4.

Dhara, S., Das, P. P., Uppaluri, R., & Purkait, M. K. (2023). Phosphorus recovery from municipal wastewater treatment plants. In Maulin P. Shah (Ed.), *Development in Wastewater
Treatment Research and Processes* (pp. 49−72). Elsevier. Available from https://doi.org/
10.1016/B978-0-323-95684-0.00014-2.

Dhara, S., Das, P. P., Uppaluri, R., & Purkait, M. K. (2023). Biological approach for energy
self-sufficiency of municipal wastewater treatment plants. In Mika Sillanpää, Ali Khadir, &
Khum Gurung (Eds.), *Resource Recovery in Municipal Waste Waters* (pp. 235−260).
Elsevier. Available from https://doi.org/10.1016/B978-0-323-99348-7.00006-0.

Dhara, S., Samanta, N. S., Das, P. P., et al. (2023). Ravenna Grass-Extracted Alkaline Lignin-
Based Polysulfone Mixed Matrix Membrane (MMM) for Aqueous Cr (VI) Removal. *ACS
Applied Polymer Materials*, *5*, 6399−6411. Available from https://doi.org/10.1021/
acsapm.3c00999.

Feng, R., Chen, K., Yan, X., Hu, X., Zhang, Y., & Wu, J. (2019). Synthesis of ZSM-5 zeolite
using coal fly ash as an additive for the methanol to propylene(MTP) reaction. *MDPI, China
Catalysts*, *9*(10). Available from https://doi.org/10.3390/catal9100788, https://www.mdpi.
com/2073-4344/9/10/788/pdf.

Hailu, S. L., McCrindle, R. I., Seopela, M. P., & Combrinck, S. (2019). Speciation of major and trace
elements leached from coal fly ash and the kinetics involved. *Journal of Environmental Science
and Health - Part A Toxic/Hazardous Substances and Environmental Engineering*, *54*(12),
1186−1196. Available from https://doi.org/10.1080/10934529.2019.1636599, http://www.tandf.
co.uk/journals/titles/10934529.asp.

Holler, H., & Wirsching, U. (1985). Zeolite formation from fly ash. *Fortschritte der
Mineralogie*, *63*, 21−43.

Hollman, G. G., Steenbruggen, G., & Janssen-Jurkovičová, M. (1999). Two-step process for the
synthesis of zeolites from coal fly ash. *Fuel*, *78*(10), 1225−1230. Available from https://doi.
org/10.1016/S0016-2361(99)00030-7.

Inada, M., Tsujimoto, H., Eguchi, Y., Enomoto, N., & Hojo, J. (2005). Microwave-assisted zeolite synthesis from coal fly ash in hydrothermal process. *Fuel*, *84*(12−13), 1482−1486.
Available from https://doi.org/10.1016/j.fuel.2005.020.002.

Jansen, J. C., Arafat, A., Barakat, A. K., Bekkum, H., et al. (1992). *Molecular sieves*
(pp. 507−521). Van Nostrand Reinhold.

Jin, H., Liu, Y., Wang, C., Lei, X., Guo, M., Cheng, F., et al. (2018). Two-step modification
towards enhancing the adsorption capacity of fly ash for both inorganic Cu(II) and organic
methylene blue from aqueous solution. *Environmental Science and Pollution Research*, *25*
(36), 36449−36461. Available from https://doi.org/10.1007/s11356-018-3585-7, http://www.
springerlink.com/content/0944-1344.

Ju, T., Jiang, J., Meng, Y., Yan, F., Xu, Y., Gao, Y., et al. (2020). An investigation of the
effect of ultrasonic waves on the efficiency of silicon extraction from coal fly ash.
*Ultrasonics Sonochemistry*, *60*, 104765. Available from https://doi.org/10.1016/j.
ultsonch.2019.104765.

Kleinhans, U., Wieland, C., Frandsen, F. J., & Spliethoff, H. (2018). Ash formation and deposi-
tion in coal and biomass fired combustion systems: Progress and challenges in the field of
ash particle sticking and rebound behavior. *Progress in Energy and Combustion Science*, *68*,
65−168. Available from https://doi.org/10.1016/j.pecs.2018.020.001.

Koshy, N., & Singh, D. N. (2016). Fly ash zeolites for water treatment applications. *Journal of
Environmental Chemical Engineering*, *4*(2), 1460−1472. Available from https://doi.org/
10.1016/j.jece.2016.020.002, http://www.journals.elsevier.com/journal-of-environmental-chemi-
cal-engineering/.

Li, Y., Gao, X., & Wu, H. (2013). Further investigation into the formation mechanism of ash cenospheres from an Australian coal-fired power station. *Energy and Fuels*, *27*(2), 811−815. Available from https://doi.org/10.1021/ef3020553.

Liu, P., Yang, L., Wang, Q., Wan, B., Ma, Q., Chen, H., et al. (2020). Speciation transformation of rare earth elements (REEs) during heating and implications for REE behaviors during coal combustion. *International Journal of Coal Geology*, *219*. Available from https://doi.org/10.1016/j.coal.2019.103371, http://www.sciencedirect.com/science/journal/01665162.

C. Ltd. (2019). *Coal fly ash industry analysis report of 2019—Research on current scales and developing trends*. Insights & Info Consulting Ltd.

McCarthy, M. J., Zheng, L., Dhir, R. K., & Tella, G. (2018). Dry-processing of long-term wet-stored fly ash for use as an addition in concrete. *Cement and Concrete Composites*, *92*, 205−215. Available from http://www.sciencedirect.com/science/journal/09589465, https://doi.org/10.1016/j.cemconcomp.2017.10.004.

Messina., Ferone., Colangelo, F., Roviello., & Cioffi. (2018). Alkali activated waste fly ash as sustainable composite: Influence of curing and pozzolanic admixtures on the early-age physico-mechanical properties and residual strength after exposure at elevated temperature. *Composites Part B: Engineering*, *132*, 161−169. Available from https://doi.org/10.1016/j.compositesb.2017.080.012.

Mondal, P., Samanta, N. S., Kumar, A., & Purkait, M. K. (2020). Recovery of $H_2SO_4$ from wastewater in the presence of NaCl and $KHCO_3$ through pH responsive polysulfone membrane: Optimization approach. *Polymer Testing*, *86*. Available from https://www.journals.elsevier.com/polymer-testing, https://doi.org/10.1016/j.polymertesting.2020.106463.

Murayama, N., Takahashi, T., Shuku, K., Lee, H. H., & Shibata, J. (2008). Effect of reaction temperature on hydrothermal syntheses of potassium type zeolites from coal fly ash. *International Journal of Mineral Processing*, *87*(3−4), 129−133. Available from https://doi.org/10.1016/j.minpro.2008.030.001.

Ngu, L., Wu, H., & Zhang, D.-K. (2007). Charaterization of ash cenospheres in fly ash from Australian power stations. *Energy Fuel*, *21*, 3437−3445. Available from https://doi.org/10.1021/ef700340k.

Ojumu, T. V., Du Plessis, P. W., & Petrik, L. F. (2016). *Synthesis of zeolite A from coal fly ash using ultrasonic treatment - A replacement for fusion step*, . *Ultrasonics Sonochemistry* (31, pp. 342−349). Available from http://www.elsevier.com/inca/publications/store/5/2/5/4/5/1, https://doi.org/10.1016/j.ultsonch.2016.01.016.

Palmer, B. (2015). *Coal ash, fly ash, bottom ash, and boiler slag*. NRDC.

Pan, J., Nie, T., Vaziri Hassas, B., Rezaee, M., Wen, Z., & Zhou, C. (2020). Recovery of rare earth elements from coal fly ash by integrated physical separation and acid leaching. *Chemosphere*, *248*, 126112. Available from https://doi.org/10.1016/j.chemosphere.2020.126112.

Panek, R., Wdowin, M., Franus, W., Czarna, D., Stevens, L. A., Deng, H., … Snape, C. E. (2017). Fly ash-derived MCM-41 as a low-cost silica support for polyethyleneimine in post-combustion CO2 capture. *Journal of $CO_2$ Utilization*, *22*, 81−90. Available from http://www.journals.elsevier.com/journal-of-co2-utilization/, https://doi.org/10.1016/j.jcou.2017.09.015.

Park, J., Dattatraya Saratale, G., Cho, S.-K., & Bae, S. (2020). Synergistic effect of Cu loading on Fe sites of fly ash for enhanced catalytic reduction of nitrophenol. *Science of The Total Environment*, *705*, 134544. Available from https://doi.org/10.1016/j.scitotenv.2019.134544.

Purkait, M. K., Sinha, M. K., Mondal, P., & Singh, R. (2018a). *Photoresponsive membranes*, *Interface Science and Technology* (25). Elsevier B.V. Available from http://www.elsevier.com/wps/find/bookdescription.cws_home/BS_3710/description#description, https://doi.org/10.1016/B978-0-12-813961-5.00004-8.

Purkait, M. K., Sinha, M. K., Mondal, P., & Singh, R. (2018b). *pH-responsive membranes*, . *Interface Science and Technology* (25). Elsevier B.V. Available from: http://www.elsevier. com/wps/find/bookdescription.cws_home/BS_3710/description#description. Available from 10.1016/B978-0-12-813961-5.00002-4.

Querol, X., Alastuey, A., López-Soler, A., Plana, F., Andrés, J. M., Juan, R., ... Ruiz, C. R. (1997). A fast method for recycling fly ash: Microwave-assisted zeolite synthesis. *Environmental Science and Technology*, *31*(9), 2527−2533. Available from https://doi.org/ 10.1021/es960937t.

Sakthivel, T., Reid, D. L., Goldstein, I., Hench, L., & Seal, S. (2013). Hydrophobic high surface area zeolites derived from fly ash for oil spill remediation. *Environmental Science and Technology*, *47*(11), 5843−5850. Available from https://doi.org/10.1021/es3048174.

Samanta, N. S., Banerjee, S., Mondal, P., Anweshan, A., Bora, U., & Purkait, M. K. (2021). Preparation and characterization of zeolite from waste Linz-Donawitz (LD) process slag of steel industry for removal of Fe3 + from drinking water. *Advanced Powder Technology*, *32* (9), 3372−3387. Available from https://doi.org/10.1016/j.apt.2021.070.023, http://www.else-vier.com.

Samanta, N. S., Das, P. P., Mondal, P., Changmai, M., & Purkait, M. K. (2022). Critical review on the synthesis and advancement of industrial and biomass waste-based zeolites and their applications in gas adsorption and biomedical studies. *Journal of the Indian Chemical Society*, *99*(11). Available from https://doi.org/10.1016/j.jics.2022.100761. Available from: https://www.sciencedirect.com/journal/journal-of-the-indian-chemical-society.

Samanta, N. S., Das, P. P., Mondal, P., Bora, U., & Purkait, M. K. (2022). Physico-chemical and adsorption study of hydrothermally treated zeolite A and FAU-type zeolite X prepared from LD (Linz−Donawitz) slag of the steel industry. *International Journal of Environmental Analytical Chemistry*. Available from http://www.tandf.co.uk/journals/titles/03067319.asp, https://doi.org/10.1080/03067319.2022.2079082.

Samanta, N. S., Anweshan, A., Mondal, P., Bora, U., & Purkait, M. K. (2023). Synthesis of pre-cipitated calcium carbonate from LD-slag using $CO_2$. *Materials Today Communications*, *36*. Available from http://www.journals.elsevier.com/materials-today-communications/, https:// doi.org/10.1016/j.mtcomm.2023.106588.

Song, M., Lin, S., & Takahashi, F. (2020). Coal fly ash amendment to mitigate soil water evapo-ration in arid/semi-arid area: An approach using simple drying focusing on sieve size and temperature. *Resources, Conservation and Recycling*, *156*. Available from http://www.else-vier.com/locate/resconrec, https://doi.org/10.1016/j.resconrec.2020.104726.

Sharma, M., Das, P. P., et al. (2023). Polyurethane Foams as Packing and Insulating Materials. In Ram K. Gupta (Ed.), *Polyurethanes: Preparation, Properties, and Applications* (pp. 83−99). American Chemical Society. Available from https://doi.org/10.1021/bk-2023-1454.ch004.

Shekhar Samanta, N., Das, P. P., Dhara, S., & Purkait, M. K. (2023). An overview of precious metal recovery from steel industry slag: Recovery strategy and utilization. *Industrial and Engineering Chemistry Research*, *62*(23), 9006−9031. Available from https://doi.org/ 10.1021/acs.iecr.3c00604.

Sōmiya, S., & Roy, R. (2000). Hydrothermal synthesis of fine oxide powders. *Bulletin of Materials Science*, *23*(6), 453−460. Available from https://doi.org/10.1007/BF02903883.

Sontakke, A. D., Das, P. P., et al. (2021). Thin-film composite nanofiltration hollow fiber mem-branes toward textile industry effluent treatment and environmental remediation applica-tions. *Emergent materials*, *5*, 1409−1427. Available from https://doi.org/10.1007/s42247-021-00261-y.

Sun, Q., Cai, C., Zhang, S., Tian, S., Li, B., Xia, Y., & Sun, Q. (2019). Study of localized deformation in geopolymer cemented coal gangue-fly ash backfill based on the digital speckle correlation method. *Construction and Building Materials*, *215*, 321−331. Available from https://doi.org/10.1016/j.conbuildmat.2019.040.208.

Supelano, G. I., Gómez Cuaspud, J. A., Moreno-Aldana, L. C., Ortiz, C., Trujillo, C. A., Palacio, C. A., ... Mejía Gómez, J. A. (2020). Synthesis of magnetic zeolites from recycled fly ash for adsorption of methylene blue. *Fuel*, *263*. Available from http://www.journals.elsevier.com/fuel/, https://doi.org/10.1016/j.fuel.2019.116800.

Valeev, D., Kunilova, I., Alpatov, A., Varnavskaya, A., & Ju, D. (2019). Magnetite and carbon extraction from coal fly ash using magnetic separation and flotation methods. *Minerals*, *9* (5). Available from https://doi.org/10.3390/min9050320, https://www.mdpi.com/2075-163X/9/5/320/pdf.

Visa, M. (2016). Synthesis and characterization of new zeolite materials obtained from fly ash for heavy metals removal in advanced wastewater treatment. *Powder Technology*, *294*, 338−347. Available from http://www.elsevier.com/locate/powtec, https://doi.org/10.1016/j.powtec.2016.02.019.

Wang, J., Zhang, Y., Wang, T., Xu, H., & Pan, W. P. (2020). Effect of modified fly ash injection on As, Se, and Pb emissions in coal-fired power plant. *Chemical Engineering Journal*, *380*. Available from http://www.elsevier.com/inca/publications/store/6/0/1/2/7/3/index.htt, https://doi.org/10.1016/j.cej.2019.122561.

Wang, Z., Dai, S., Zou, J., French, D., & Graham, I. T. (2019). Rare earth elements and yttrium in coal ash from the Luzhou power plant in Sichuan, southwest China: Concentration, characterization and optimized extraction. *International Journal of Coal Geology*, *203*, 1−14. Available from http://www.sciencedirect.com/science/journal/01665162, https://doi.org/10.1016/j.coal.2019.01.001.

Wirsching, U. (1975). Experimente zum Einfluß des Gesteinsglas-Chemismus auf die Zeolithbildung durch hydrothermale Umwandlung. *Contributions to Mineralogy and Petrology*, *49*(2), 117−124. Available from https://doi.org/10.1007/BF00373855.

Woolard, C. D., Strong, J., & Erasmus, C. R. (2002). Evaluation of the use of modified coal ash as a potential sorbent for organic waste streams. *Applied Geochemistry*, *17*(8), 1159−1164. Available from https://doi.org/10.1016/S0883-2927(02)00057-4.

Wu, H., & Li, Y. (2019). Ash cenosphere fragmentation during pulverised pyrite combustion: Importance of cooling. *Proceedings of the Combustion Institute*, *37*(3), 2773−2780. Available from https://doi.org/10.1016/j.proci.2018.070.007, http://www.sciencedirect.com/science/journal/15407489.

Yan, F., Jiang, J., Li, K., Liu, N., Chen, X., Gao, Y., & Tian, S. (2017). Green synthesis of nanosilica from coal fly ash and its stabilizing effect on CaO Sorbents for $CO_2$ capture. *Environmental Science & Technology*, *51*(13), 7606−7615. Available from https://doi.org/10.1021/acs.est.7b00320.

Yanguang, C., Tingting, X., Hongjing, H., Xinhui, W., Qiqi, W., Hongwei, H., ... Hua, S. (2015). Research development of zeolites preparation from coal fly ash by microwave-hydrothermal synthesis. *Chemical Industry and Engineering Progress*, *34*, 2916−2924. Available from https://doi.org/10.16085/j.issn.1000-6613.2015.08.002.

# Chapter 5

# Synthesis of zeolite from biomass fly ash

## 5.1 Introduction

Zeolite is an aluminosilicate mineral that is crystalline in nature and contains metals like Ti, Sn, and Zn as well as two integrated oxygen and aluminum atoms. It has an organized cage-like structure made of aluminum silicate and is a porous crystalline material. Although oxygen atoms are located in the corners of the four-sided framework in the zeolite structure, aluminum or silicon atoms are located in the center. The ions and molecules in the zeolite lattice are thought to have weak links, making it possible to remove them with little risk of compromising the zeolite's structural integrity (Das, Mondal, & Purkait, et al., 2022; Schulman, Wu, & Liu, 2020; Sontakke, Deepti, Samanta, & Purkait, et al., 2023). Zeolites can be used in a variety of processes, including ion exchange, filtration, adsorption, separation, and catalysis because of their chemical constituents, negatively charged surface, and interior porosity characteristics. Some of the most significant properties of zeolites include pore channel, pore volume, pore diameter, and cation exchange capacity (CEC). Its ion exchange characteristics are also significant for plant nutrition due to its permeability and CEC. Zeolites can be incorporated into fertilizers, and when applied to soil, they have two effects: they prolong the action of fertilizers and prevent the leaching of nutrients. Natural and artificial zeolites are two different forms of zeolite. Over the course of millions of years, natural zeolites have been created either from the ash of an active volcano or from sea salts (Chakraborty, Gautam, Das, & Hazarika, 2019; Das, Duarah, & Purkait, 2023; Khaleque et al., 2020; Mondal, Samanta, Meghnani, & Purkait, 2019; Mondal, Samanta, Kumar, & Purkait, 2020; Sontakke, Das, Mondal, & Purkait, al., 2022). Contrarily, synthetic zeolites are created under hydrothermal treatment settings (90°C−250°C) in an alkaline atmosphere and can be either large or tiny in size. In addition, zeolites can be divided into three groups based on the Si/Al molar ratio: (1) high silica contained zeolites: Si/Al $>$ 5 for instance, ZSM-5 (MFI) and zeolite B (BEA); (2) intermediate silica contained zeolites: 2 $<$ Si/Al $<$ 5 viz. zeolite Y (FAU), mordenite (MOR), faujasite (FAU), and chabazite (CHA); (3) zeolites with low silica contained: Si/Al $<$ 2 like natrolite

Waste-based Zeolite. DOI: https://doi.org/10.1016/B978-0-443-22316-7.00005-X
**115**

(NAT), phillipsite (PHI), cancrinite (CAN), zeolite X (FAU), and analcime (ANA) (Das, Anweshan, & Purkait, 2021; Das, Anweshan, Mondal, et al., 2021; Das, Mondal, et al., 2021; Das, Sharma, & Purkait, al., 2022; Samanta et al., 2021; Volli & Purkait, 2015). Zeolites are grouped according to their modified structure during the production process. Ten different kinds of naturally occurring zeolites, such as heulandite, chabazite, laumontite, analcime, ferrierite, and erionite, and more than 250 different types of synthetic zeolites have been studied thus far (Crémoux et al., 2019). Every variation has one or more zeolites; for example, faujasite contains two distinct kinds of zeolite minerals, zeolite X and zeolite Y, each of which has unique physical and chemical properties. Due to this, the crystalline structure and chemical makeup of zeolites determine how they should be used based on the intended physicochemical attributes. With an increase in the Si/Al molar ratio, other zeolite qualities including thermal stability, electrical phenomenon, and hydrophobicity also increase. Zeolites are biocompatible, and nonhazardous materials that possess special properties viz. molecular sieve structure, water adsorbent, and ionic exchangeability, which leads to their widespread uses in a variety of medical, gas purification, and wastewater treatment fields (Bharti, Das, & Purkait, 2023; Das, Deepti, & Purkait, 2023; Das, Dhara, & Purkait, 2023a, 2023b, 2024; Das, Sontakke, & Purkait, 2023; Das, Sontakke, Samanta, & Purkait, 2023; Dhara, Samanta, & Das, 2023; Samanta, Das, Dhara, & Purkait, 2023; Sharma & Das, 2021, 2022a, 2022b). Also, zeolites made from biomass and industrial waste, which are Si and Al-enriched, provide greater benefits than zeolites that are already on the market. Additionally, it was mentioned that waste-derived zeolite materials are less expensive rather than their traditional counterparts and that industrial solid waste that is Al/Si-rich may result in the valorization of byproducts (Bhandari, Volli, & Purkait, 2015; Wibowo, Sutisna Rokhmat, Murniati, & Khairurrijal Abdullah, 2017). To lower operating costs, many studies have recently concentrated on the use of inexpensive raw materials for zeolite production (Chakraborty, Das, & Mondal, 2023; Das & Mondal, 2023; Dhara, Das, Uppaluri, & Purkait, 2023a, 2023b; Sharma & Das, 2023). Therefore using waste with high Si and Al content to produce various types of zeolites, for example, sugarcane bagasse fly ash (SBFA), rice husk ash, palm oil mill fly ash (POMFA), bamboo leaf, and asbestos waste, may be thought of as a cost-reduction measure (Samanta, Das, Mondal, Changmai, & Purkait, 2022).

This chapter primarily focuses on the zeolite-like materials synthesis from various ash generated from various biomass viz. rice husk ash, bagasse fly ash, POMFA, and bamboo leaf. Also, the characterization of each of the aforementioned biomass sources has been critically discussed to enhance the fabrication of different kinds of zeolites. Further, recent advancements, various challenges, and possible solutions associated with zeolite synthesis have been covered in detail.

## 5.2 Outline of zeolite synthesis

Zeolites are aluminosilicate and crystalline silicate minerals having a relatively uniform and open microporous structure (<2 nm). The framework of the zeolite is a three-dimensional $AlO_4$ and $SiO_4$ network. As shown in Fig. 5.1, the tetrahedral configuration is associated with the shared oxygen atoms, creating voids, and attached by either pore holes or rings, thus determining the shape and size of zeolites. The zeolitic structure has a number of physicochemical properties, including porosity, surface area, CEC, and bulk density, which make it more desirable for a variety of applications. Zeolites have a great capacity for adsorption, which is one of their figure most significant chemical properties. A sufficient concentration of Si and Al is present in coal fly ash (CFA), which is considered the primary feedstock for the production of zeolite. The varieties of zeolite formation and their concentration are greatly influenced by the interaction of variables including activation solution-to-fly ash ratio, crystallization duration, temperature, pressure, solution alkalinity, and aging period (Bukhari, Behin, Kazemian, & Rohani, 2014).

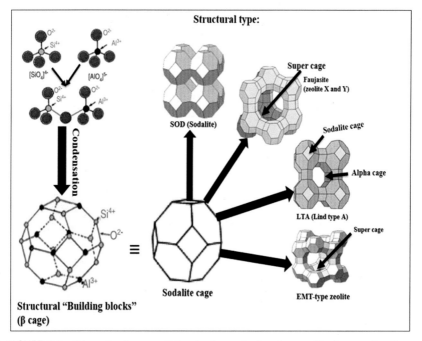

**FIGURE 5.1** Schematic diagram of the development of various zeolite frameworks. *From Samanta, N. S., Das, P. P., Mondal, P., Changmai, M., & Purkait, M. K. (2022). Critical review on the synthesis and advancement of industrial and biomass waste-based zeolites and their applications in gas adsorption and biomedical studies. Journal of the Indian Chemical Society, 99(11). https://doi.org/10.1016/j.jics.2022.100761.*

## 5.2.1 Rice husk ash

The rice husk (RH), which is removed from the rice grains after milling, is the topmost layer of the paddy grain. Due to changes in the climate, kind of paddy, geographic conditions, and crop year, it has been discovered that the chemical composition of RH varies from one sample to another (Anweshan, Das, Dhara, & Purkait, 2023; Das, Samanta, Dhara, & Purkait, 2023; Das, Sontakke, & Purkait, 2023). Poor in nutrients, RH is hardly ever used as animal food. It has historically been used as a fertilizer addition, as a cooking fuel, in stock breeding carpets, and for paving or landfill applications. The manufacturing of biochar, as well as composites and partition boards, are further advantageous uses for RH. A key component of improving soil is biochar. RH has enormous potential as a source of bioenergy, an alternative renewable energy source (Yaghoubi, Allahyari, Firouzi, Damalas, & Marzban, 2019). However, at this, very few amounts of RH are utilized for energy generation and other purposes like composting and silica manufacturing. For the industrial manufacture of rice husk ash (RHA), fluidized bed combustion is typically used. Only a few boiler manufacturers have perfected the technology necessary to produce superior-quality RHA in their boilers. But as technology develops, certain initiatives might result in high-quality ash. Depending on the temperature, three different forms of RHA—amorphous, partly crystalline, and crystalline—are created throughout the confinement of RH. The RH typically burns off between 2 and 5 hours at temperatures between 400°C and 1000°C. RH is burned at a regulated temperature of less than 800°C, producing ash primarily made of amorphous silica. The silica ($SiO_2$) will get sticky and the particles may aggregate and have a tendency to stay in the apparatus if the temperature is just a little bit too high. The formation of unwanted crystalline silica in unacceptable amounts will also occur at higher temperatures. The optimal combustion process temperature is between 500°C and 800°C to prepare amorphous silica ash with a high amorphous silica content and low carbon content. The speed at which combustion occurs increases with temperature. As the temperature rises over 850°C, the $SiO_2$ in the ash produces a cluster and then changes to crystal silica powder (Krishnarao, Subrahmanyam, & Jagadish Kumar, 2001; Pode, 2016).

### 5.2.1.1 Characterization of RHA

Mineralogical characterization: Cristobalite and amorphous material were found in the RHA using X-ray diffraction (XRD) analysis. A silica polymorph called cristobalite is created at elevated temperatures (1470°C—1723°C). At lower temperature (800°C), cristobalite can manifest as a metastable phase in the presence of alkali cations. This may be a result of the multiple temperature gradients present in the grate-firing method employed in this investigation to acquire the RHA. Potassium zeolites were discovered in all samples, along with amorphous material in varying proportions. The supplemental material contains the diffractograms. The zeolites found in potassium were known as merlinoite and chabazite-K (Gomes Flores et al., 2021). In a prior study, the same kinds of potassium zeolites were discovered, and the conditions for their synthesis were $100°C-150°C$, $24-72$ hours, and $3-5$ mol/L for temperature, duration, and KOH content, respectively (Gomes Flores et al., 2021). As chabazite-K and merlinoite are competing phases, increasing the $K^+$ concentration during synthesis can encourage merlinoite crystallization (Flores, Schneider, Marcilio, Ferret, & Oliveira, al., 2017).

Morphology: The scanning electron microscopy (SEM) picture of the RHA is displayed in Fig. 5.2. Observable surface pores and a "spongy" quality are present in this substance. Its skeleton structure preserves the shape of a RH; Kordatos and their research team have discovered a morphology that is similar to this (Kordatos et al., 2008). The SEM pictures of the materials that were synthesized are displayed in Fig. 5.3. In all samples, merlinoite is seen. According to earlier studies that solely used pure precursors for the synthesis, merlinoite is identified by bundles of needle-like crystals with square cross sections, either in the form of a transparent solution or a gel. It is possible to spot several walnut-shaped agglomerates, which most likely match the previously proposed chabazite zeolites. Utilizing kaolin and CFA as the starting materials, Che and their research group recognized this

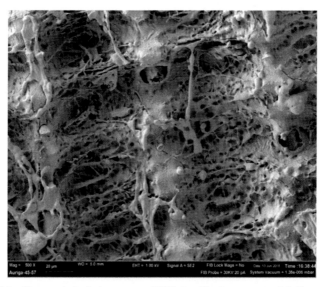

**FIGURE 5.2** The morphological structure of RHA is utilized as raw material. *From Flores, C. G., Schneider, H., Dornelles, J. S., Gomes, L. B., Marcilio, N. R., & Melo, P. J. (2021). Synthesis of potassium zeolite from rice husk ash as a silicon source.* Cleaner Engineering and Technology, 4, *100201. https://doi.org/10.1016/j.clet.2021.100201.*

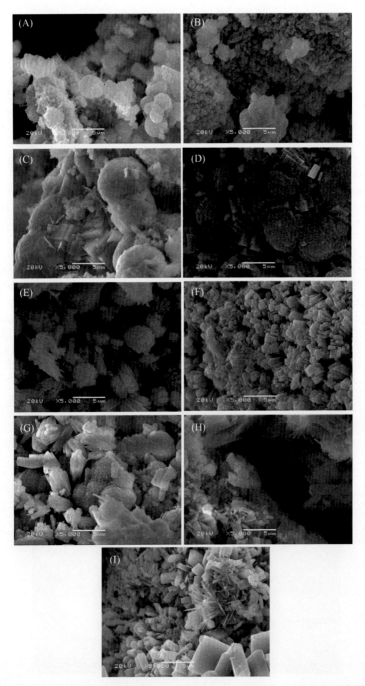

**FIGURE 5.3** Morphological structure of the fabricated zeolites: (A) Z1, (B) Z2, (C) Z3, (D) Z4, (E) Z5, (F) Z6, (G) Z7, (H) Z8, and (I) Z9. *From Flores, C. G., Schneider, H., Dornelles, J. S., Gomes, L. B., Marcilio, N. R., & Melo, P. J. (2021). Synthesis of potassium zeolite from rice husk ash as a silicon source.* Cleaner Engineering and Technology, 4, *100201. https://doi. org/10.1016/j.clet.2021.100201.*

potassium chabazite form (Che, Du, Zhu, Fang, & Wang, 2019; Che, Fang, Li, Chen, & Du, 2019). A similar shape for chabazite was also achieved using a fluoride-route synthesis (Liu et al., 2014). The crystals of analcime are spherical icositetrahedrons. When zeolite synthesis was carried out using only pure precursors or discarded materials, the same shape was discovered (namely, RHA). As sodalite can have several morphologies, identifying it can be challenging.

Specific surface area: Fig. 5.4 displays the $N_2$ isotherms for the processed samples. The synthetic material is mesoporous since the samples showed hysteresis in the isotherms. Additionally, every hysteresis loop was of type H3, showing the presence of slit-shaped holes connected to plate-like agglomerates of particles. The fact that RHA displayed a higher specific surface area compared to the synthetic material may be due to the grate-firing combustion method that was utilized to burn the RH. This process runs at different temperatures, which results in reduced combustion efficacy and more residual carbon in the ash. The Brunauer—Emmett—Teller (BET) surface area of the ash is increased by residual carbon because it appears as small particles with a high surface area. As the fabrication temperature rises or the fabrication period lengthens, there is a tendency for the specific surface area to decrease (Gomes Flores et al., 2021). This might be because samples created at lower temperatures have a higher proportion of amorphous materials (i.e., less crystallinity) (or shorter times). Sánchez-Hernández, López-Delgado, Padilla, Galindo, and López-Andrés (2016) demonstrated the synthesis of sodalite and analcime from aluminum slag with

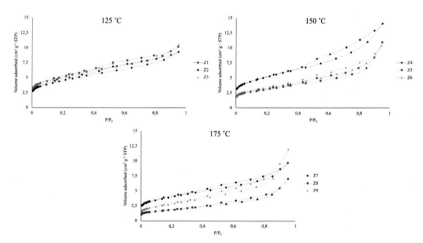

**FIGURE 5.4**  Nitrogen adsorption−desorption isotherms for zeolites prepared at 125°C, 150°C, and 175°C. *From Flores, C. G., Schneider, H., Dornelles, J. S., Gomes, L. B., Marcilio, N. R., & Melo, P. J. (2021). Synthesis of potassium zeolite from rice husk ash as a silicon source. Cleaner Engineering and Technology, 4, 100201. https://doi.org/10.1016/j.clet.2021.100201.*

specific surface areas of 4.6 and 15.5 m$^2$/g, respectively. For merlinoite, a similar pattern of behavior is predicted.

### 5.2.1.2 Several types of zeolite synthesis from RHA

The sodium version of zeolite Y (NaY) was fabricated by synthesizing silica from RH. By using XRD, SEM, and BET, the samples as they were produced were analyzed. When made in two steps, NaY zeolite exhibits good crystallinity and is appropriate for use in a variety of applications. The manufacturing of zeolite particles makes good use of RHA (NaY and NaP). After maximizing the various fabrication parameters (24 hours of crystallization time and 110°C), the physicochemical properties of the produced zeolite were characterized. According to SEM results, the produced NaY zeolite has a consistent distribution of particle sizes. The results from the particle size distribution revealed the zeolite Y's homogeneity. The presence of zeolite P caused the surface area of the final sample from the one-step route to be less than that of the two steps, according to BET data. A mixed phase including NaY and sodium-form zeolite P resulted from a longer crystallization period in the two-step procedure (NaP). Zeolite P was produced using a one-step synthetic process that was researched in addition. For the fabrication of zeolite NaY with RHS, 24 hours of crystallization time was ideal (Mohamed, Mkhalid, & Barakat, 2015). Using RHA as the silicon source, it is also possible to produce potassium zeolites without templates by hydrothermal alkaline treatment, providing a previously unexplored use for this waste material. Additionally, this research has demonstrated that it is feasible to conduct a synthesis using tap water and commercially available chemicals. This investigation also confirmed the creation of zeolites with faster reaction times. Additionally, by improving the synthesis's operational parameters, pure zeolite might be produced. In light of the industrial-scale zeolite production, these findings are pertinent. The potassium zeolites created in the present study have the potential to be used as plant fertilizers. The ideal synthesis conditions for this application were 125°C and 8 hours of treatment time, respectively (Gomes Flores et al., 2021). Amorphous rice husk-derived silica might be employed as a precursor material for the hydrothermal process of synthesizing H-Beta at 135°C, followed by three days of crystallization. It was demonstrated that only the pure phase of Beta was present in the products with gel Si/Al ratios of 8−20. A mixed phase of ZSM-12 and Beta was visible in samples with gel Si/Al ratios of 50−200. With increasing Si/Al ratios, the ZSM-12 phase became more important and was nearly pure at a ratio of 200. The analysis of H-Beta with a gel silica/alumina ratio of 13 revealed a spherical shape with approximately 1.5 m-sized particles. The BET area and acid amount were found to be 670 m$^2$/g and 1.17 mmol/g, respectively (Loiha, Prayoonpokarach, Songsiriritthigun, & Wittayakun, 2009).

### 5.2.2 Sugarcane bagasse fly ash

Bagasse waste is generated in enormous quantities by the sugar industry. An innovative method of environmental rehabilitation involves burning bagasse garbage to produce bagasse fly ash (BFA) and reuses of that BFA to create value-added materials. Si and Al are present in ash particles, making them good starting points for the production of zeolites. BFA is more desirable than CFA since it has more components and only Si and Al are extracted during the alkaline treatment process, which produces high-quality zeolite. Purnomo, Salim, and Hinode (2012) investigated the synthesis of Si and Al-enrich low-cost BFA with sodium hydroxide (NaOH) fusion to produce very pure and crystalline zeolite types—A and X. The Si/Al ratio is modified during the zeolite-making process as per the desired zeolite type and is processed using the traditional hydrothermal technique. Zeolite mineral dissolves in a very alkaline solution because low temperature produces particles of higher purity and smaller size, which improves porosity and ion exchange capacity. Sugarcane waste was used by Oliveira, Cunha, and Ruotolo (2019) to create zeolite A, a cheap adsorbent for removing heavy metals. For the synthesis of zeolite, they chose the hydrothermal approach, using NaOH as the mineralizing agent. The calcination duration and temperature increased the produced zeolite mineral's ion exchange capacity. When the analysis was run for an 8-hour calcination period at 800°C, all the carbon constituents were eliminated from the fly ash. The research also showed that zeolite structure can be changed to improve $Cu^{2+}$ absorption capability (142 mg/g) by including aluminum isopropoxide.

#### 5.2.2.1 Characterization of SBFA

Compositional analyses: First, it was established how much fuel was lost upon igniting (LOI) in sugarcane bagasse ash (SCBA) samples. LOI has long been used as a metric to represent the amount of unburned carbon contained in combustion ashes. As a pretreatment for X-ray fluorescence (XRF) tests used for the elemental analysis or to repair erroneous XRF data, LOI is also used. The types of oxide and elements in the SCBA are extremely constant (Chancey, Stutzman, Juenger, & Fowler, 2010). SCBA has a concentration of 87.3% ($SiO_2$, $Al_2O_3$, and $Fe_2O_3$). $SiO_2$ (76.16%) makes up the majority of SCBA's chemical makeup. In the event that the SCBA's high $SiO_2$ content reveals that it is primarily amorphous, this material will undoubtedly exhibit pozzolanic reactivity. Its low CaO level (2.32%), however, suggests that the SCBA's hydraulic reactivity may be minimal. The high $K_2O$ content of the SCBA (3.97%) results in a high equivalent $Na_2O$ concentration of up to 2.8%. This means that, as a potential supplementary cementitious material (SCM), the SCBA must be employed with caution or at relatively low replacement levels; otherwise, it could put concrete at a rather high danger of an alkali-aggregate reaction (ASTM C1778-19b, 2019). An LOI of 2.46%

applies to the SCBA. Unburned carbon may control the LOI, which may have a negative impact on concrete's performance (especially fresh-state properties) (Zhang, Liao, Kumar, Zhang, & Ma, 2020).

Mineralogical characterization: Quantitative XRD tests were used to detect and measure the phases in the CFA and SCBA, and Fig. 5.5 shows the representative diffraction peaks for every phase. Quartz (ICSD 27831) and anorthite ($CaAl_2Si_2O_8$, ICSD 23922) are the two primary crystalline phases seen in the SCBA diffraction pattern, and a very little amount of cristobalite ($SiO_2$, ICSD 35536) is also discernible. This suggests that Ca could penetrate the $SiO_2/Al_2O_3$ tetrahedral lattice more easily, leading to the creation of anorthite rather than mullite during the formation of SCBA, but the precise process underlying this needs to be investigated (Yan, Guo, Ma, Zhao, & Cheng, 2018). According to the available literature, these phases may exist, but their existence cannot be confirmed with certainty due to peak overlapping or their small numbers. In Fig. 5.5, both the samples exhibit humps in the 2θ range of 20−40 degrees, indicating their amorphous nature. The relative proportions of crystalline and amorphous phases in SCBA are obtained by the Rietveld refinement technique because Rietveld analysis is an established methodology that is frequently employed to assess a full-component quantitative phase analysis from the XRD pattern. According to the literature, $SiO_2$ and $Al_2O_3$ make up the majority of the noncrystalline phase in the SCBA (60.6% of the ash and 77.2% of the amorphous phase, respectively) (Zhang et al., 2020).

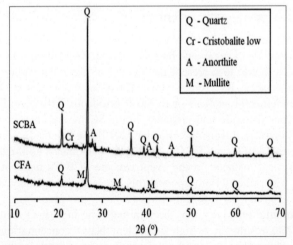

**FIGURE 5.5** XRD profile of CFA and SCBA. *From Zhang, P., Liao, W., Kumar, A., Zhang, Q., & Ma, H. (2020). Characterization of sugarcane bagasse ash as a potential supplementary cementitious material: Comparison with coal combustion fly ash. Journal of Cleaner Production, 277, 123834. https://doi.org/10.1016/j.jclepro.2020.123834.*

Morphology: Fig. 5.6 shows the morphologies of the SCBA grains. The ashes' particle sizes range from submicron to more than 20 μm. The majority of SCBA particles have unusual forms and sizes, such as prismatic and fibrous. This discrepancy can be explained by the relatively low combustion temperature attained during SCBA ignition. Because there is no ball-bearing effect, as seen in the grain morphology, the SCBA may not be able to enhance the rheological character of fresh-state cement-based materials when used as an SCM (Xu & Shi, 2018).

### 5.2.2.2 Zeolite preparation from SBFA

Using BFA as the silica source, crystalline zeolite Na-A and Na-X have been effectively synthesized by using alkali fusion, extraction with residue removal, and hydrothermal process. The produced zeolites have an ion exchange capacity that is equivalent to commercial goods and very high porosity. It was simpler to manufacture zeolites with higher purity and smaller particle sizes at low zeolitization temperatures. The low treatment temperature, however, necessitated a longer curing period. So, in this study, it is possible to avoid the occurrence of sodalite's low porosity and ion exchange

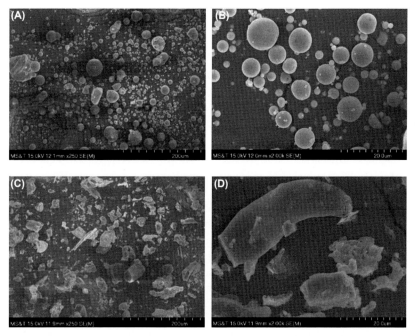

**FIGURE 5.6** Morphologies of the ash grains: (A, B) CFA and (C, D) SCBA. *From Zhang, P., Liao, W., Kumar, A., Zhang, Q., & Ma, H. (2020). Characterization of sugarcane bagasse ash as a potential supplementary cementitious material: Comparison with coal combustion fly ash.* Journal of Cleaner Production, 277, 123834. *https://doi.org/10.1016/j.jclepro.2020.123834.*

capability, which are typical of this frequent zeolitization procedure. In addition to the synthesized zeolite's crystalline purity, it has been discovered that particle size has a major impact on its ion exchange capacity and porosity. Meanwhile, as the molecular weight of the additional salts increases (NaBr > NaNO$_3$ > NaCl), the impacts of anions on the zeolitization rate improvement become more pronounced (Purnomo et al., 2012). In a different study, BFA, a free precursor produced in significant quantities in sugar refineries and ethanol facilities, was effectively used to manufacture zeolite Na-A. The proposed synthesis method uses BFA waste instead of commercial chemicals that contain Al and Si sources, making it sustainable and clean. Following the removal of carbon through calcination and hydrothermal treatment of the produced ash, zeolitic materials are created. Copper sorption from a contaminated solution verified the as-synthesized materials' ion exchangeability. An additional supply of aluminum was mixed into the synthetic medium, which resulted in a significant and sustained improvement in the ion exchange capacity of the synthetic zeolite up to 142 mg/g. At the same temperature, the concentration of copper, and adsorbent dose of this amazing ion exchange capability is significantly greater than that of commercial polymeric commercial resins (46.6 mg/g). In addition to its remarkable ion exchange capabilities, this product comes from a sustainable source and finds a use for the substantial amounts of biomass waste produced by a significant industrial sector (Oliveira et al., 2019).

### 5.2.3  Palm oil mill fly ash

#### 5.2.3.1  Characterization of POMFA

The palm oil industry generates a large amount of solid POMFA, which comprises nearly 39% silica. The generated amorphous silica (SAS) can be produced via alkaline extraction of fly ash followed by sol−gel precipitation. POMFA is made up of 30% carbon, 39% silica, and other inorganic compounds. Where silica and carbon are the primary components, moreover, it has been stated that around 60% of silica can be recovered from POMFA (Jha & Singh, 2012). The characterization of POMFA is illustrated in the subsequent section.

Particle size and compositional analysis: The particle size of POMFA is an important factor for zeolite and geopolymer-like sorbent materials. As shown in Fig. 5.7, the particle size distribution of the palm oil fuel ash (PF) sample was precise to less than 150 μm. According to Assi, Anay, Leaphart, Soltangharaei, and Ziehl (2018), small particles have higher mechanical strengths because fewer microcracks and air pockets are formed, which leads to the formation of denser microstructure. In an experiment, Opiso et al. (2021) synthesized geopolymer material from POMFA where the particle size of the starting material was achieved below 150 μm. The chemical

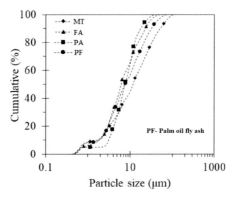

**FIGURE 5.7** Particle size analysis of PF. *From Opiso, E. M., Tabelin, C. B., Maestre, C. V., Aseniero, J. P. J., Park, I., & Villacorte-Tabelin, M. (2021). Synthesis and characterization of coal fly ash and palm oil fuel ash modified artisanal and small-scale gold mine (ASGM) tailings based geopolymer using sugar mill lime sludge as Ca-based activator.* Heliyon, 7(4). https://doi. org/10.1016/j.heliyon.2021.e06654.

composition and physical characteristics of POMFA were obtained from XRF, and other techniques are illustrated in Tables 5.1 and 5.2, respectively.

Functional group analysis: The functional groups present in palm oil fuel ash were identified by Fourier infrared spectroscopic analysis and depicted in Fig. 5.8. As reported by Opiso et al. (2021), the IR peak band between 1500 and 450/cm corresponds to the existence of silicon (Si) and aluminum (Al) minerals, respectively, in the produced ash sample. The alumina-silicate broad peak appears within the range 1150 and 950/cm attributed to the asymmetric vibrational stretching of Al−O−Si or Si−O−Al and the absorption peak exhibited at 776/cm is ascribed to the tetrahedral vibrational mode of silicates ions. Additionally, the peak that arises at 693/cm represents the asymmetrical bending of Si−O. In addition, the adsorption band at 693/cm and 874/cm could be attributed to the stretching of Al-oxyhydroxides/oxides bonds (Park et al., 2018) (Fig. 5.8B). Moreover, as can be seen in Fig. 5.8A, the vibrational stretching of the Al−O−H and Si−O−Si bond present in the PF sample was found at the region of 2510 and 3620/cm, respectively. The study also revealed the existence of quartz in the diagnostic region, 820−740/cm, which also makes evidence of the asymmetrical vibration of the Si−O bond.

Mineralogical characterization: The XRD pattern of PF, as shown in Fig. 5.9, reveals that this type of waste ash comprises relatively abundant aluminum-silicate minerals that may be used as a precursor material for the preparation of zeolite-like material.

### 5.2.3.2 Different kinds of zeolite synthesis using POMFA

POMFA comprises an adequate amount of silica ($SiO_2$), which may be used as a solid additive or as a source for the zeolite preparation. In a zeolite preparation

**TABLE 5.1** Chemical constituents of palm oil fly ash.

| Constituents | SiO$_2$ | Fe$_2$O$_3$ | Al$_2$O$_3$ | SO$_3$ | CaO | Na$_2$O | Hg (mg/kg) | Zn (mg/kg) | Cd (mg/kg) | Pb (mg/kg) | Cu (mg/kg) | As (µg/kg) | Si/Al |
|---|---|---|---|---|---|---|---|---|---|---|---|---|---|
| Concentration (wt.%) | 46.2 | 5.47 | 10.4 | 0.67 | 7.4 | 7.28 | 0.076 | 1265 | - | 0.052 | 64 | 0.0001 | 4.4 |

*Source:* From Opiso, E. M., Tabelin, C. B., Maestre, C. V., Aseniero, J. P. J., Park, I., & Villacorte-Tabelin, M. (2021). Synthesis and characterization of coal fly ash and palm oil fuel ash modified artisanal and small-scale gold mine (ASGM) tailings based geopolymer using sugar mill lime sludge as Ca-based activator. *Heliyon, 7*(4). https://doi.org/10.1016/j.heliyon.2021.e06654.

**TABLE 5.2** Physical properties of palm oil fly ash.

| Physical characteristics | Color | SG | SSA | D[4, 3] | D[3, 2] | D(v, 0.1) | D(v, 0.5) | D(v, 0.9) |
|---|---|---|---|---|---|---|---|---|
| POMFA | Black | 2.2 | 3.41 | 20.06 | 3.49 | 3.23 | 18.46 | 38.27 |

SG, specific gravity; SSA, specific surface area; D[4, 3], volume moment mean, μm; D[3, 2], surface area moment mean, (μm)$^2$; D(v, 0.1), 10% of the particle by volume below this size; D(v, 0.5), 50% of the particle by volume below this size; D(v, 0.9), 90% of the particle by volume below this size.
Source: From Bashar, I. I., Alengaram, U. J., Jumaat, M. Z., & Islam, A. (2014). The effect of variation of molarity of alkali activator and fine aggregate content on the compressive strength of the fly ash: Palm oil fuel ash based geopolymer mortar. Advances in Materials Science and Engineering, 2014. https://doi.org/10.1155/2014/245473.

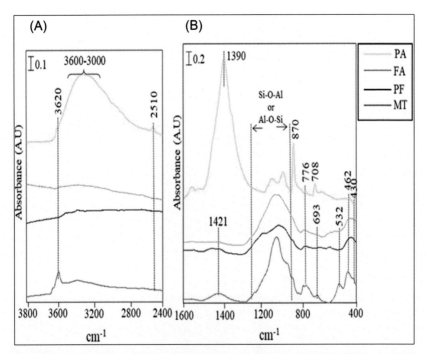

**FIGURE 5.8** Deconvoluted FTIR spectra of palm oil fly ash: (A) 2400−4000/cm and (B) 400−1600/cm. *From Opiso, E. M., Tabelin, C. B., Maestre, C. V., Aseniero, J. P. J., Park, I., & Villacorte-Tabelin, M. (2021). Synthesis and characterization of coal fly ash and palm oil fuel ash modified artisanal and small-scale gold mine (ASGM) tailings based geopolymer using sugar mill lime sludge as Ca-based activator.* Heliyon, 7(4). https://doi.org/10.1016/j.heli-yon.2021.e06654.

study, Kongnoo, Tontisirin, Worathanakul, and Phalakornkule (2017) opted for traditional alkaline fusion-assisted hydrothermal treatment to generate 13X-zeolite from POMFA. The influences of fusion time, $Al_2O_3$ and $NaOH/SiO_2$ molar ratios, acid activation, and temperature on $CO_2$ sorption capacity were investigated in this work. The $CO_2$ adsorption capacity and BET surface area of the synthesized zeolite mineral were estimated to be 223 mg/g and 643 $m^2$/g, respectively. Further investigation of this study also stated that the crystallinity of zeolite-type 13X was reduced by 17% after acid treatment, but the FAU-like framework was shown to be stable. In a recent study, Davina, Utama, Saputra, and Bahri (2019) fabricated POMFA-based zeolite-type P1 through the alkaline extraction method assisted with the precipitation of sol−gel, followed by a two-stage technique. In the initial step, amorphous silica (SAS) powder was generated from POMFA via a basic leaching route using NaOH as the basic source, followed by precipitation on a sol−gel. In the final stage, SAS was utilized in a hydrothermal technique to generate zeolite P (maintaining the Si/Al = 5, 8, and 11, respectively) over an 8-hour period under 200 kPa pressure and 120°C temperature. The selected Si/Al ratio

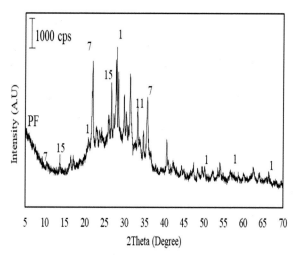

**FIGURE 5.9** XRD pattern of palm oil fuel ash. *From Opiso, E. M., Tabelin, C. B., Maestre, C. V., Aseniero, J. P. J., Park, I., & Villacorte-Tabelin, M. (2021). Synthesis and characterization of coal fly ash and palm oil fuel ash modified artisanal and small-scale gold mine (ASGM) tailings based geopolymer using sugar mill lime sludge as Ca-based activator.* Heliyon, 7(4). *https://doi.org/10.1016/j.heliyon.2021.e06654.*

has a remarkable impact on zeolite preparation; as the Si/Al ratio rises, the peak intensity increases. This result suggests that a greater Si/Al ratio accelerates the synthesis process significantly. The study also revealed that several zeolite types were obtained as the study progressed with a constant Si/Al ratio of 8 and variable NaOH concentrations of 1 N, 2 N, and 3 N, respectively. Zeolite-type P1 was found to develop at 1 N NaOH concentration, whereas sodium tecto-hexaaluminohexasilicate carbonate and analcime types of zeolite were found at a concentration of 2 N and 3 N, respectively. In a different study, Kishimoto (2000) opted for synthetic POMFA as a precursor material for the synthesis of zeolite-like minerals. Ultrasound-irradiation energy was employed in zeolite preparation with an irradiation duration of 140 min and keeping a KOH/POMFA ratio of 1.5:1−2.5:1. The study achieved blended zeolite, which was a blended of zeolite A, X, and P, respectively. The findings also revealed that synthesis parameters have a significant impact on zeolite topology, infrared and phases, diffraction, and chemical substituent, which makes the zeolite most favorable for versatile applications.

## 5.2.4 Bamboo leaf

### 5.2.4.1 Preparation of bamboo leaf ash

Literature on zeolite mineral synthesis from BLA is very scarce. In a bamboo leaf ash-based zeolite synthesis (Ng et al., 2017), BLA was first prepared to obtain high-content $SiO_2$, which was utilized for zeolite synthesis. In this typical experiment, a large amount of bamboo leaf was collected from a bamboo factory directly. To prepare the ash, only 45 g bamboo leaves are taken and treated with distilled water for washing and then cut into small pieces (ca. 3 cm). The fresh bamboo leaves were used in the nitric acid treatment, which was conducted at an ambient temperature and acid concentration of 1.5 M for 15 hours while stirring at 90 rpm. The processed bamboo leaves were filtered out and washed with deionized water for several cycles until pH reached 7. Before getting into fine pieces the obtained leaves were dried at room temperature. The leaves were heated at a rate of 1.0 C/min for 6 hours while being burned at 600°C. After combustion, a yield of 31.7% pure white ash was generated. For the purposes of comparison, bamboo leaves that had not been subjected to initial acid treatment were also burned under similar circumstances.

### 5.2.4.2 Characterization of bamboo leaf and ash

Composition analysis: The elemental and quantitative analysis of BLA was carried out through the XRF analyzer. Table 5.3 illustrates the XRF data of the bamboo leaves-derived ashes. The ash from the bamboo leaves that had not previously been treated with nitric acid showed a significant concentration of inorganic impurities, including CaO (5.96%), $SO_3$ (3.96%), MgO (2.03%), $K_2O$ (8.00%), and $P_2O_5$ (2.13%). The inorganic contaminants were isolated from the bamboo leaves after acid leaching. Because of this, the inorganic constituents drastically decreased to 0.26%, while bamboo leaf ash's (BLA's) $SiO_2$ content increased to 99.37%. Additionally, the residue of black carbon was not found in BLA, as evident from the XRF analysis (Table 5.3).

*bdl—below detection limit

Thermogravimetric analysis: The thermogravimetric analysis (TGA) and XRD analysis of BLA have been shown in Fig. 5.10. Basically, bamboo leaves comprise both types of components: organic and inorganic. The existence of noncombustible minerals in bamboo leaves and the thermal disintegration profile of organic constituents were investigated through TGA. Fig. 5.10A shows the bamboo leaf thermogravimetric and differential thermogravimetric (TG/DTG) profiles. There are three primary processes in weight loss: the first step (160°C) is caused by water absorption, the second step (160°C–380°C) is caused by cellulose and hemicellulose decomposition, and the third step is induced by lignin decomposition (Ng et al., 2017). According to the TG/DTG results, bamboo leaves had a substantially larger percentage of noncombustible ash (31.7 wt.%) as compared with RH (20 wt. %) (Zakharov, Belyakov, & Tsvigunov, 1993).

Phase analysis: The phase analysis of BLA was investigated by XRD pattern, as depicted in Fig. 5.10B. As can be seen, only one broad peak appeared at $2\theta = 23$ degrees, which ascribed the presence of amorphous silica in the produced BLA (Ng et al., 2017).

**TABLE 5.3** XRF analysis of bamboo leaves ashes with and without nitric acid treatment.

| Samples | $SiO_2$ | $K_2O$ | CaO | $SO_3$ | $P_2O_5$ | MgO | $Fe_2O_3$ | $Al_2O_3$ | $Na_2O$ |
|---|---|---|---|---|---|---|---|---|---|
| Before treatment | 77.09 | 8.00 | 5.96 | 3.96 | 2.13 | 2.03 | 0.15 | 0.14 | 0.46 |
| After treatment | 99.37 | 0.03 | 0.07 | *bdl | 0.26 | 0.04 | 0.04 | 0.06 | 0.04 |

Source: From Ng, E. P., Chow, J. H., Mukti, R. R., Muraza, O., Ling, T. C., & Wong, K. L. (2017). Hydrothermal synthesis of zeolite a from bamboo leaf biomass and its catalytic activity in cyanoethylation of methanol under autogenic pressure and air conditions. Materials Chemistry and Physics, 201, 78–85. https://doi.org/10.1016/j.matchemphys.2017.08.044.

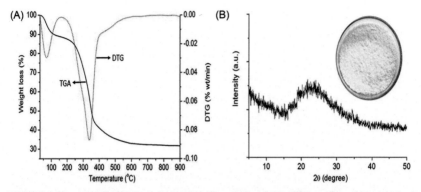

**FIGURE 5.10** (A) TGA/DTG analysis of bamboo leaf and (B) XRD profile of bamboo leaf ash after combustion. Inset: Physical appearance of BLA. *From Ng, E. P., Chow, J. H., Mukti, R. R., Muraza, O., Ling, T. C., & Wong, K. L. (2017). Hydrothermal synthesis of zeolite a from bamboo leaf biomass and its catalytic activity in cyanoethylation of methanol under autogenic pressure and air conditions.* Materials Chemistry and Physics, 201, *78–85. https://doi.org/ 10.1016/j.matchemphys.2017.08.044.*

### 5.2.4.3 Zeolite synthesis from bamboo leaf biomass ash

It is well known that zeolite is a mineral of aluminosilicates, which mainly comprises alumina and silica in their structural framework. From Table 5.3, it can also be seen that bamboo leaves generated ash contained a very high amount of silica (77.09%). Therefore these ashes could be employed as impactful resources for zeolite-like material synthesis. In this context, Ng et al. (2017) utilized bamboo leaves extracted from 99% pure silica to prepare zeolite A. Before zeolite synthesis acid treatment was carried out to extract the unwanted material that leads to better zeolite crystallinity. This study focused on zeolite A generation by changing the process conditions like heating temperature, initial gel molar constituents, and crystallization duration with the BLA precursor. The research showed that the synthesis variables remarkably enhanced the shape, size, and, crystallinity, of the pre-pared A-type zeolite. The zeolitization of precursor solution was carried out for different times 1, 3, 5, and 7 hours, respectively and hydrothermal treat-ment of 7 hours showed zeolite formation with substantial crystallinity, according to their findings (Ng et al., 2017). The four distinct field emission scanning electron microscopy (FESEM) images of the zeolite minerals pro-duced at different hydrothermal conditions have been shown in Fig. 5.11. The XRD results obtained from the study revealed that when the hydrogel was kept at 70°C heat treatment, the final product was amorphous and the data was in good agreement with the FESEM analysis, as illustrated in Fig. 5.11A. When the heating temperature was raised to 80°C, the crystalline structure of zeolite A predominant (Fig. 5.11B). At 90°C, fully solid crystal-line structures were produced, as shown by the XRD profile and SEM

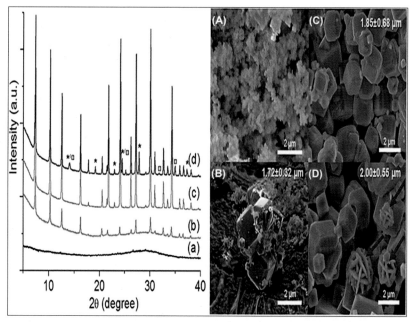

**FIGURE 5.11** XRD profiles and FESEM pictures of as-synthesized zeolite material at different crystalline temperatures: (A) 70°C, (B) 80°C, (C) 90°C, and (D) 100°C for 7 hours. The * and □ signs in the XRD profile indicate that hydroxysodalite and cancrinite dense phases exist in the prepared samples. The arrows noticed in the FESEM image signify the cocrystallization of sodalite and cancrinite phases, respectively. *From Ng, E. P., Chow, J. H., Mukti, R. R., Muraza, O., Ling, T. C., & Wong, K. L. (2017). Hydrothermal synthesis of zeolite a from bamboo leaf biomass and its catalytic activity in cyanoethylation of methanol under autogenic pressure and air conditions.* Materials Chemistry and Physics, 201, *78−85. https://doi.org/10.1016/j.matchemphys.2017.08.044.*

morphology (Fig. 5.11C). The same morphology of zeolite-type A has also been found by Samanta et al. (2021) and Samanta, Das, Mondal, Bora, and Purkait (2022), taking steel industry solid waste as a starting material. As the investigation proceeded further, the XRD and FESEM data revealed that at crystallization temperatures beyond 100°C, the dense phases of hydroxysodalite and cancrinite phases were detected (Fig. 5.11D). The lepispheric shape intergrown on top of the zeolite A cubic crystals was captured and shown in Fig. 5.11D. The hydrogel $SiO_2/Al_2O_3$ ratio effect on zeolite A crystallinity was also examined. The study showed that the crystalline phase of zeolite A was present in all samples; however, no other zeolite phases, for instance, NaX, cancrinite, and hydroxysodalite were observed when hydrogel ratio was maintained within a range of 1−2.5. This study also showed that zeolite A produced from BLA, as a precursor, had magnificent catalytic performance in the solvent-free cyanoethylation of methanol under autogenic pressure and moisture-tolerant conditions, with approximately 82% reactant conversion and 100% product selectivity even reused up to 10 successive reaction cycles. The results achieved from XRD and FESEM analysis imply that the crystalline nature of zeolite A was temperature-dependent and had a thermally activated mechanism. In a different research work, Setiadji et al. (2018) prepared zeolite-type ZSM-11 by utilizing bamboo leaf-extracted silica as the starting material. The high-purity silica was isolated from bamboo leaves by leaching using NaOH as a basic source. In this typical experiment, bamboo leaves were cleaned with distilled water thoroughly and kept in the sunlight for 2 days to evaporate the water content. At 650°C for 5 hours, the dried leaves were calcined to ash. The generated ash was introduced in the aging chamber containing 1 M NaOH and kept in the reaction temperature at 85°C for 1 hour. The obtained solution was then filtered and rinsed

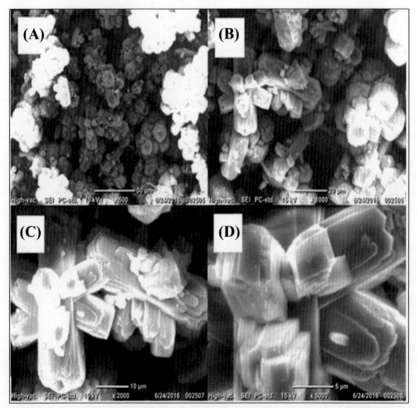

**FIGURE 5.12** Morphological analysis of synthetic ZSM-11 sample: (A) 500×, (B) 1000×, (C) 2000×, and (D) 5000× magnifications. *From Setiadji, S., Nurazizah, I. S., Sundari, C. D. D., Darmalaksana, W., Nurbaeti, D. F., Novianti, I., Azizah, T. B. N., Abdurrahman, D., & Ivansyah, A. L. (2018). Synthesis of zeolite ZSM-11 using bamboo leaf as silica source. In IOP conference series: Materials science and engineering (Vol. 434, No. 1). Institute of Physics Publishing. https://doi.org/10.1088/1757-899X/434/1/012084.*

with distilled water several times and 3 M $H_2SO_4$ was added drop-wise to the filtrate until a gel-type product was produced. The aging process was conducted with the gel material for 24 hours. The aged sample was then dried at 110°C for 24 hours to achieve the final product, that is, $SiO_2$. Once the silica was produced then it was allowed for zeolite crystallization through conventional hydrothermal technique at 170°C for 2 days. The FESEM images of the synthesized zeolite sample were recorded with four different magnification ranges, as shown in Fig. 5.12. According to the result, coffin and hexagonal-shaped zeolite morphology were obtained after the hydrothermal process.

Fig. 5.13 shows the X-ray diffractogram of the produced zeolite sample ZSM-11. The standard peaks appeared at $2\theta = 7.94$ degrees, 8.82 degrees, 14.82 degrees, 23.16 degrees, 23.99 degrees, 29.82 degrees, and 45.23 degrees indicates the existence of ZSM-11 in the prepared solid sample.

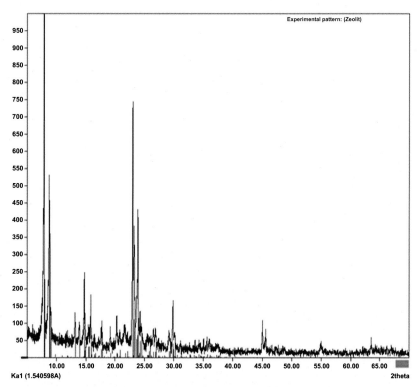

**FIGURE 5.13**   X-ray diffractogram of synthetic ZSM-11 zeolite. *From Setiadji, S., Nurazizah, I. S., Sundari, C. D. D., Darmalaksana, W., Nurbaeti, D. F., Novianti, I., Azizah, T. B. N., Abdurrahman, D., & Ivansyah, A. L. (2018). Synthesis of zeolite ZSM-11 using bamboo leaf as silica source. In* IOP conference series: Materials science and engineering *(Vol. 434, No. 1). Institute of Physics Publishing. https://doi.org/10.1088/1757-899X/434/1/012084.*

According to the findings, the significant sharp peak at 7.852 degrees was ascribed to the perfect crystallization of the zeolite phase (Setiadji et al., 2018). The XRD result reveals a higher degree of zeolite crystallinity with the above-mentioned reaction conditions.

## 5.3 Conclusion and future outlook

In the upcoming years, biomass-derived adsorbents will have a significant place in water treatment applications. The ash produced from different biomass sources such as bagasse, RH, palm oil, and bamboo leaf is used as a potential source for zeolite synthesis.

Several characteristics such as morphology, phase analysis, thermal activity, and compositional analysis of different biomass ash have been elucidated. It was seen that biomass ash is amorphous in nature, comprising several oxides along with a high amount of silica and alumina, which are often used as precursor materials in zeolite synthesis.

Typical type zeolite formations, like zeolite ZSM-11, zeolite 13X, zeolite A and X, and zeolite P1, taking biomass ash as starting material have been addressed in this chapter. The effect of various physical parameters, for instance, hydrothermal reaction temperature, time, silica-alumina ratio, and so on, was examined to achieve the highest zeolite production with good crystallinity. Several characterizations of the final product (zeolite) were discussed. Cubical, lepispheric, coffin, and hexagonal-shaped zeolite morphology was observed in the studies. Apart from this, BET analysis revealed a high surface area of biomass-ash-derived zeolites that should be beneficial for gas and heavy metal adsorption as well as environmental protection. POMFA-based zeolite X had a high surface area (643 $m^2$/g) and showed a higher affinity toward $CO_2$ capture (223 mg/g). RHA-derived zeolite-type A was found to be a potential sorbent for $Cu^{2+}$ adsorption from contaminated water.

Biomass-generated ash is useful for various types of zeolite synthesis; however, the presence of other constituents like $K_2O$, CaO, and MgO inhibits the zeolitization process and reduces the zeolite crystallinity. Hence, the reduction of toxic components from ash is an utmost challenging task. Therefore it is suggested that leaching treatment before zeolitzation can remove such detrimental components from the ash and can produce fresh zeolite having high adsorption capability. It is also recommended to the researchers find a unique technique that can reduce the zeolite production cost and can fulfill the demand, rather than using commercially available zeolites.

The direct utilization of zeolite-composite like sorbent for adsorption applications is very significant and reported in a few studies. The zeolite generated from biomass ash might be used to create composite materials for hazardous gas and metal adsorption.

# References

Anweshan., Das, Pranjal P., Dhara, S., & Purkait, Mihir K. (2023). Nanosensors in food science and technology. In A. Husen, & K. S. Siddiqi (Eds.), *Advances in smart nanomaterials and their applications* (pp. 247−272). Elsevier. Available from http://doi.org/10.1016/B978-0-323-99546-7.00015-X.

Assi, L., Anay, R., Leaphart, D., Soltangharaei, V., & Ziehl, P. (2018). Understanding early geopolymerization process of fly ash-based geopolymer paste using pattern recognition. *Journal of Materials in Civil Engineering*, *30*(6). Available from https://doi.org/10.1061/(ASCE) MT.1943-5533.0002270, http://ascelibrary.org/mto/resource/1/jmcee7/.

Bhandari, R., Volli, V., & Purkait, M. K. (2015). Preparation and characterization of fly ash based mesoporous catalyst for transesterification of soybean oil. *Journal of Environmental Chemical Engineering*, *3*(2), 906−914. Available from https://doi.org/10.1016/j.jece.2015.040.008, http://www.journals.elsevier.com/journal-of-environmental-chemical-engineering/.

Bharti, M., Das, Pranjal P., & Purkait, Mihir K. (2023). A review on the treatment of water and wastewater by electrocoagulation process: Advances and emerging applications. *Journal of Environmental Chemical Engineering*, *11*, 111558. Available from https://doi.org/10.1016/j.jece.2023.111558.

Bukhari, S. S., Behin, J., Kazemian, H., & Rohani, S. (2014). A comparative study using direct hydrothermal and indirect fusion methods to produce zeolites from coal fly ash utilizing single-mode microwave energy. *Journal of Materials Science*, *49*(24), 8261−8271. Available from https://doi.org/10.1007/s10853-014-8535-2, http://www.springer.com/journal/10853.

Chakraborty, S., Das, Pranjal P., & Mondal, P. (2023). Recent advances in membrane technology for the recovery and reuse of valuable resources. In M. Sillanpaa, A. Khadir, & K. Gurung (Eds.), *Resource recovery in industrial waste waters* (pp. 695−719). Elsevier. Available from http://doi.org/10.1016/B978-0-323-95327-6.00028-2.

Chakraborty, S., Gautam, S. P., Das, P. P., & Hazarika, M. K. (2019). Instant controlled pressure drop (DIC) treatment for improving process performance and milled rice quality. *Journal of The Institution of Engineers (India): Series A*, *100*(4), 683−695. Available from https://doi.org/10.1007/s40030-019-00403-w, http://www.springer.com/engineering/civil + engineering/journal/40030.

Chancey, R. T., Stutzman, P., Juenger, M. C. G., & Fowler, D. W. (2010). Comprehensive phase characterization of crystalline and amorphous phases of a Class F fly ash. *Cement and Concrete Research*, *40*(1), 146−156. Available from https://doi.org/10.1016/j.cemconres.2009.080.029, 00088846.

Che, S., Du, T., Zhu, S., Fang, X., & Wang, Y. (2019). Eco-friendly synthesis of kaolin-based chabazite for $CO_2$ capture. *Journal of the Ceramic Society of Japan*, *127*(9), 606−611. Available from https://doi.org/10.2109/jcersj2.19056, https://www.jstage.jst.go.jp/article/jcersj2/127/9/127_19056/_pdf/-char/en.

Che, S., Fang, X., Li, S., Chen, X., & Du, T. (2019). Modification of potassium chabazites derived from fly ash by dosing extra cations: Promoted $CO_2$ adsorption capacities and fine-tuned frameworks. Wiley-VCH Verlag, China. *Zeitschrift fur Anorganische und Allgemeine Chemie.*, *645*(24), 1365−1371. Available from https://doi.org/10.1002/zaac.201900196, http://onlinelibrary.wiley.com/journal/10.1002/(ISSN)1521-3749.

Crémoux, T., Batonneau-Gener, I., Moissette, A., Paillaud, J. L., Hureau, M., Ligner, E., ... Nouali, H. (2019). Influence of framework Si/Al ratio and topology on electron transfers in zeolites. *Physical*

*Chemistry Chemical Physics*, *21*(27), 14892–14903. Available from https://doi.org/10.1039/c9cp01166h, http://pubs.rsc.org/en/journals/journal/cp.

Das, P. P., Anweshan, A., Mondal, P., Sinha, A., Biswas, P., Sarkar, S., & Purkait, M. K. (2021). Integrated ozonation assisted electrocoagulation process for the removal of cyanide from steel industry wastewater. *Chemosphere*, *263*. Available from https://doi.org/10.1016/j.chemosphere.2020.128370, http://www.elsevier.com/locate/chemosphere.

Das, P. P., Mondal, P., Anweshan., Sinha, A., Biswas, P., Sarkar, S., & Purkait, M. K. (2021). Treatment of steel plant generated biological oxidation treated (BOT) wastewater by hybrid process. *Separation and Purification Technology*, *258*. Available from https://doi.org/10.1016/j.seppur.2020.118013, http://www.journals.elsevier.com/separation-and-purification-technology/.

Das, P. P., Anweshan, A., & Purkait, M. K. (2021). Treatment of cold rolling mill (CRM) effluent of steel industry. *Separation and Purification Technology*, *274*. Available from https://doi.org/10.1016/j.seppur.2021.119083, http://www.journals.elsevier.com/separation-and-purification-technology/.

Das, P. P., Sharma, M., & Purkait, M. K. (2022). Recent progress on electrocoagulation process for wastewater treatment: A review. *Separation and Purification Technology*, *292*. Available from https://doi.org/10.1016/j.seppur.2022.121058, http://www.journals.elsevier.com/separation-and-purification-technology/.

Das, Pranjal P., & Mondal, P. (2023). Membrane-assisted potable water reuses applications: Benefits and drawbacks. In M. Sillanpaa, A. Khadir, & K. Gurung (Eds.), *Resource recovery in drinking water treatment* (pp. 289–309). Elsevier. Available from http://doi.org/10.1016/B978-0-323-99344-9.00014-1.

Das, Pranjal P., Mondal, P., & Purkait, M. K. (2022). *Recent advances in synthesis of iron nanoparticles via green route and their application in biofuel production. Green nano solution for bioenergy production enhancement. Clean energy production Technologies* (pp. 79–104). Springer Science and Business Media LLC. Available from 10.1007/978-981-16-9356-4_4.

Das, Pranjal P., Deepti., & Purkait, Mihir K. (2023). Industrial wastewater to biohydrogen production via potential bio-refinery route. In Maulin P. Shah (Ed.), *Biorefinery for water and wastewater treatment* (pp. 159–179). Springer. Available from http://doi.org/10.1007/978-3-031-20822-5_8.

Das, Pranjal P., Dhara, S., & Purkait, Mihir K. (2023). The anaerobic ammonium oxidation process: Inhibition, challenges and opportunities. In Maulin P. Shah (Ed.), *Ammonia oxidizing bacteria: Applications in industrial wastewater treatment* (pp. 56–82). Royal Society of Chemistry. Available from http://doi.org/10.1039/BK9781837671960-00056.

Das, Pranjal P., Dhara, S., & Purkait, Mihir K. (2023). Hybrid electrocoagulation and ozonation techniques for industrial wastewater treatment. In Maulin P. Shah (Ed.), *Sustainable industrial wastewater treatment and pollution control* (pp. 107–128). Springer. Available from http://doi.org/10.1007/978-981-99-2560-5_6.

Das, Pranjal P., Dhara, S., & Purkait, Mihir K. (2024). Ozone-based oxidation processes for the removal of pharmaceutical products from wastewater. In Maulin P. Shah, & Pooja Ghosh (Eds.), *Development in wastewater treatment research and processes* (pp. 287–308). Elsevier. Available from http://doi.org/10.1016/B978-0-443-19207-4.00003-3.

Das, Pranjal P., Duarah, P., & Purkait, M. K. (2023). *Fundamentals of food roasting process. High-temperature processing of food products* (pp. 103–130). Elsevier BV. Available from 10.1016/b978-0-12-818618-3.00005-7.

Davina, V., Utama, P.S., Saputra, E., & Bahri, S. (2019). Zeolite Na-P1 derived from palm oil mill fly ash: Synthesis and characterization. In *Journal of physics: Conference series* (Vol. 1351, No. 1). Institute of Physics Publishing, Indonesia. 10.1088/1742-6596/1351/1/012103 17426596 http://iopscience.iop.org/journal/1742-6596.

Das, Pranjal P., Samanta, N. S., Dhara, S., & Purkait, Mihir K. (2023). Biofuel production from algal biomass. In Maulin P. Shah (Ed.), *Green approach to alternative fuel for a sustainable future* (pp. 167−179). Elsevier. Available from http://doi.org/10.1016/B978-0-12-824318-3.00009-6.

Das, Pranjal P., Sontakke, A. D., & Purkait, Mihir K. (2023). Electrocoagulation process for wastewater treatment: Applications, challenges, and prospects. In Maulin P. Shah (Ed.), *Development in wastewater treatment research and processes* (pp. 23−48). Elsevier. Available from http://doi.org/10.1016/B978-0-323-95684-0.00015-4.

Das, Pranjal P., Sontakke, A. D., & Purkait, Mihir K. (2023). Rice straw for biofuel production. In Maulin P. Shah (Ed.), *Green approach to alternative fuel for a sustainable future* (pp. 153−166). Elsevier. Available from http://doi.org/10.1016/B978-0-12-824318-3.00034-5.

Das, Pranjal P., Sontakke, A. D., Samanta, N. S., & Purkait, Mihir K. (2023). Emerging contaminants in wastewater: Eco-toxicity and sustainability assessment. In Maulin P. Shah (Ed.), *Industrial wastewater reuse* (pp. 63−87). Springer. Available from http://doi.org/10.1007/978-981-99-2489-9_4.

Dhara, S., Das, Pranjal P., Uppaluri, R., & Purkait, Mihir K. (2023). Phosphorus recovery from municipal wastewater treatment plants. In Maulin P. Shah (Ed.), *Development in wastewater treatment research and processes* (pp. 49−72). Elsevier. Available from http://doi.org/10.1016/B978-0-323-95684-0.00014-2.

Dhara, S., Das, Pranjal P., Uppaluri, R., & Purkait, Mihir K. (2023). Biological approach for energy self-sufficiency of municipal wastewater treatment plants. In M. Sillanpaa, A. Khadir, & K. Gurung (Eds.), *Resource recovery in municipal waste waters* (pp. 235−260). Elsevier. Available from http://doi.org/10.1016/B978-0-323-99348-7.00006-0.

Dhara, S., Samanta, N. S., Das, Pranjal P., Uppaluri, R. V. S., & Purkait, M. K. (2023). Ravenna grass-extracted alkaline lignin-based polysulfone Mixed Matrix Membrane (MMM) for aqueous Cr(VI) removal. *Applied Polymer Materials*, 5, 6399−6411. Available from https://doi.org/10.1021/acsapm.3c00999.

Flores, C. G., Schneider, H., Marcilio, N. R., Ferret, L., & Oliveira, J. C. P. (2017). Potassic zeolites from Brazilian coal ash for use as a fertilizer in agriculture. *Waste Management*, 70, 263−271. Available from https://doi.org/10.1016/j.wasman.2017.080.039, http://www.elsevier.com/locate/wasman.

Gomes Flores, C., Schneider, H., Dornelles, J. S., Gomes, L. B., Marcilio, N. R., & Melo, P. J. (2021). Synthesis of potassium zeolite from rice husk ash as a silicon source. *Cleaner Engineering and Technology*, 4, 100201. Available from https://doi.org/10.1016/j.clet.2021.100201, 26667908.

Jha, B., & Singh, D. N. (2012). ChemInform abstract: A review on synthesis, characterization and industrial applications of flyash zeolites. *ChemInform*, 43(25), 09317597. Available from https://doi.org/10.1002/chin.201225227.

Khaleque, A., Alam, M. M., Hoque, M., Mondal, S., Haider, J. B., Xu, B., ... Moni, M. A. (2020). Zeolite synthesis from low-cost materials and environmental applications: A review. *Environmental Advances*, 2. Available from https://doi.org/10.1016/j.envadv.2020.100019, http://www.journals.elsevier.com/environmental-advances.

Kishimoto, T. (2000). Environmental research. *NTT Review*, 12(5), 36−41.

Kongnoo, A., Tontisirin, S., Worathanakul, P., & Phalakornkule, C. (2017). Surface characteristics and $CO_2$ adsorption capacities of acid-activated zeolite 13X prepared from palm oil mill fly ash. *Fuel*, *193*, 385−394. Available from https://doi.org/10.1016/j.fuel.2016.120.087, http://www.journals.elsevier.com/fuel/.

Kordatos, K., Gavela, S., Ntziouni, A., Pistiolas, K. N., Kyritsi, A., & Kasselouri-Rigopoulou, V. (2008). Synthesis of highly siliceous ZSM-5 zeolite using silica from rice husk ash. *Microporous and Mesoporous Materials*, *115*(1-2), 189−196. Available from https://doi.org/ 10.1016/j.micromeso.2007.120.032, 13871811.

Krishnarao, R. V., Subrahmanyam, J., & Jagadish Kumar, T. (2001). Studies on the formation of black particles in rice husk silica ash. *Journal of the European Ceramic Society*, *21*(1), 99−104. Available from https://doi.org/10.1016/S0955-2219(00)00170-9, 09552219.

Liu, B., Zheng, Y., Hu, N., Gui, T., Li, Y., Zhang, F., . . . Kita, H. (2014). Synthesis of low-silica CHA zeolite chabazite in fluoride media without organic structural directing agents and zeolites. *Microporous and Mesoporous Materials*, *196*, 270−276. Available from https://doi.org/10.1016/j.micromeso.2014.050.019, 13871811.

Loiha, S., Prayoonpokarach, S., Songsiriritthigun, P., & Wittayakun, J. (2009). Synthesis of zeolite beta with pretreated rice husk silica and its transformation to ZSM-12. *Materials Chemistry and Physics*, *115*(2-3), 637−640. Available from https://doi.org/10.1016/j.matchemphys.2009.010.031, 02540584.

Mohamed, R. M., Mkhalid, I. A., & Barakat, M. A. (2015). Rice husk ash as a renewable source for the production of zeolite NaY and its characterization. *Arabian Journal of Chemistry*, *8* (1), 48−53. Available from https://doi.org/10.1016/j.arabjc.2012.120.013, http://colleges.ksu. edu.sa/Arabic%20Colleges/CollegeOfScience/ChemicalDept/AJC/default.aspx. (ScienceDirect http://www.sciencedirect.com/science/journal/18785352).

Mondal, P., Samanta, N. S., Meghnani, V., & Purkait, M. K. (2019). Selective glucose permeability in presence of various salts through tunable pore size of pH responsive PVDF-co-HFP membrane. *Separation and Purification Technology*, *221*, 249−260. Available from https://doi.org/10.1016/j.seppur.2019.040.001, http://www.journals.elsevier.com/separation-and-purification-technology/.

Mondal, P., Samanta, N. S., Kumar, A., & Purkait, M. K. (2020). Recovery of $H_2SO_4$ from wastewater in the presence of NaCl and $KHCO_3$ through pH responsive polysulfone membrane: Optimization approach. *Polymer Testing*, *86*. Available from https://doi.org/10.1016/j. polymertesting.2020.106463, https://www.journals.elsevier.com/polymer-testing.

Ng, E. P., Chow, J. H., Mukti, R. R., Muraza, O., Ling, T. C., & Wong, K. L. (2017). Hydrothermal synthesis of zeolite a from bamboo leaf biomass and its catalytic activity in cyanoethylation of methanol under autogenic pressure and air conditions. *Materials Chemistry and Physics*, *201*, 78−85. Available from https://doi.org/10.1016/j.matchemphys.2017.080.044, http://www.journals.elsevier.com/materials-chemistry-and-physics/.

Oliveira, J. A., Cunha, F. A., & Ruotolo, L. A. M. (2019). Synthesis of zeolite from sugarcane bagasse fly ash and its application as a low-cost adsorbent to remove heavy metals. *Journal of Cleaner Production*, *229*, 956−963. Available from https://doi.org/10.1016/j.jclepro.2019.050.069, https://www.journals.elsevier.com/journal-of-cleaner-production.

Opiso, E. M., Tabelin, C. B., Maestre, C. V., Aseniero, J. P. J., Park, I., & Villacorte-Tabelin, M. (2021). Synthesis and characterization of coal fly ash and palm oil fuel ash modified artisanal and small-scale gold mine (ASGM) tailings based geopolymer using sugar mill lime sludge as Ca-based activator. *Heliyon*, *7*(4). Available from https://doi.org/10.1016/j.heliyon.2021.e06654, http://www.journals.elsevier.com/heliyon/.

Park, I., Tabelin, C. B., Seno, K., Jeon, S., Ito, M., & Hiroyoshi, N. (2018). Simultaneous suppression of acid mine drainage formation and arsenic release by Carrier-microencapsulation using aluminum-catecholate complexes. *Chemosphere*, *205*, 414−425. Available from https://doi.org/10.1016/j.chemosphere.2018.040.088, http://www.elsevier.com/locate/chemosphere.

Pode, R. (2016). Potential applications of rice husk ash waste from rice husk biomass power plant. *Renewable and Sustainable Energy Reviews*, *53*, 1468−1485. Available from https://doi.org/10.1016/j.rser.2015.090.051, https://www.journals.elsevier.com/renewable-and-sustainable-energy-reviews.

Purnomo, C. W., Salim, C., & Hinode, H. (2012). Synthesis of pure Na-X and Na-A zeolite from bagasse fly ash. *Microporous and Mesoporous Materials*, *162*, 6−13. Available from https://doi.org/10.1016/j.micromeso.2012.060.007, 13871811.

Samanta, N. S., Banerjee, S., Mondal, P., Anweshan., Bora, U., & Purkait, M. K. (2021). Preparation and characterization of zeolite from waste Linz-Donawitz (LD) process slag of steel industry for removal of Fe3 + from drinking water. *Advanced Powder Technology*, *32* (9), 3372−3387. Available from https://doi.org/10.1016/j.apt.2021.070.023, http://www.elsevier.com.

Samanta, N. S., Das, P. P., Mondal, P., Changmai, M., & Purkait, M. K. (2022). Critical review on the synthesis and advancement of industrial and biomass waste-based zeolites and their applications in gas adsorption and biomedical studies. *Journal of the Indian Chemical Society*, *99*(11). Available from https://doi.org/10.1016/j.jics.2022.100761, https://www.sciencedirect.com/journal/journal-of-the-indian-chemical-society.

Samanta, N. S., Das, Pranjal P., Dhara, S., & Purkait, Mihir K. (2023). An overview of precious metal recovery from steel industry slag: Recovery strategy and utilization. *Industrial & Engineering Chemistry Research*, *62*, 9006−9031. Available from https://doi.org/10.1021/acs.iecr.3c00604.

Samanta, N. S., Das, P. P., Mondal, P., Bora, U., & Purkait, M. K. (2022). Physico-chemical and adsorption study of hydrothermally treated zeolite A and FAU-type zeolite X prepared from LD (Linz−Donawitz) slag of the steel industry. *International Journal of Environmental Analytical Chemistry*. Available from https://doi.org/10.1080/03067319.2022.2079082, http://www.tandf.co.uk/journals/titles/03067319.asp.

Schulman, E., Wu, W., & Liu, D. (2020). Two-dimensional zeolite materials: Structural and acidity properties. *Materials*, *13*(8). Available from https://doi.org/10.3390/MA13081822, https://www.mdpi.com/1996-1944/13/8/1822.

Setiadji, S., Nurazizah, I.S., Sundari, C.D. D., Darmalaksana, W., Nurbaeti, D.F., Novianti, I., ...Ivansyah, A.L. (2018). Synthesis of zeolite ZSM-11 using bamboo leaf as silica source. In *IOP conference series: Materials science and engineering* (Vol. 434, No. 1). Institute of Physics Publishing, Indonesia. https://doi.org/10.1088/1757-899X/434/1/012084 1757899X. https://iopscience.iop.org/journal/1757-899X.

Sharma, M., Das, Pranjal P., Sood, T., Chakraborty, A., & Purkait, M. K. (2021). Ameliorated polyvinylidene fluoride based proton exchange membrane impregnated with graphene oxide, and cellulose acetate obtained from sugarcane bagasse for application in microbial fuel cell. *Journal of Environmental Chemical Engineering*, *9*, 106681. Available from https://doi.org/10.1016/j.jece.2021.106681.

Sharma, M., Das, Pranjal P., Sood, T., Chakraborty, A., & Purkait, M. K. (2022). Reduced graphene oxide incorporated polyvinylidene fluoride/cellulose acetate proton exchange membrane for energy extraction using microbial fuel cells. *Journal of Electroanalytical Chemistry*, *907*, 115890. Available from https://doi.org/10.1016/j.jelechem.2021.115890.

Sharma, M., Das, Pranjal P., Chakraborty, A., & Purkait, M. K. (2022). Clean energy from salinity gradients using pressure retarded osmosis and reverse electrodialysis: A review. *Sustainable Energy Technologies and Assessments, 49*, 101687. Available from https://doi.org/10.1016/j.seta.2021.101687.

Sharma, M., Das, Pranjal P., Kumar, S., & Purkait, M. K. (2023). Polyurethane foams as packing and insulating materials. In Ram K. Gupta (Ed.), *Polyurethanes: Preparation, properties, and applications* (pp. 83–99). American Chemical Society. Available from http://doi.org/10.1021/bk-2023-1454.ch004.

Sontakke, A. D., Das, P. P., Mondal, P., & Purkait, M. K. (2022). Thin-film composite nanofiltration hollow fiber membranes toward textile industry effluent treatment and environmental remediation applications: Review. *Emergent Materials, 5*(5), 1409–1427. Available from https://doi.org/10.1007/s42247-021-00261-y, https://www.springer.com/journal/42247.

Sontakke, A. D., Deepti., Samanta, N. S., & Purkait, M. K. (2023). *Smart nanomaterials in the medical industry* (pp. 23–50). Elsevier BV. Available from 10.1016/b978-0-323-99546-7.00025-2.

Sánchez-Hernández, R., López-Delgado, A., Padilla, I., Galindo, R., & López-Andrés, S. (2016). One-step synthesis of NaP1, SOD and ANA from a hazardous aluminum solid waste. *Microporous and Mesoporous Materials, 226*, 267–277. Available from https://doi.org/10.1016/j.micromeso.2016.010.037, http://www.elsevier.com/inca/publications/store/6/0/0/7/6/0.

Volli, V., & Purkait, M. K. (2015). Selective preparation of zeolite X and A from flyash and its use as catalyst for biodiesel production. *Journal of Hazardous Materials, 297*, 101–111. Available from https://doi.org/10.1016/j.jhazmat.2015.040.066, http://www.elsevier.com/locate/jhazmat.

Wibowo E. Sutisna Rokhmat M. Murniati R. Khairurrijal Abdullah M. 2017, 170, pp. 8-13 Indonesia utilization of natural zeolite as sorbent material for seawater desalination, Procedia Engineering..

Xu, G., & Shi, X. (2018). Characteristics and applications of fly ash as a sustainable construction material: A state-of-the-art review. *Resources, Conservation and Recycling, 136*, 95–109. Available from https://doi.org/10.1016/j.resconrec.2018.040.010, http://www.elsevier.com/locate/resconrec.

Yaghoubi, H., Allahyari, M. S., Firouzi, S., Damalas, C. A., & Marzban, S. (2019). Identifying sustainable options for rice husk valorization using the analytic hierarchy process. *Outlook on Agriculture, 48*(2), 117–125. Available from https://doi.org/10.1177/0030727018821384, https://journals.sagepub.com/home/OAG.

Yan, K., Guo, Y., Ma, Z., Zhao, Z., & Cheng, F. (2018). Quantitative analysis of crystalline and amorphous phases in pulverized coal fly ash based on the Rietveld method. *Journal of Non-Crystalline Solids, 483*, 37–42. Available from https://doi.org/10.1016/j.jnoncrysol.2017.120.043, http://www.journals.elsevier.com/journal-of-non-crystalline-solids/.

Zakharov, A. I., Belyakov, A. V., & Tsvigunov, A. N. (1993). Forms of extraction of silicon compounds in rice husks. *Glass and Ceramics, 50*(9-10), 420–425. Available from https://doi.org/10.1007/bf00683590, 0361-7610.

Zhang, P., Liao, W., Kumar, A., Zhang, Q., & Ma, H. (2020). Characterization of sugarcane bagasse ash as a potential supplementary cementitious material: Comparison with coal combustion fly ash. *Journal of Cleaner Production, 277*123834. Available from https://doi.org/10.1016/j.jclepro.2020.123834, 09596526.

# Chapter 6

# Utilization of municipal solid waste fly ash for zeolite preparation

## 6.1 Introduction

Municipal solid waste incineration has emerged significantly as a viable approach to mitigate waste volume and mass while simultaneously facilitating energy recovery. Consequently, it is progressively assuming an indispensable role as a disposal option. A significant quantity of municipal solid waste incinerator fly ash (MSWI FA) is produced through the process of incineration, constituting around 3%−15% of the initial waste mass. Based on the Chinese Standard GB 18485-2014, MSWI FA is categorized as hazardous waste due to its significant levels of toxicity attributed to dioxins and heavy metals. Insufficient treatment of MSWI FA may result in the infiltration of elevated levels of heavy metals into the soil and groundwater, posing a potential risk to human health (Lee & van Deventer, 2002; Okonji, Dominic, Pernitsky, & Achari, 2020; Samadi-Maybodi & Masoomeh Pourali, 2013). Based on the Chinese regulatory framework outlined in GB 16889, it is permissible to transport MSWI FA to sanitary landfills subsequent to undergoing appropriate stabilization procedures. Nevertheless, the escalating fabrication of MSWI FA coupled with the limited availability of land resources has resulted in exorbitant disposal expenses for landfills (Chang & Wey, 2006; Cristelo et al., 2020; Phua, Giannis, Dong, Lisak, & Ng, 2019; Zhang et al., 2021). Consequently, it is imperative to investigate ecologically sustainable reusable approaches for MSWI FA. At present, conventional methods for treating MSWI FA in an environmentally safe manner encompass solidification/stabilization techniques, separation methods, as well as thermal treatment approaches. The S/S technique, which stands for solidification/stabilization, is a commonly employed method for treating municipal solid waste incinerator fly ash (Mondal, Samanta, Meghnani, & Purkait, 2019; Samanta, Mondal, & Purkait, 2023). This approach utilizes process equipment that is straightforward and user-friendly in nature. Nevertheless, the extended release of heavy metals through leaching exhibits an inherent instability, while the process of cement solidification yields a

Waste-based Zeolite. DOI: https://doi.org/10.1016/B978-0-443-22316-7.00006-1
**145**

substantial increase in volume, hence posing challenges in terms of waste management. The separation of heavy metals and soluble salts from MSWI FA can be achieved by a low-temperature separation method, resulting in enhanced quality and uniformity of the treated MSWI FA. However, the extensive use of this technology is constrained by the issue of secondary pollution through wastewater and the requirement for harsh reaction conditions (Hu et al., 2013; Montañés, Sánchez-Tovar, & Roux, 2014; Nagib & Inoue, 2000; Tang, Erzat, & Liu, 2014). There are several other methods used for the treatment of water and wastewater, such as electrocoagulation (Changmai et al., 2022; Das, Sontakke, Purkait, & Shah, 2023a; Das, Anweshan, & Purkait, 2021; Das, Anweshan, Mondal, et al., 2021; Das, Dhara, & Purkait, 2023; Das, Mondal, et al., 2021; Das, Sharma, & Purkait, 2022), membrane (Chakraborty et al., 2023; Das, Duarah, Purkait, & Jafari, 2023; Dhara, Shekhar Samanta, Das, Uppaluri, & Purkait, 2023; Sharma, Das, Sood, Chakraborty, & Purkait, 2021; Sharma, Das, Chakraborty, & Purkait, 2022; Sharma, Das, Sood, Chakraborty, & Purkait, 2022; Sontakke, Das, Mondal, & Purkait, 2022), adsorption (Chakraborty, Gautam, Das, & Hazarika, 2019; Das, Mondal, Sillanpää, Khadir, & Gurung, 2023; Samanta, Das, & Mondal, 2022; Samanta, Das, Mondal, Changmai, & Purkait, 2022; Samanta, Das, et al., 2023; Sharma, Das, Kumar, & Purkait, 2023; Shekhar Samanta, Das, Dhara, & Purkait, 2023), nanotechnology (Anweshan et al., 2023; Das, Samanta, Dhara, Purkait, & Shah, 2023; Das, Mondal, & Purkait, 2022; Duarah, Das, & Purkait, 2023; Sharma, Das, Purkait, Husen, & Siddiqi, 2023), and biological treatments (Das, Deepti, & Purkait, 2023; Das, Sontakke, Samanta, & Purkait, 2023; Das, Sontakke, Purkait, & Shah, 2023b; Dhara, Das, Uppaluri, Purkait, & Shah, 2023; Dhara, Das, Uppaluri, Purkait, Sillanpää, et al., 2023; Sharma, Das, Chakraborty, et al., 2023).

Thermal remediation procedures offer several benefits, including a significant reduction in volume, a low leaching rate for heavy metals, and the capability to degrade dioxins. Nevertheless, thermal approaches exhibit a notable drawback in terms of their substantial energy requirements, the following necessity for flue gas treatment to mitigate the presence of volatile heavy metals like Pb, Cd, and Zn, and the associated financial burden of such treatment processes. Hence, there is an urgent requirement for a treatment approach that combines high efficiency, energy conservation, and cost-effectiveness (Behin, Bukhari, Kazemian, & Rohani, 2016; Cardoso, Paprocki, Ferret, Azevedo, & Pires, 2015; Chindaprasirt & Rattanasak, 2017). In recent years, there has been an increasing focus on the hydrothermal treatment method as a potentially effective strategy for managing several types of solid wastes (Aldahri, Behin, Kazemian, & Rohani, 2016; Anuwattana & Khummongkol, 2009; Chen et al., 2020; Ferreira, Ribeiro, & Ottosen, 2003; Zhang et al., 2022). Aluminosilicate zeolite has been effectively synthesized from solid waste materials, exhibiting notable characteristics such as a significant specific surface area and porosity. The primary components of MSWI FA consist of oxides, which are produced as a result of the highly oxidizing conditions within the incinerator. Notably, circulating fluidized bed (CFB) MSWI fly ash exhibits elevated concentrations of calcium oxide (CaO), silicon dioxide ($SiO_2$), aluminum oxide ($Al_2O_3$), and iron oxide ($Fe_2O_3$). The alkalinity of MSWI FA originates from its elevated calcium oxide (CaO) content, rendering it conducive for zeolite formation through the involvement of hydroxide ions ($OH^-$). Several research studies have employed $SiO_2$ and $Al_2O_3$ present in MSWI FA to produce aluminosilicate zeolites, such as gismondine, gmelinite, tobermorite, and katoite, using thermal techniques. Synthetic zeolites possess the ability to stabilize heavy metals within the resulting hydrothermal products, hence mitigating the transfer of heavy metals into the liquid phase (Dhara, Samanta, Uppaluri, & Purkait, 2023; Samanta, Anweshan, Mondal, Bora, & Purkait, 2023; Samanta et al., 2021; Sharma, Samanta, et al., 2023; Sontakke et al., 2023). Additionally, these zeolites have a noteworthy adsorption capacity for $Cs^+$, $NH_4^+$, and $Cu^{2+}$. Previous research has demonstrated that zeolite P exhibits notable efficacy in water softening and can further serve as a viable catalyst for the production of eco-friendly detergents. Furthermore, it has also been employed for the purpose of gas separation and the removal of radioactive waste (Das, Dhara, & Purkait, 2024). Previous studies have indicated that the Si/Al molar ratio plays a crucial role in the production of zeolite P. Nevertheless, the available sources of Si and Al in MSWI FA are restricted, with Si often present in a crystalline form, such as quartz. Therefore the inclusion of silica−alumina additives is important to facilitate the formation of zeolite during the hydrothermal process (Fan, Zhang, Zhu, & Liu, 2008; Ren et al., 2020; Shi et al., 2017; Yang & Yang, 1998).

The synthesis of zeolites using conventional hydrothermal methods at elevated temperatures is known to be energy-intensive. Typically, this process requires heating at temperatures ranging from 90°C to 200°C for a duration spanning from several hours to several weeks. The utilization of microwave and ultrasonic energy is employed as an alternate approach for the synthesis of zeolites. Microwave heating (MH) is classified as a form of bulk heating, characterized by several advantageous features such as uniform heating, rapid heating rate, selective heating, and energy efficiency (Abdullahi, Harun, & Othman, 2017; Bukhari, Behin, Kazemian, & Rohani, 2015; Pal, Das, Das, & Bandyopadhyay, 2013; Sivalingam & Sen, 2019; Zubowa et al., 2008). The application of ultrasonic irradiation has been found to expedite heterogeneous processes involving liquid and solid reactants, leading to a decrease in both the time required for crystallization and the temperature at which crystallization occurs. Nevertheless, the temperature range for ultrasonic treatment typically falls between 50°C and 95°C. This temperature range is not optimal for facilitating $SiO_2$ dissolution in MSWI FA when it is subjected to an alkaline solution. As a result, ultrasonic treatment is commonly employed in conjunction with alkaline fusion and standard hydrothermal treatment methods. The microwave hydrothermal process has the capability to achieve elevated temperatures ranging from 100°C to 200°C. This elevated temperature range has the potential to expedite the usage of $SiO_2$ in MSWI FA (Inada, Eguchi, Enomoto, & Hojo, 2005; Qiu et al., 2019; Querol et al., 1997).

This chapter focuses on various techniques utilized for the preparation of zeolites from municipal solid waste. The techniques, especially the alkali fusion−assisted hydrothermal technique and microwave-assisted hydrothermal method, have been discussed and summarized. The operating parameters of alkali fusion−assisted hydrothermal technique, viz. temperature, mineralizer concentration, and solid-to-liquid ratio, as well as microwave-assisted hydrothermal method, viz. treatment time, NaOH concentration, and $Na_2SiO_3$ dosage, have been covered in detail. Further, the challenges and future recommendations on zeolite preparation from municipal solid waste have also been included toward the end of the chapter.

## 6.2 Various techniques to prepare zeolite from municipal solid waste

### 6.2.1 Alkali fusion followed by hydrothermal method

#### 6.2.1.1 Effects of operating temperature

Fig. 6.1 illustrates the significant impact of operating temperature on the Brunauer−Emmett−Teller (BET) surface areas and average pore radii of the manufactured products. The particular surface areas obtained at operating

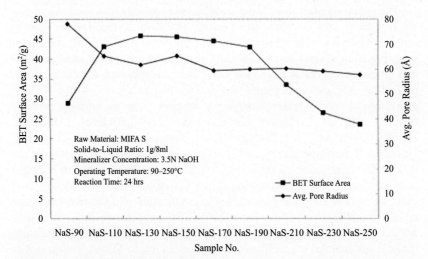

**FIGURE 6.1** **Effects of the operating temperature on the cation exchange capacity of synthesized products originated from MIFA S** . *From Yang, G. C. C., & Yang, T.-Y. (1998). Journal of Hazardous Materials, 62(1), 75−89. https://doi.org/10.1016/S0304-3894(98)00163-0.*

temperatures ranging from 110°C to 190°C were around 45 m$^2$/g. However, the highest BET surface area was achieved at an operating temperature of 130°C. The operating temperature below 110°C may have been insufficient for achieving optimal conversion. Consequently, a somewhat limited BET surface area was acquired. Conversely, when the operating temperature exceeds 190°C, a significant decrease in the BET surface area is seen. The occurrence of distinct mineral species in the synthesized product at the critical temperature of 190°C can be attributed to this phenomenon. With the exception of the pore radius at 90°C, the average pore radii observed across all tested temperatures fall within the range of 57−65 Å. According to the data presented in Fig. 6.2, it can be observed that the cation exchange capacity (CEC) of the synthesized products exhibited an upward trend with increasing operating temperature, reaching its peak at 190°C. Subsequently, the CEC value exhibited a negative correlation with rising temperatures. The X-ray diffraction (XRD) study results indicate that the principal mineral species observed at temperatures of 190°C and 250°C are identical. The disparity in CEC values between these two instances may be attributed to variations in the proportions of distinct zeolite-like substances (Yang & Yang, 1998).

### 6.2.1.2 Effects of mineralizer concentration

The influence of the concentration of the mineralizer on the BET surface area and average pore radius is depicted in Fig. 6.3. Typically, insufficient alkalinity in a mineralizer hinders the synthesis process, resulting in inadequate development of the zeolite structure. The 1 N NaOH solution yielded a synthesized product with low crystallinity, leading to a significantly reduced BET surface area.

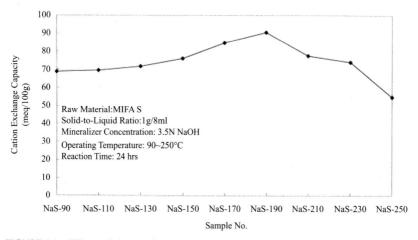

**FIGURE 6.2** **Effects of the operating temperature on the cation exchange capacity of synthesized products originated from MIFA S** . *From Yang, G. C. C., & Yang, T.-Y. (1998).* Journal of Hazardous Materials, 62(1), 75−89. https://doi.org/10.1016/S0304-3894(98)00163-0.

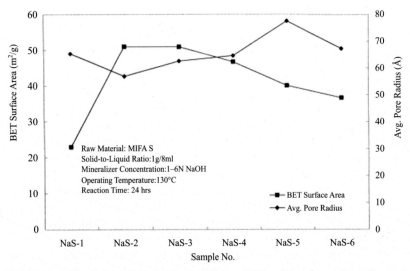

**FIGURE 6.3**    **Effects of mineralizer concentration on BET surface area and average pore radius of synthesized products originated from MIFA S** . *From Yang, G. C. C., & Yang, T.-Y. (1998).* Journal of Hazardous Materials, *62(1), 75−89. https://doi.org/10.1016/S0304-3894(98) 00163-0.*

Consequently, the synthesized goods exhibited higher values of BET surface area. A reduction in the specific surface area was seen when a concentration exceeding 4 N was employed. There is a hypothesis suggesting that a transition in zeolite type occurs within the vicinity of this particular concentration of mineralizer. The forthcoming presentation of XRD findings reveals that gismondine has undergone a conversion into different zeolite species. Nevertheless, the observed alteration of the zeolite type did not result in a substantial increase in the average pore radius. Similarly, the concentration of the mineralizer had an impact on the CEC values of the synthesized products, as shown in Fig. 6.4. A CEC of 50.61 meq/100 g was found when a solution of 1 N NaOH was employed. When the concentration of the mineralizer was kept within the range of 2−4 N, a CEC value of approximately 64 meq/100 g could be achieved. The CEC value experienced a decrease following the transition to a different zeolite type. The synthesized products obtained in this investigation exhibit a much lower CEC value compared to commercially available zeolites. Typically, commercial zeolites exhibit CEC values ranging from 200 to 300 meq/100 g. However, certain zeolites can possess exceptionally high CEC values, reaching up to 920 meq/100 g (Yang & Yang, 1998).

### 6.2.1.3    Effects of solid-to-liquid ratio

Fig. 6.5 illustrates the impact of the solid-to-liquid ratio, specifically the municipal incinerator fly ash (MIFA)/mineralizer ratio, on the BET surface area and average pore radius of MIFA S subsequent to its manufacture.

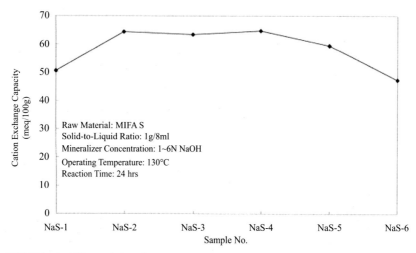

**FIGURE 6.4    Effects of the mineralizer concentration on cation exchange capacity of synthesized products originated from MIFA S** . *From Yang, G. C. C., & Yang, T.-Y. (1998).* Journal of Hazardous Materials, 62(1), 75–89. https://doi.org/10.1016/S0304-3894(98)00163-0.

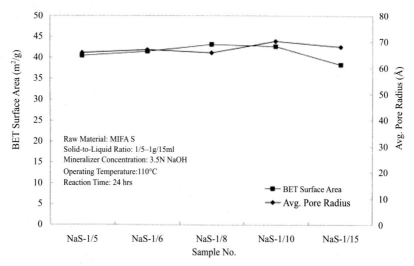

**FIGURE 6.5    Effects of the solid-to-liquid on BET surface area and average pore radius of synthesized products originated from MIFA S** . *From Yang, G. C. C., & Yang, T.-Y. (1998).* Journal of Hazardous Materials, 62(1), 75–89. https://doi.org/10.1016/S0304-3894(98)00163-0.

Typically, the fluctuation in BET surface area is negligible, falling within the range of 38–43 $m^2$/g. In a similar vein, the average pore radii observed in the synthesized products exhibited a range of 66–70 Å. Nevertheless, the synthesized products exhibit higher values of BET surface area and average

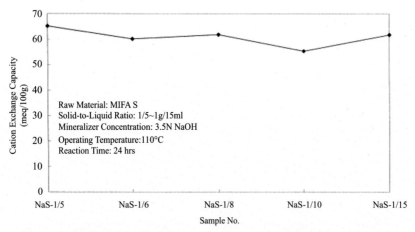

**FIGURE 6.6** **Effects of the solid-to-liquid ratio on the cation exchange capacity of synthe-sized products originated from MIFA S** . *From Yang, G. C. C., & Yang, T.-Y. (1998).* Journal of Hazardous Materials, *62(1), 75−89. https://doi.org/10.1016/S0304-3894(98)00163-0.*

pore radius compared to the initial MIFA S, specifically measuring at 2.54 m$^2$/g and 45.85 Å, respectively. It is noteworthy to mention that a con-centration of 3.5 N was employed as the mineralizer concentration for inves-tigating the effects of MIFA/mineralizer in the investigation. A mineralizer concentration of 3.5 N is sufficient to facilitate the conversion of MIFA S into materials with zeolite-like characteristics. When the concentration of a mineralizer exceeds the threshold value required for the formation of a zeo-lite structure, the influence of the MIFA-to-mineralizer ratio on specific sur-face area and average pore radius is deemed to be negligible. Fig. 6.6 also demonstrated a comparable impact of the ratio between solid and liquid phases on the CEC value (Yang & Yang, 1998).

## 6.2.2 Microwave-assisted hydrothermal method

### 6.2.2.1 Effect of microwave-assisted hydrothermal treatment time

Fig. 6.7A displays the XRD diffractograms obtained from the products syn-thesized at various MH durations ranging from 0.5 to 2 hours. The intensity of the peak observed in quartz exhibits a declining pattern, whereas the peak intensity observed in zeolite Na-P1 demonstrates a clear upward trend as the MH time increases. The crystallinity percentages of samples S8, S9, S10, and S11 are 37.33%, 38.06%, 44.28%, and 51.68%, respectively. These values exhibit a peak at 2 hours. The reason behind the enhanced dissolving rate of SiO$_2$ and Al$_2$O$_3$ in MSWI FA is attributed to the prolonged MH time. Additionally, the crystallinity of zeolite Na-P1 exhibits a gradual initial

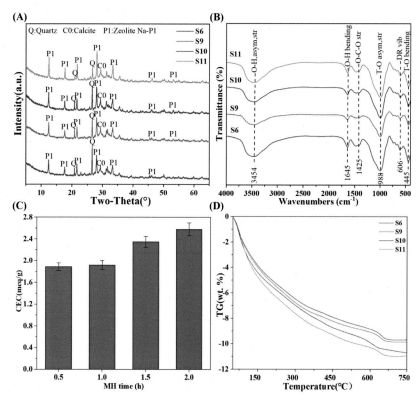

**FIGURE 6.7** The characteristic results of products generated with distinct MH times: (A) XRD diffractograms; (B) FTIR spectra; (C) CEC value; (D) TG curves . *From Zhou, Q., Jiang, X., Qiu, Q., Zhao, Y., & Long, L. (2023)*. Science of the Total Environment, 855, 158741. https://doi.org/10.1016/j.scitotenv.2022.158741.

increase followed by a quick acceleration (Fan et al., 2008). The expansion of zeolite Na-P1 crystallinity in the product is facilitated by an extended period of MH. The identification of a distinct zeolite structure is observed with an augmentation in MH time. When the MH duration is increased from 1.5 hours to 2.0 hours, a cluster of zeolite Na-P1 with a cactus/cabbage form is detected. This observation aligns with the architectures of zeolite synthesized by Pal et al. (2013) and Pal et al. (2013). Fig. 6.7B displays the Fourier Transform Infrared (FTIR) spectra of the synthetic compounds S6, S9, S10, and S11. The vibrational frequency of zeolite Na-P1, as determined by Dr spectroscopy, is found to be 606/cm. The T-O bending vibration of the $TO_4$ moiety in zeolite Na-P1 has a characteristic peak at a wavenumber of 445/cm. The obtained results align with the existing body of literature and provide confirmation of the effective fabrication of zeolite Na-P1. Fig. 6.7C illustrates the variation in the CEC for the different products produced at various measurement hours. The CEC exhibits an increase as the residence time of the material in the MH process is extended from 0.5 to 2 hours. At the 2-hour mark, the CEC reaches its peak value of 2.58 meq/g, which is roughly 133 times greater than the CEC of MSWI FA. The reason for prolonging the MH duration is to increase the dissolution of aluminum (Al) and silicon (Si), which subsequently promotes the crystallinity of zeolite Na-P1 in the resulting product. The CEC of the products produced after MH treatment for 0.5 and 1.0 hours exhibits a marginal rise from 1.88 to 1.92 meq/g. The acceleration of the formation of the CEC becomes evident when the MH time surpasses 1.5 hours. The increase in the CEC is found to be only 9.79% when the measurement horizon duration is extended from 1.5 to 2 hours. The observed pattern of change in the CEC aligns with the trend observed in the crystallinity of Na-P1 zeolite, suggesting a positive correlation between CEC and the yield of zeolite Na-P1 (Kumar & Jena, 2022; Liu et al., 2018; Novembre, Gimeno, & Del Vecchio, 2021). In relation to the objective of energy conservation, it is not advisable to prolong the MH duration any further. Based on the thermogravimetric (TG) curves depicted in Fig. 6.7D, the mass-loss rates for samples S6, S9, S10, and S11 are determined to be 4.87%, 5.06%, 5.29%, and 5.89%, respectively, during the first stage. In the subsequent stage, the corresponding mass-loss rates for the same samples are found to be 3.02%, 3.14%, 3.31%, and 3.39%, respectively. The observed trend of increasing mass loss suggests that there is a positive correlation between the duration of MH and the synthetic yield of zeolite Na-P1. Based on the aforementioned study and taking into account the energy usage, a maintenance and handling period of 2 hours was chosen.

### 6.2.2.2    Effects of NaOH concentration

Fig. 6.8A displays the XRD patterns obtained from the products synthesized using NaOH solutions of varying concentrations. In the absence of NaOH, the synthesis of the product did not yield any discernible zeolite crystal phase. The introduction of NaOH only results in the formation of zeolite Na-P1. The relationship between the strength of the peaks associated with zeolite Na-P1 crystals and the concentration of NaOH does not exhibit a linear trend. The degrees of crystallinity for S12, S13, S11, S14, and S15 are recorded as 0%, 36.43%, 34.63%, 51.68%, and 30.60%, respectively. The zeolite Na-P1 exhibits the maximum level of crystallinity in the product S11, which is manufactured using a 1 M NaOH solution. Insufficient dissolution of Si and Al from MSWI FA results in a limited synthetic yield of zeolite when the concentration of NaOH is below 1 M. Although the solubility of $Al_2O_3$ is comparatively lower than that of $SiO_2$ in alkaline solutions, the dissolution of Al is enhanced when exposed to NaOH solutions with concentrations exceeding 1 M. Hence, it can be observed that the molar ratio of Si to Al decreases as the concentration of NaOH increases from 1.5 to 2 M. This

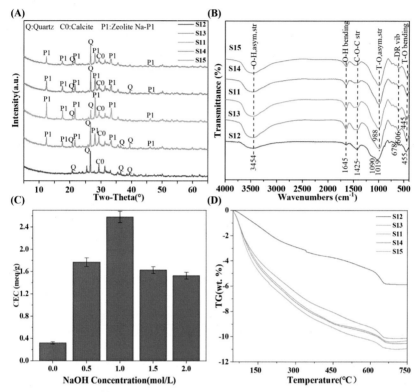

**FIGURE 6.8** **The characteristic results of products generated with distinct NaOH concentrations: (A) XRD diffractograms; (B) FTIR spectra; (C) CEC; and (D) TG curves** . *From Zhou, Q., Jiang, X., Qiu, Q., Zhao, Y., & Long, L. (2023).* Science of the Total Environment, *855, 158741. https://doi.org/10.1016/j.scitotenv.2022.158741.*

drop in ratio is unfavorable for the production of zeolite Na-P1. The absence of NaOH in the MH process results in the lack of zeolite structure. The presence of zeolite Na-P1 with diamond-like characteristics is very likely at a NaOH concentration of 0.5 M. This finding aligns with the morphology previously reported by Huo et al. in 2012 (Huo et al., 2012). The zeolite Na-P1 structure may be characterized by its resemblance to cabbage or cactus. This particular zeolite is obtained by the synthesis process involving 1 M and 1.5 M NaOH solutions. A conglomeration of zeolite Na-P1 exhibiting wool ball and cabbage/cactus morphologies is observed in a 2 M NaOH solution. This phenomenon has been documented in multiple research investigations as well. The structures and morphologies of the synthesized products exhibit notable differences when varying the concentration of NaOH, as compared to the structures and morphologies of MSWI FA. Fig. 6.8B displays the FTIR spectra corresponding to products S11−S15. The FTIR spectra of S12 have similarities to that of MSWI FA. Notably, the presence of bands at

1090, 1019, and 678/cm is indicative of quartz, suggesting the absence of zeolite formation. The presence of zeolite Na-P1 is confirmed by the observation of characteristic bands at approximately 988 and 606/cm for products S11, S13, S14, and S15 (Tang, Ge, Wang, He, & Cui, 2015). Fig. 6.8C illustrates a distinct and pronounced increase followed by a subsequent decrease in the CEC as the concentration of NaOH is incrementally raised from 0 to 2 M. The highest CEC value for product S11 is observed to be 2.58 meq/g when the concentration of NaOH reaches 1 M. The CEC of this material closely resembles that of zeolite Na-P1 (3.00 meq/g) produced from coal fly ash by Miki Inada et al. in 2005 (Inada et al., 2005). This discovery represents a significant advancement in the production of zeolites using MSWI FA as a starting material. The decrease in the CEC of the product may be observed as the concentration of NaOH is raised. This decrease can be attributed to a corresponding decline in the crystallinity of zeolite Na-P1. The presence of a concentrated NaOH solution ( $>1$ M) impedes the enhancement of the CEC during the synthesis of zeolite. Fig. 6.8D displays the TG curves corresponding to S11−S15. The initial stage of S11−S15 exhibits mass-loss rates of 5.90%, 1.94%, 4.86%, 5.49%, and 5.33% correspondingly. These rates align with the shifting pattern observed in the CEC, which peaks at a NaOH concentration of 1 M. The mass-loss rates of samples S11−S15 at temperatures ranging from 200°C to 450°C exhibit values of 3.39%, 1.80%, 3.20%, 3.46%, and 3.67%. These rates demonstrate an upward trend as the concentration of NaOH solution increases, aligning with the findings reported by Huo et al. (2012) on the synthesized product (Huo et al., 2012).

### 6.2.2.3 Effect of $Na_2SiO_3$ dosage

Fig. 6.9A displays the XRD diffractograms obtained from the products containing varying amounts of $Na_2SiO_3$. Upon the introduction of $Na_2SiO_3$, the characteristic peaks associated with sodalite (hydrated) and katoite are no longer observable, but the peaks belonging to zeolite Na-P1 and zeolite Na-P2 become evident. The inclusion of $Na_2SiO_3$ serves to control the molar ratio of Si/Al, which represents the proportion of Si and Al present in the reaction system. The presence of hydrated sodalite is exclusively found when 20 wt.% $Na_2SiO_3$ (with a Si/Al molar ratio of 1.30) is added. However, when the $Na_2SiO_3$ concentration is increased to 30 wt.% (with a Si/Al molar ratio of 1.44), peaks corresponding to zeolite Na-P1 are detected. Nevertheless, the maximum quantity of $Na_2SiO_3$ incorporated is not achieved, resulting in the coexistence of zeolite Na-P1 and Na-P2 when 40 wt.% $Na_2SiO_3$ is added (with a Si/Al molar ratio of 1.58). However, upon further raising the $Na_2SiO_3$ dosage to 50 wt.% (with a Si/Al molar ratio of 1.72), only medium silica zeolite Na-P2 was discovered. The crystallinity percentages for S6, S7, and S8 are 37.33%, 36.56%, and 36.07%, respectively. The substitution of low silica zeolite Na-P1 with medium silica

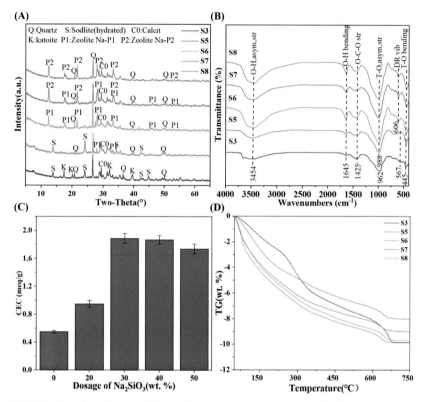

**FIGURE 6.9** **The characteristic results of products generated with distinct dosage of Na₂SiO₃: (A) XRD diffractograms; (B) FTIR spectra; (C) CEC value; (D) TG curves** . *From Zhou, Q., Jiang, X., Qiu, Q., Zhao, Y., & Long, L. (2023).* Science of the Total Environment, 855, *158741. https://doi.org/10.1016/j.scitotenv.2022.158741.*

zeolite Na-P2 is observed as the dosage of Na₂SiO₃ increases. Additionally, the introduction of Na₂SiO₃ in amounts exceeding 30 wt.% leads to a reduction in the crystallinity of the zeolite. The spherical and irregularly shaped crystals are indicative of zeolite particle growth. This growth becomes more apparent with an increase in the quantity of Na₂SiO₃ supplied. As the dose of Na₂SiO₃ increases, the morphology of the crystal transitions from spherical to irregular, characterized by rough edges. The FTIR spectra corresponding to the products are depicted in Fig. 6.9B. The product S5 has notable vibrational frequencies at 3454, 1645, 1425, 988, and 445/cm, which can be attributed to the presence of sodalite (hydrated). This compound has been consistently identified in multiple instances. The vibrational mode of double rings (Dr) within the zeolite Na-P1 and Na-P2 frameworks has been detected with a wavenumber of 606/cm. The disappearance of the band at 567/cm is observed as the concentration of Na₂SiO₃ increases, suggesting that the addition of $Na_2SiO_3$ enhances the use of $Al_2O_3$ in MSWI FA. The T-O bending vibration exhibits a characteristic frequency of 445/cm (Alkan, Hopa, Yilmaz, & Güler, 2005; Azizi, Ghasemi, & Amiripour, 2014; Bohra, Kundu, & Naskar, 2013; Zong, Zhao, Chen, Liu, & Cang, 2020). Fig. 6.9C illustrates the values corresponding to the CEC of S3, S5, S6, S7, and S8. It clearly demonstrates a discernible upward trajectory in growth as the $Na_2SiO_3$ content is progressively increased from 0 to 30 wt.%. On the other hand, it can be observed that the CEC of products containing a dosage of $Na_2SiO_3$ more than 30 wt.% demonstrates a gradual decline due to a reduction in the crystalline nature of said goods. The CEC of the product reaches its highest value of 1.89 meq/g when 30 wt.% of $Na_2SiO_3$ is added. The molar ratio of silicon to aluminum (Si/Al) in the product is 1.44, which is very similar to the ratio of 1.50 used in the production of zeolite Na-P1 from coal fly ash. Fig. 6.9D displays the TG curves corresponding to samples S3, S5, S6, S7, and S8. The mass-loss rates of S3, S5, S6, S7, and S8 throughout the temperature range of 50°C−200°C are recorded as 2.32%, 3.64%, 4.87%, 4.66%, and 5.24% correspondingly. The reason for the decreased mass-loss rate of S7 compared to S6 can be attributed to the formation of zeolite Na-P2. The production of zeolite Na-P2 requires a greater amount of silicon and aluminum sources compared to the production of zeolite Na-P1. The quantity of synthetic zeolite present in S7 is comparatively lower than that found in S6, mostly attributed to the simultaneous presence of zeolite Na-P1 and Na-P2 in S7. The sole zeolite identified in S8 is zeolite Na-P2, which contains a greater quantity of zeolitic water compared to other zeolites present in S8 (Li et al., 2012). Consequently, the combined amount of adsorbed water and zeolitic water in S8 surpasses that found in S6. Consequently, the rate of mass loss for S8 is greater in comparison to S6. The rates of mass loss at temperatures ranging from 200°C to 450°C are 4.54%, 2.57%, 3.02%, 2.76%, and 2.95%, respectively. The significant mass loss observed in S3 can be attributed to the desorption of hydroxyl (OH) groups present in katoite. The observed mass loss of S6 has a greater magnitude compared to S5, S7, and S8 within the temperature range of 200°C−450°C. This discrepancy can be attributed to the fact that S6 possesses the highest quantity of synthetic zeolite, as indicated by the CEC measurements. The findings of this study indicate that the dosage of $Na_2SiO_3$ is a crucial factor in the synthesis of zeolites, with a concentration of 30 wt.% $Na_2SiO_3$ being identified as the best condition for enhancing the CEC. Consequently, a concentration of 30 wt.% of $Na_2SiO_3$ was used for the remaining experimental procedures.

## 6.3 Challenges and future recommendation

The production of zeolites from municipal solid waste (MSW) presents several challenges, primarily due to the complex and diverse nature of MSW composition. There exist numerous notable concerns associated with the synthesis of zeolite from MSW.

1. MSW composition variability: The composition of MSW displays considerable variety, spanning a wide range of organic and inorganic constituents. Therefore the task of obtaining consistent and replicable zeolite products from MSW poses a significant difficulty (Sun et al., 2011).

2. Contaminants and impurities: The inclusion of pollutants and foreign substances in MSW can present difficulties during the synthesis of zeolite and negatively affect the overall excellence of the final output. The presence of contaminants in a given sample might encompass a range of substances, such as heavy metals, organic pollutants, and a diverse array of impurities. The elimination of these contaminants presents a significant challenge.

3. Source separation: The method of source separation plays a vital role in the attainment of superior-grade raw materials for the synthesis of zeolites. The process entails efficiently segregating and categorizing diverse constituents of MSW. The deployment of effective separation systems may result in considerable expenses and necessitate a substantial workforce (Sharifikolouei et al., 2021).

4. Energy and resource intensity: Traditional zeolite synthesis procedures are known for their high energy and resource requirements, which are attributed to the use of elevated temperatures, prolonged reaction times, and the incorporation of chemicals or reagents. The problem lies in the development of synthesis procedures that effectively optimize energy consumption and resource utilization.

5. Environmental impact: The minimization of environmental impact holds significant importance in the process of synthesizing zeolite from MSW. This involves the mitigation of energy use and emissions, together with the efficient handling of waste byproducts.

6. Scale-up issues: The complexities involved in the process of upscaling zeolite synthesis from a small-scale laboratory setting to a large-scale industrial output can be sophisticated. The challenge of ensuring consistent product quality and improving operational efficiency in the context of increased production volumes is a significant barrier (Nowak et al., 2010).

7. Zeolite product purity: The achievement of high degrees of purity in zeolite products is of great significance in various applications. The challenge of mitigating contaminants and pollutants in the final product presents notable complexities.

8. Regulatory compliance: The incorporation of environmental regulations and guidelines in the handling of MSW and potentially hazardous waste substances may increase the complexity and costs of the zeolite manufacturing procedure (Villaescusa, Wheatley, Bull, Lightfoot, & Morris, 2001).

The scientific exploration of synthesizing zeolites from MSW is an ongoing area of interest since it presents potential solutions to challenges pertaining to waste management and materials synthesis. The subsequent recommendations and approaches presented aim to improve the zeolite synthesis process from MSW.

1. Advanced waste separation and sorting: The potential for enhancing the quality of raw materials utilized in zeolite synthesis, as well as improving efficiency and cost reduction, can be realized through the progress made in waste separation and sorting procedures. The possibility for enhancing source separation within waste management systems exists through the integration of automation, artificial intelligence, and robots.

2. Waste-to-energy integration: This study aims to analyze the prospective amalgamation of waste-to-energy technologies with the production of zeolite from MSW. This possesses the capacity to meet the energy demands for synthesis while also reducing the amount of waste that needs to be disposed of.

3. Contaminant removal: Propose innovative approaches for the removal of contaminants and impurities from MSW feedstock, hence improving the overall quality of the zeolite output. The implementation of adsorbents or chemical treatments may be necessary to selectively target and remove particular contaminants.

4. Technological innovations: The utilization of technological innovations, such as nanotechnology, 3D printing, and advanced materials science, can be utilized to enhance the techniques of zeolite synthesis from MSW.

5. Life cycle analysis: To comprehensively evaluate the environmental and economic ramifications of zeolite production from MSW in comparison to conventional synthesis methods, it is imperative to carry out meticulous life cycle studies. The provided information possesses the capacity to boost understanding and assist improvements in processes.

6. Regulatory support: To comprehensively evaluate the environmental and economic ramifications of zeolite production from MSW in comparison to conventional synthesis methods, it is imperative to carry out meticulous life cycle studies. The provided information possesses the capacity to boost understanding and assist improvements in processes.

7. Demonstration plants: It is advisable to incorporate demonstration facilities as a means to successfully showcase the feasibility and benefits of producing zeolite from MSW. These projects have the capacity to serve as evidence of a concept's viability and elicit attention from prospective investors.

## 6.4 Conclusion

Both microwave and alkali fusion—assisted hydrothermal methods are regarded as highly promising and adaptable approaches for the synthesis of zeolites. These approaches demonstrate distinct advantages and have the ability to address specific difficulties related to conventional zeolite synthesis. The reactivity of the raw materials, specifically MSW, is augmented by the use of the alkali fusion process. This process facilitates the creation of an appropriate precursor for the synthesis of zeolite. This approach offers a customized mineral composition that acts as a fundamental basis for the subsequent hydrothermal treatment. In contrast, the hydrothermal method enables the controlled formation of zeolites through the manipulation of key variables, including temperature, pressure, and duration. This technique provides a means to achieve precise crystallization conditions. The aforementioned combination facilitates the development of zeolite structures that possess distinct characteristics and advantageous attributes. The incorporation of microwave help into the hydrothermal process represents a notable technological achievement. The utilization of microwave irradiation expedites the rate of reaction kinetics, resulting in reduced reaction durations and enhanced energy utilization. This not only conforms to ecological principles but also boosts the overall productivity of zeolite synthesis. The phenomenon of selective heating caused by microwaves plays a role in achieving a more even dispersion of energy, hence facilitating the process of homogenous nucleation and crystallization. Furthermore, the use of MSW as a primary element in these methodologies underscores the dedication to sustainable and environmentally conscious approaches. Through the process of repurposing waste materials, these strategies effectively mitigate environmental impact while simultaneously fostering the principles of the circular economy by converting waste into a useful resource.

The combination of the alkali fusion—assisted hydrothermal technique and the integration of microwave assistance offers a comprehensive and effective method for the synthesis of zeolites. The aforementioned methodologies demonstrate the capacity for achieving sustainable and scalable zeolite synthesis, hence creating opportunities for additional investigation and utilization in several domains, including catalysis, adsorption, and environmental remediation. The ongoing pursuit of environmentally friendly and highly efficient technologies has led to the emergence of methodologies that have great potential for the advancement of zeolite synthesis and its wide-ranging applications.

## 6.5 AI Disclosure

During the preparation of this work, the author(s) used Quillbot to improve the English and for smooth writing. After using this tool/service, the author(s) reviewed and edited the content as needed and take(s) full responsibility for the content of the publication.

# References

Abdullahi, T., Harun, Z., & Othman, M. H. D. (2017). A review on sustainable synthesis of zeolite from kaolinite resources via hydrothermal process. *Advanced Powder Technology, 28*(8), 1827−1840. Available from https://doi.org/10.1016/j.apt.2017.040.028. Available from, https://www.sciencedirect.com/science/article/pii/S0921883117301930.

Aldahri, T., Behin, J., Kazemian, H., & Rohani, S. (2016). Synthesis of zeolite Na-P from coal fly ash by thermo-sonochemical treatment. *Fuel, 182,* 494−501. Available from https://doi.org/10.1016/j.fuel.2016.060.019. Available from, https://www.sciencedirect.com/science/article/pii/S0016236116304720.

Alkan, M., Hopa, Ç., Yilmaz, Z., & Güler, H. 1 (2005). The effect of alkali concentration and solid/liquid ratio on the hydrothermal synthesis of zeolite NaA from natural kaolinite. *Microporous and Mesoporous Materials, 86*(1), 176−184. Available from https://doi.org/10.1016/j.micromeso.2005.070.008. Available from, https://www.sciencedirect.com/science/article/pii/S138718110500291X.

Anuwattana, R., & Khummongkol, P. (2009). Conventional hydrothermal synthesis of Na-A zeolite from cupola slag and aluminum sludge. *Journal of Hazardous Materials, 166*(1), 227−232. Available from https://doi.org/10.1016/j.jhazmat.2008.110.020. Available from, https://www.sciencedirect.com/science/article/pii/S0304389408016786.

Anweshan., Das, P. P., Dhara, S., Purkait, M. K., Husen, A., & Siddiqi, K. S. (2023). *Chapter 12 - Nanosensors in food science and technology micro and nano technologies* (pp. 247−272). Elsevier Available from. Available from https://www.sciencedirect.com/science/article/pii/B978032399546700015X, 10.1016/B978-0-323-99546-7.00015-X.

Azizi, S. N., Ghasemi, S., & Amiripour, F. (2014). A new attitude to environment: Preparation of an efficient electrocatalyst for methanol oxidation based on Ni-doped P zeolite nanoparticles synthesized from stem sweep ash. *Electrochimica Acta, 137,* 395−403. Available from https://doi.org/10.1016/j.electacta.2014.050.158. Available from, https://www.sciencedirect.com/science/article/pii/S0013468614011736.

Behin, J., Bukhari, S. S., Kazemian, H., & Rohani, S. (2016). Developing a zero liquid discharge process for zeolitization of coal fly ash to synthetic NaP zeolite. *Fuel, 171,* 195−202. Available from https://doi.org/10.1016/j.fuel.2015.120.073. Available from, https://www.sciencedirect.com/science/article/pii/S0016236115013393.

Bohra, S., Kundu, D., & Naskar, M. K. (2013). Synthesis of cashew nut-like zeolite NaP powders using agro-waste material as silica source. *Materials Letters, 106,* 182−185. Available from https://doi.org/10.1016/j.matlet.2013.040.080. Available from, https://www.sciencedirect.com/science/article/pii/S0167577X13005545.

Bukhari, S. S., Behin, J., Kazemian, H., & Rohani, S. (2015). Conversion of coal fly ash to zeolite utilizing microwave and ultrasound energies: A review. *Fuel, 140,* 250−266. Available from https://doi.org/10.1016/j.fuel.2014.090.077. Available from, https://www.sciencedirect.com/science/article/pii/S0016236114009466.

Cardoso, A. M., Paprocki, A., Ferret, L. S., Azevedo, C. M. N., & Pires, M. (2015). Synthesis of zeolite Na-P1 under mild conditions using Brazilian coal fly ash and its application in wastewater treatment. *Fuel, 139,* 59−67. Available from https://doi.org/10.1016/j.fuel.2014.080.016. Available from, https://www.sciencedirect.com/science/article/pii/S0016236114007856.

Chakraborty, S., Gautam, S. P., Das, P. P., & Hazarika, M. K. (2019). Instant controlled pressure drop (DIC) treatment for improving process performance and milled rice quality. *Journal of The Institution of Engineers (India): Series A, 100*(4), 683−695. Available from https://doi.

org/10.1007/s40030-019-00403-w. Available from, https://doi.org/10.1007/s40030-019-00403-w.

Chakraborty, S., Das, P. P., Mondal, P., Sillanpää, M., Khadir, A., & Gurung, K. (2023). *34 - Recent advances in membrane technology for the recovery and reuse of valuable resources* (pp. 695−719). Elsevier Available from. Available from https://www.sciencedirect.com/science/article/pii/B9780323953276000282, 10.1016/B978-0-323-95327-6.00028-2.

Chang, F.-Y., & Wey, M.-Y. (2006). Comparison of the characteristics of bottom and fly ashes generated from various incineration processes. *Journal of Hazardous Materials, 138*(3), 594−603. Available from https://doi.org/10.1016/j.jhazmat.2006.050.099. Available from, https://www.sciencedirect.com/science/article/pii/S0304389406006212.

Changmai, M., Das, P. P., Mondal, P., Pasawan, M., Sinha, A., Biswas, P., ... Purkait, M. K. (2022). Hybrid electrocoagulation−microfiltration technique for treatment of nanofiltration rejected steel industry effluent. *International Journal of Environmental Analytical Chemistry, 102*(1), 62−83. Available from https://doi.org/10.1080/03067319.2020.1715381. Available from, https://doi.org/10.1080/03067319.2020.1715381.

Chen, Y., Armutlulu, A., Sun, W., Jiang, W., Jiang, X., Lai, B., & Xie, R. (2020). Ultrafast removal of Cu(II) by a novel hierarchically structured faujasite-type zeolite fabricated from lithium silica fume. *Science of The Total Environment, 714*136724. Available from https://doi.org/10.1016/j.scitotenv.2020.136724. Available from, https://www.sciencedirect.com/science/article/pii/S0048969720302345.

Chindaprasirt, P., & Rattanasak, U. (2017). Characterization of the high-calcium fly ash geopolymer mortar with hot-weather curing systems for sustainable application. *Advanced Powder Technology, 28*(9), 2317−2324. Available from https://doi.org/10.1016/j.apt.2017.060.013. Available from, https://www.sciencedirect.com/science/article/pii/S0921883117302571.

Cristelo, N., Segadães, L., Coelho, J., Chaves, B., Sousa, N. R., & de Lurdes Lopes, M. (2020). Recycling municipal solid waste incineration slag and fly ash as precursors in low-range alkaline cements. *Waste Management, 104*, 60−73. Available from https://doi.org/10.1016/j.wasman.2020.010.013. Available from, https://www.sciencedirect.com/science/article/pii/S0956053X20300131.

Das, P. P., Deepti., & Purkait, M. K. (2023). *Industrial wastewater to biohydrogen production via potential bio-refinery route* (pp. 159−179). Cham: Springer International Publishing Available from. Available from https://doi.org/10.1007/978-3-031-20822-5_8, 10.1007/978-3-031-20822-5_8.

Das, P. P., Mondal, P., Anweshan., Sinha, A., Biswas, P., Sarkar, S., & Purkait, M. K. (2021). Treatment of steel plant generated biological oxidation treated (BOT) wastewater by hybrid process. *Separation and Purification Technology, 258*118013. Available from https://doi.org/10.1016/j.seppur.2020.118013. Available from, https://www.sciencedirect.com/science/article/pii/S1383586620324862.

Das, P. P., Anweshan., Mondal, P., Sinha, A., Biswas, P., Sarkar, S., & Purkait, M. K. (2021). Integrated ozonation assisted electrocoagulation process for the removal of cyanide from steel industry wastewater. *Chemosphere, 263*128370. Available from https://doi.org/10.1016/j.chemosphere.2020.128370. Available from, https://www.sciencedirect.com/science/article/pii/S0045653520325650.

Das, P. P., Anweshan., & Purkait, M. K. (2021). Treatment of cold rolling mill (CRM) effluent of steel industry. *Separation and Purification Technology, 274*119083. Available from https:-//doi.org/10.1016/j.seppur.2021.119083. Available from, https://www.sciencedirect.com/science/article/pii/S1383586621007930.

Das, P. P., Mondal, P., & Purkait, M. K. (2022). *Recent advances in synthesis of iron nanoparticles via green route and their application in biofuel production* (pp. 79–104). Singapore: Springer Nature Singapore Available from. Available from https://doi.org/10.1007/978-981-16-9356-4_4, 10.1007/978-981-16-9356-4_4.

Das, P. P., Sharma, M., & Purkait, M. K. (2022). Recent progress on electrocoagulation process for wastewater treatment: A review. *Separation and Purification Technology, 292*121058. Available from https://doi.org/10.1016/j.seppur.2022.121058. Available from, https://www.sciencedirect.com/science/article/pii/S1383586622006153.

Das, P. P., Sontakke, A. D., Purkait, M. K., & Shah, M. P. (2023a). *Chapter 2 - Electrocoagulation process for wastewater treatment: applications, challenges, and prospects* (pp. 23–48). Elsevier Available from. Available from https://www.sciencedirect.com/science/article/pii/B9780323956840000154, 10.1016/B978-0-323-95684-0.00015-4.

Das, P. P., Samanta, N. S., Dhara, S., Purkait, M. K., & Shah, M. P. (2023). Chapter 13 - Biofuel production from algal biomass (pp. 167–179). Elsevier Available from. Available from https://www.sciencedirect.com/science/article/pii/B9780128243183000096, 10.1016/B978-0-12-824318-3.00009-6.

Das, P. P., Mondal, P., Sillanpää, M., Khadir, A., & Gurung, K. (2023). *14 -* Membrane-assisted potable water reuses applications: benefits and drawbacks (pp. 289–309). Elsevier Available from. Available from https://www.sciencedirect.com/science/article/pii/B9780323993449000141, 10.1016/B978-0-323-99344-9.00014-1.

Das, P. P., Sontakke, A. D., Samanta, N. S., & Purkait, M. K. (2023). *Emerging contaminants in wastewater: eco-toxicity and sustainability assessment* (pp. 63–87). Singapore: Springer Nature Singapore Available from. Available from https://doi.org/10.1007/978-981-99-2489-9_4, 10.1007/978-981-99-2489-9_4.

Das, P. P., Sontakke, A. D., Purkait, M. K., & Shah, M. P. (2023b). *Chapter 12 - Rice straw for biofuel production* (pp. 153–166). Elsevier Available from. Available from https://www.sciencedirect.com/science/article/pii/B9780128243183000345, 10.1016/B978-0-12-824318-3.00034-5.

Das, P. P., Dhara, S., & Purkait, M. K. (2023). *Hybrid electrocoagulation and ozonation techniques for industrial wastewater treatment* (pp. 107–128). Singapore: Springer Nature Singapore Available from. Available from https://doi.org/10.1007/978-981-99-2560-5_6, 10.1007/978-981-99-2560-5_6.

Das, P. P., Duarah, P., Purkait, M. K., & Jafari, S. M. (2023). *5 - Fundamentals of food roasting process unit operations and processing equipment in the food industry* (pp. 103–130). Woodhead Publishing Available from. Available from https://www.sciencedirect.com/science/article/pii/B9780128186183000057, 10.1016/B978-0-12-818618-3.00005-7.

Dhara, S., Shekhar Samanta, N., Das, P. P., Uppaluri, R. V. S., & Purkait, M. K. (2023). Ravenna grass-extracted alkaline lignin-based polysulfone mixed matrix membrane (MMM) for aqueous Cr(VI) removal. *ACS Applied Polymer Materials, 5*(8), 6399–6411. Available from https://doi.org/10.1021/acsapm.3c00999. Available from, https://doi.org/10.1021/acsapm.3c00999.

Dhara, S., Das, P. P., Uppaluri, R., Purkait, M. K., Sillanpää, M., Khadir, A., & Gurung, K. (2023). *12 -* Biological approach for energy self-sufficiency of municipal wastewater treatment plants (pp. 235–260). Elsevier Available from. Available from https://www.sciencedirect.com/science/article/pii/B9780323993487000060, 10.1016/B978-0-323-99348-7.00006-0.

Das, P. P., Dhara, S., & Purkait, M. K. (2024). Ozone-based oxidation processes for the removal of pharmaceutical products from wastewater. In Maulin P. Shah, & Pooja Ghosh (Eds.),

*Development in wastewater treatment research and processes* (pp. 287−308). Elsevier. Available from https://doi.org/10.1016/B978-0-443-19207-4.00003-3.

Dhara, S., Das, P. P., Uppaluri, R., Purkait, M. K., & Shah, M. P. (2023). *Chapter 3 - Phosphorus recovery from municipal wastewater treatment plants* (pp. 49−72). Elsevier Available from. Available from https://www.sciencedirect.com/science/article/pii/B9780323956840000142, 10.1016/B978-0-323-95684-0.00014-2.

Dhara, S., Samanta, N. S., Uppaluri, R., & Purkait, M. K. (2023). High-purity alkaline lignin extraction from Saccharum ravannae and optimization of lignin recovery through response surface methodology. *International Journal of Biological Macromolecules*, *234*123594. Available from https://doi.org/10.1016/j.ijbiomac.2023.123594. Available from, https://www.sciencedirect.com/science/article/pii/S0141813023004877.

Duarah, P., Das, P. P., & Purkait, M. K. (2023). *Technological advancement in the synthesis and application of nanocatalysts* (pp. 191−214). Singapore: Springer Nature Singapore Available from. Available from https://doi.org/10.1007/978-981-99-3292-4_10, 10.1007/978-981-99-3292-4_10.

Fan, Y., Zhang, F.-S., Zhu, J., & Liu, Z. (2008). Effective utilization of waste ash from MSW and coal co-combustion power plant—Zeolite synthesis. *Journal of Hazardous Materials*, *153*(1), 382−388. Available from https://doi.org/10.1016/j.jhazmat.2007.080.061. Available from, https://www.sciencedirect.com/science/article/pii/S0304389407012368.

Ferreira, C., Ribeiro, A., & Ottosen, L. (2003). Possible applications for municipal solid waste fly ash. *Journal of Hazardous Materials*, *96*(2), 201−216. Available from https://doi.org/10.1016/S0304-3894(02)00201-7. Available from, https://www.sciencedirect.com/science/article/pii/S0304389402002017.

Hu, H., Luo, G., Liu, H., Qiao, Y., Xu, M., & Yao, H. (2013). Fate of chromium during thermal treatment of municipal solid waste incineration (MSWI) fly ash. *Proceedings of the Combustion Institute*, *34*(2), 2795−2801. Available from https://doi.org/10.1016/j.proci.2012.060.181. Available from, https://www.sciencedirect.com/science/article/pii/S1540748912002891.

Huo, Z., Xu, X., Lü, Z., Song, J., He, M., Li, Z., . . . Yan, L. (2012). Synthesis of zeolite NaP with controllable morphologies. *Microporous and Mesoporous Materials*, *158*, 137−140. Available from https://doi.org/10.1016/j.micromeso.2012.030.026. Available from, https://www.sciencedirect.com/science/article/pii/S1387181112001758.

Inada, M., Eguchi, Y., Enomoto, N., & Hojo, J. (2005). Synthesis of zeolite from coal fly ashes with different silica−alumina composition. *Fuel*, *84*(2), 299−304. Available from https://doi.org/10.1016/j.fuel.2004.080.012. Available from, https://www.sciencedirect.com/science/article/pii/S0016236104002479.

Kumar, M. M., & Jena, H. (2022). Direct single-step synthesis of phase pure zeolite Na−P1, hydroxy sodalite and analcime from coal fly ash and assessment of their Cs + and Sr2 + removal efficiencies. *Microporous and Mesoporous Materials*, *333*, 1387−1811. Available from https://doi.org/10.1016/j.micromeso.2022.111738. Available from, https://www.sciencedirect.com/science/article/pii/S1387181122000609.

Lee, W. K. W., & van Deventer, J. S. J. (2002). The effects of inorganic salt contamination on the strength and durability of geopolymers. *Colloids and Surfaces A: Physicochemical and Engineering Aspects*, *211*(2), 115−126. Available from https://doi.org/10.1016/S0927-7757(02)00239-X. Available from, https://www.sciencedirect.com/science/article/pii/S092777570200239X.

Li, Q., Xu, H., Li, F., Li, P., Shen, L., & Zhai, J. (2012). Synthesis of geopolymer composites from blends of CFBC fly and bottom ashes. *Fuel*, *97*, 366−372. Available from https://doi.

org/10.1016/j.fuel.2012.020.059. Available from, https://www.sciencedirect.com/science/article/pii/S0016236112001962.

Liu, Y., Yan, C., Zhao, J., Zhang, Z., Wang, H., Zhou, S., & Wu, L. (2018). Synthesis of zeolite P1 from fly ash under solvent-free conditions for ammonium removal from water. *Journal of Cleaner Production*, *202*, 11−22. Available from https://doi.org/10.1016/j.jclepro.2018.080.128. Available from, https://www.sciencedirect.com/science/article/pii/S0959652618324703.

Mondal, P., Samanta, N. S., Meghnani, V., & Purkait, M. K. (2019). Selective glucose permeability in presence of various salts through tunable pore size of pH responsive PVDF-co-HFP membrane. *Separation and Purification Technology*, *221*, 249−260. Available from https://doi.org/10.1016/j.seppur.2019.040.001. Available from, https://www.sciencedirect.com/science/article/pii/S1383586618346252.

Montañés, M. T., Sánchez-Tovar, R., & Roux, M. S. (2014). The effectiveness of the stabilization/solidification process on the leachability and toxicity of the tannery sludge chromium. *Journal of Environmental Management*, *143*, 71−79. Available from https://doi.org/10.1016/j.jenvman.2014.040.026. Available from, https://www.sciencedirect.com/science/article/pii/S0301479714002151.

Nagib, S., & Inoue, K. (2000). Recovery of lead and zinc from fly ash generated from municipal incineration plants by means of acid and/or alkaline leaching. *Hydrometallurgy*, *56*(3), 269−292. Available from https://doi.org/10.1016/S0304-386X(00)00073-6. Available from, https://www.sciencedirect.com/science/article/pii/S0304386X00000736.

Novembre, D., Gimeno, D., & Del Vecchio, A. (2021). Synthesis and characterization of Na-P1 (GIS) zeolite using a kaolinitic rock. *Scientific Reports*, *11*(1)4872. Available from https://doi.org/10.1038/s41598-021-84383-7. Available from, https://doi.org/10.1038/s41598-021-84383-7.

Nowak, B., Pessl, A., Aschenbrenner, P., Szentannai, P., Mattenberger, H., Rechberger, H., ... Winter, F. (2010). Heavy metal removal from municipal solid waste fly ash by chlorination and thermal treatment. *Journal of Hazardous Materials*, *179*(1), 323−331. Available from https://doi.org/10.1016/j.jhazmat.2010.030.008. Available from, https://www.sciencedirect.com/science/article/pii/S0304389410003146.

Okonji, S. O., Dominic, J. A., Pernitsky, D., & Achari, G. (2020). Removal and recovery of selenium species from wastewater: Adsorption kinetics and co-precipitation mechanisms. *Journal of Water Process Engineering*, *38*101666. Available from https://doi.org/10.1016/j.jwpe.2020.101666. Available from, https://www.sciencedirect.com/science/article/pii/S2214714420305444.

Pal, P., Das, J. K., Das, N., & Bandyopadhyay, S. (2013). Synthesis of NaP zeolite at room temperature and short crystallization time by sonochemical method. *Ultrasonics Sonochemistry*, *20*(1), 314−321. Available from https://doi.org/10.1016/j.ultsonch.2012.070.012. Available from, https://www.sciencedirect.com/science/article/pii/S1350417712001514.

Phua, Z., Giannis, A., Dong, Z.-L., Lisak, G., & Ng, W. J. (2019). Characteristics of incineration ash for sustainable treatment and reutilization. *Environmental Science and Pollution Research*, *26*(17), 16974−16997. Available from https://doi.org/10.1007/s11356-019-05217-8. Available from, https://doi.org/10.1007/s11356-019-05217-8.

Qiu, Q., Jiang, X., Lv, G., Chen, Z., Lu, S., Ni, M., ... Cao, J. (2019). Adsorption of copper ions by fly ash modified through microwave-assisted hydrothermal process. *Journal of Material Cycles and Waste Management*, *21*(3), 469−477. Available from https://doi.org/10.1007/s10163-018-0806-6. Available from, https://doi.org/10.1007/s10163-018-0806-6.

Querol, X., Alastuey, A., López-Soler, A., Plana, F., Andrés, J. M., Juan, R., . . . Ruiz, C. R. (1997). A fast method for recycling fly ash: Microwave-assisted zeolite synthesis. *Environmental Science & Technology*, *31*(9), 2527−2533. Available from https://doi.org/10.1021/es960937t. Available from, https://doi.org/10.1021/es960937t.

Ren, X., Liu, S., Qu, R., Xiao, L., Hu, P., Song, H., . . . Gao, X. (2020). Synthesis and characterization of single-phase submicron zeolite Y from coal fly ash and its potential application for acetone adsorption. *Microporous and Mesoporous Materials*, *295*109940. Available from https://doi.org/10.1016/j.micromeso.2019.109940. Available from, https://www.sciencedirect.com/science/article/pii/S1387181119307991.

Samadi-Maybodi, A., & Masoomeh Pourali, S. (2013). Microwave-assisted aging synthesis of bismuth modified zeolite-P microspheres via BiOCl nanoflake transformation. *Microporous and Mesoporous Materials*, *167*, 127−132. Available from https://doi.org/10.1016/j.micromeso.2012.020.012. Available from, https://www.sciencedirect.com/science/article/pii/S1387181112000728.

Samanta, N., Das, P. P., & Mondal, P. (2022). Physico-chemical and adsorption study of hydrothermally treated zeolite A and FAU-type zeolite X prepared from LD (Linz−Donawitz) slag of the steel industry. *International Journal of Environmental Analytical Chemistry*, *13*, 1−23. Available from https://doi.org/10.1080/03067319.2022.2079082.

Samanta, N. S., Banerjee, S., Mondal, P., Anweshan., Bora, U., & Purkait, M. K. (2021). Preparation and characterization of zeolite from waste Linz-Donawitz (LD) process slag of steel industry for removal of Fe3 + from drinking water. *Advanced Powder Technology*, *32* (9), 3372−3387. Available from https://doi.org/10.1016/j.apt.2021.070.023. Available from, https://www.sciencedirect.com/science/article/pii/S0921883121003605.

Samanta, N. S., Das, P. P., Mondal, P., Changmai, M., & Purkait, M. K. (2022). Critical review on the synthesis and advancement of industrial and biomass waste-based zeolites and their applications in gas adsorption and biomedical studies. *Journal of the Indian Chemical Society*, *99*(11)100761. Available from https://doi.org/10.1016/j.jics.2022.100761. Available from, https://www.sciencedirect.com/science/article/pii/S001945222200423X.

Samanta, N. S., Das, P. P., Sharma, M., Purkait, M. K., Sillanpää, M., Khadir, A., & Gurung, K. (2023). *12 - Recycle of water treatment plant sludge and its utilization for wastewater treatment* (pp. 239−264). Elsevier Available from. Available from https://www.sciencedirect.com/science/article/pii/B9780323993449000104, 10.1016/B978-0-323-99344-9.00010-4.

Samanta, N. S., Mondal, P., & Purkait, M. K. (2023). *Nanofiltration technique for the treatment of industrial wastewater*. Singapore: Springer Nature Singapore Available from. Available from https://doi.org/10.1007/978-981-99-3292-4_9, 10.1007/978-981-99-3292-4_9.

Samanta, N. S., Anweshan., Mondal, P., Bora, U., & Purkait, M. K. (2023). Synthesis of precipitated calcium carbonate from LD-slag using $CO_2$. *Materials Today Communications*, *36*106588. Available from https://doi.org/10.1016/j.mtcomm.2023.106588. Available from, https://www.sciencedirect.com/science/article/pii/S2352492823012795.

Sharifikolouei, E., Baino, F., Salvo, M., Tommasi, T., Pirone, R., Fino, D., & Ferraris, M. (2021). Vitrification of municipal solid waste incineration fly ash: An approach to find the successful batch compositions. *Ceramics International*, *47*(6), 7738−7744. Available from https://doi.org/10.1016/j.ceramint.2020.110.118. Available from, https://www.sciencedirect.com/science/article/pii/S0272884220334490.

Sharma, M., Das, P. P., Sood, T., Chakraborty, A., & Purkait, M. K. (2021). Ameliorated polyvinylidene fluoride based proton exchange membrane impregnated with graphene oxide, and cellulose acetate obtained from sugarcane bagasse for application in microbial fuel cell. *Journal of Environmental Chemical Engineering*, *9*(6)106681. Available from https://doi.

org/10.1016/j.jece.2021.106681. Available from, https://www.sciencedirect.com/science/article/pii/S2213343721016584.

Sharma, M., Das, P. P., Sood, T., Chakraborty, A., & Purkait, M. K. (2022). Reduced graphene oxide incorporated polyvinylidene fluoride/cellulose acetate proton exchange membrane for energy extraction using microbial fuel cells. *Journal of Electroanalytical Chemistry*, *907*115890. Available from https://doi.org/10.1016/j.jelechem.2021.115890. Available from, https://www.sciencedirect.com/science/article/pii/S1572665721009164.

Sharma, M., Das, P. P., Chakraborty, A., & Purkait, M. K. (2022). Clean energy from salinity gradients using pressure retarded osmosis and reverse electrodialysis: A review. *Sustainable Energy Technologies and Assessments*, *49*101687. Available from https://doi.org/10.1016/j.seta.2021.101687. Available from, https://www.sciencedirect.com/science/article/pii/S2213138821007013.

Sharma, M., Das, P. P., Purkait, M. K., Husen, A., & Siddiqi, K. S. (2023). *Chapter 16 - Energy storage properties of nanomaterials micro and nano technologies* (pp. 337−350). Elsevier Available from. Available from https://www.sciencedirect.com/science/article/pii/B9780323995467000057, 10.1016/B978-0-323-99546-7.00005-7.

Sharma, M., Samanta, N. S., Chakraborty, A., Purkait, M. K., Sillanpää, M., Khadir, A., & Gurung, K. (2023). *30* - Simultaneous treatment of industrial wastewater and resource recovery using microbial fuel cell (pp. 621−637). Elsevier Available from. Available from https://www.sciencedirect.com/science/article/pii/B9780323953276000026, 10.1016/B978-0-323-95327-6.00002-6.

Sharma, M., Das, P. P., Chakraborty, A., Purkait, M. K., Sillanpää, M., Khadir, A., & Gurung, K. (2023). *29* - Extraction of clean energy from industrial wastewater using bioelectrochemical process (pp. 601−620). Elsevier Available from. Available from https://www.sciencedirect.com/science/article/pii/B9780323953276000038, 10.1016/B978-0-323-95327-6.00003-8.

Sharma, M., Das, P. P., Kumar, S., & Purkait, M. K. (2023). Polyurethane foams as packing and insulating materials. *American Chemical Society, Polyurethanes: Preparation, Properties, and Applications Volume 3: Emerging Applications, 1454*. Available from https://doi.org/10.1021/bk-2023-1454.ch004. Available from, https://doi.org/10.1021/bk-2023-1454.ch004.

Shekhar Samanta, N., Das, P. P., Dhara, S., & Purkait, M. K. (2023). An overview of precious metal recovery from steel industry slag: Recovery strategy and utilization. *Industrial & Engineering Chemistry Research*, *62*(23), 9006−9031. Available from https://doi.org/10.1021/acs.iecr.3c00604. Available from, https://doi.org/10.1021/acs.iecr.3c00604.

Shi, D., Hu, C., Zhang, J., Li, P., Zhang, C., Wang, X., & Ma, H. (2017). Silicon-aluminum additives assisted hydrothermal process for stabilization of heavy metals in fly ash from MSW incineration. *Fuel Processing Technology*, *165*, 44−53. Available from https://doi.org/10.1016/j.fuproc.2017.050.007. Available from, https://www.sciencedirect.com/science/article/pii/S0378382017304162.

Sivalingam, S., & Sen, S. (2019). Valorization of coal fly ash into nanozeolite by sonication-assisted hydrothermal method. *Journal of Environmental Management*, *235*, 145−151. Available from https://doi.org/10.1016/j.jenvman.2019.010.042. Available from, https://www.sciencedirect.com/science/article/pii/S0301479719300428.

Sontakke, A. D., Das, P. P., Mondal, P., & Purkait, M. K. (2022). Thin-film composite nanofiltration hollow fiber membranes toward textile industry effluent treatment and environmental remediation applications: review. *Emergent Materials*, *5*(5), 1409−1427. Available from https://doi.org/10.1007/s42247-021-00261-y. Available from, https://doi.org/10.1007/s42247-021-00261-y.

Sontakke, A. D., Deepti., Samanta, N. S., Purkait, M. K., Husen, A., & Siddiqi, K. S. (2023). *Chapter 2 - Smart nanomaterials in the medical industry micro and nano technologies* (pp. 23−50). Elsevier Available from. Available from https://www.sciencedirect.com/science/article/pii/B9780323995467000252, 10.1016/B978-0-323-99546-7.00025-2.

Sun, Y., Zheng, J., Zou, L., Liu, Q., Zhu, P., & Qian, G. (2011). Reducing volatilization of heavy metals in phosphate-pretreated municipal solid waste incineration fly ash by forming pyromorphite-like minerals. *Environmental Implications of Alternative Materials in Construction and Treatment of Waste*, *31*(2), 325−330. Available from https://doi.org/10.1016/j.wasman.2010.100.011. Available from, https://www.sciencedirect.com/science/article/pii/S0956053X10005350.

Tang, H., Erzat, A., & Liu, Y. (2014). Recovery of soluble chloride salts from the wastewater generated during the washing process of municipal solid wastes incineration fly ash. *Environmental Technology*, *35*(22), 2863−2869. Available from https://doi.org/10.1080/09593330.2014.924568. Available from, https://doi.org/10.1080/09593330.2014.924568.

Tang, Q., Ge, Y.-y, Wang, K.-t, He, Y., & Cui, X.-m (2015). Preparation of porous P-type zeolite spheres with suspension solidification method. *Materials Letters*, *161*, 558−560. Available from https://doi.org/10.1016/j.matlet.2015.090.062. Available from, https://www.sciencedirect.com/science/article/pii/S0167577X15305681.

Villaescusa, L. A., Wheatley, P. S., Bull, I., Lightfoot, P., & Morris, R. E. (2001). The location and ordering of fluoride ions in pure silica zeolites with framework types IFR and STF; implications for the mechanism of zeolite synthesis in fluoride media. *Journal of the American Chemical Society*, *123*(36), 8797−8805. Available from https://doi.org/10.1021/ja016113f. Available from, https://doi.org/10.1021/ja016113f.

Yang, G. C. C., & Yang, T.-Y. (1998). Synthesis of zeolites from municipal incinerator fly ash. *Journal of Hazardous Materials*, *62*(1), 75−89. Available from https://doi.org/10.1016/S0304-3894(98)00163-0. Available from, https://www.sciencedirect.com/science/article/pii/S0304389498001630.

Zhang, Y., Han, H., Wang, X., Zhang, M., Chen, Y., Zhai, C., … Zhang, C. (2021). Utilization of NaP zeolite synthesized with different silicon species and NaAlO$_2$ from coal fly ash for the adsorption of Rhodamine B. *Journal of Hazardous Materials*, *415*125627. Available from https://doi.org/10.1016/j.jhazmat.2021.125627. Available from, https://www.sciencedirect.com/science/article/pii/S0304389421005902.

Zhang, Z., Wang, Y., Zhang, Y., Shen, B., Ma, J., & Liu, L. (2022). Stabilization of heavy metals in municipal solid waste incineration fly ash via hydrothermal treatment with coal fly ash. *Waste Management*, *144*, 285−293. Available from https://doi.org/10.1016/j.wasman.2022.030.022. Available from, https://www.sciencedirect.com/science/article/pii/S0956053X2200157X.

Zong, Y.-b, Zhao, C.-y, Chen, W.-h, Liu, Z.-b, & Cang, D.-q (2020). Preparation of hydrosodalite from fly ash using a hydrothermal method with a submolten salt system and study of the phase transition process. *International Journal of Minerals, Metallurgy and Materials*, *27*(1), 55−62. Available from https://doi.org/10.1007/s12613-019-1904-8. Available from, https://doi.org/10.1007/s12613-019-1904-8.

Zubowa, H.-L., Kosslick, H., Müller, D., Richter, M., Wilde, L., & Fricke, R. (2008). Crystallization of phase-pure zeolite NaP from MCM-22-type gel compositions under microwave radiation. *Microporous and Mesoporous Materials*, *109*(1), 542−548. Available from https://doi.org/10.1016/j.micromeso.2007.060.002. Available from, https://www.sciencedirect.com/science/article/pii/S1387181107003538.

# Chapter 7

# Zeolite synthesis from miscellaneous waste resources

## 7.1 Introduction

Zeolites are classified as crystalline tectoaluminosilicates, consisting of tetrahedral units of $[SiO_4]^{4-}$ and $[AlO_4]^{5-}$ that are interconnected at their vertices, resulting in a distinct three-dimensional framework structure. The arrangement of $[SiO_4]^{4-}$ and $[AlO_4]^{5-}$ tetrahedra results in the formation of pores and channels characterized by molecular-scale dimensions. The surplus of negative charges generated by $Al^{3+}$ is counterbalanced by alkali ($Na^+$, $K^+$) or alkaline earth metal ions ($Ca^{2+}$, $Mg^{2+}$) that are surrounded by water molecules situated within the interstitial regions. The physicochemical qualities that are pertinent to zeolites, such as molecular sieving, high surface area, and great thermal and chemical stability, are a direct result of their distinctive structure (Přech, 2018). The utilization of zeolites in many industries as sorbents, molecular sieves, ion exchangers in detergents, fertilizers, and catalysts in industrial processes is a well-anticipated phenomenon (Montalvo et al., 2012). Zeolites, as an adsorbent, can be used for the treatment of both water and wastewater (Millar, Couperthwaite, & Alyuz, 2016; Millar, Winnett, Thompson, & Couperthwaite, 2016; Wen, Dong, & Zeng, 2018). In addition to zeolites, there are several other methods used for the treatment of water and wastewater, such as electrocoagulation (Changmai et al., 2022; Das, Sontakke, Purkait, & Shah, 2023a; Das, Anweshan, & Purkait, 2021; Das, Anweshan, Mondal, et al., 2021; Das, Dhara, & Purkait, 2023; Das, Mondal, et al., 2021; Das, Sharma, & Purkait, 2022), membrane (Chakraborty et al., 2023; Das et al., 2023a; Das, Sontakke, Purkait, & Shah, 2023b; Das, Deepti, & Purkait, 2023; Das, Dhara, et al., 2023; Das, Duarah, Purkait, & Jafari, 2023; Das, Mondal, Sillanpää, Khadir, & Gurung, 2023; Das, Samanta, Dhara, Purkait, & Shah, 2023; Das, Sontakke, Samanta, & Purkait, 2023; Dhara, Shekhar Samanta, Das, Uppaluri, & Purkait, 2023; Sharma, Das, Sood, Chakraborty, & Purkait, 2021; Sharma, Das, Chakraborty, & Purkait, 2022; Sharma, Das, Sood, Chakraborty, & Purkait, 2022; Sontakke, Das, Mondal, & Purkait, 2022), adsorption (Chakraborty, Gautam, Das, & Hazarika, 2019; Das, Sontakke, Samanta, et al., 2023; Samanta, Das, et al., 2023; Samanta, Das, & Mondal, 2022; Samanta, Das,

Waste-based Zeolite. DOI: https://doi.org/10.1016/B978-0-443-22316-7.00007-3
**171**

Mondal, Changmai, & Purkait, 2022; Sharma, Das, Kumar, & Purkait, 2023; Shekhar Samanta, Das, Dhara, & Purkait, 2023), nanotechnology (Anweshan et al., 2023; Das, Duarah, Purkait, et al., 2023; Das, Mondal, & Purkait, 2022; Duarah, Das, & Purkait, 2023; Sharma, Das, Purkait, Husen, & Siddiqi, 2023), and biological treatments (Das et al., 2023a, 2023b; Das, Deepti, et al., 2023; Das, Dhara, et al., 2023; Das, Duarah, et al., 2023; Das, Mondal, et al., 2023; Das, Samanta, et al., 2023; Das, Sontakke, Samanta, et al., 2023; Dhara, Das, Uppaluri, Purkait, & Shah, 2023; Dhara, Das, Uppaluri, Purkait, Sillanpää, et al., 2023; Sharma, Das, Chakraborty, et al., 2023). Zeolites can be readily manufactured in laboratory settings through hydrothermal treatment of alkali-metal aluminosilicate gels that exhibit reactivity (Mondal, Samanta, Kumar, & Kumar Purkait, 2020). This process occurs within a temperature range spanning from ambient conditions to 200°C and at autogenously generated pressures (Das, Dhara, & Purkait, 2024). The duration of crystallization exhibits a range of timeframes, spanning from a few hours to several days, with the intention of enhancing the crystalline structure of the zeolitic products (Khaleque et al., 2020). The synthesis process primarily involved the combination of costly chemical reagents that consist of high-quality supplies of silicon (such as colloidal silica, alkali silicates, and silicon alkoxides) and aluminum (such as aluminum (hydr)oxides, nitrates, alkoxides, and sodium aluminate). Additional reagents are necessary for the reaction, including mineralizing agents such as sodium hydroxide (NaOH) and potassium hydroxide (KOH), as well as structure-guiding agents in the form of organic surfactants (Cundy & Cox, 2005). Consequently, numerous studies have concentrated on exploring the utilization of alternative sources for silicon (Si) and aluminum (Al) that are more affordable and readily available (Mondal, Samanta, Meghnani, & Purkait, 2019; Samanta, Mondal, & Purkait, 2023). A number of cost-effective geological silicate and aluminosilicate compounds, as well as affordable industrial by-products, have been suggested as economical options for initiating zeolite syntheses. Coal fly ashes are commonly utilized in the synthesis of zeolites due to their significant global production volumes. In addition, porcelain and glass wastes are abundant sources of silicon (Si) and aluminum (Al) that are utilized in the production of zeolites (Dhara, Samanta, Uppaluri, & Purkait, 2023; Samanta et al., 2021; Sharma, Samanta, et al., 2023; Sontakke et al., 2023). Zeolites with a low Si/Al ratio, such as linde type A (LTA), faujasite (FAU), Na-P1, and hydroxysodalite (HS), are typically the desired outcome in the synthesis of zeolites from glass wastes. This is achieved by employing suitable synthesis conditions and methodologies. Industrial raw powder glass as a precursor material can be employed for the synthesis of LTA, X, and HS zeolites. The utilization of an alkaline fusion process, followed by a subsequent hydrothermal treatment, was chosen as a suitable approach due to the inherent challenge of zeolitization of glass powder residue using hydrothermal treatment in basic conditions (Cheng, Xu, &

Liu, 2021). In their study, Terzano, D'Alessandro, Spagnuolo, Romagnoli, and Medici (2015) conducted hydrothermal treatment of municipal glass and aluminum solid wastes at a temperature of 60°C for a duration of 7 days (Terzano et al., 2015). The outcome of this process was the synthesis of a zeotype material, which exhibited a zeolite LTA content of 30%. It is important to note that NaOH was employed as the mineralizing agent in this experiment. Noviello et al. (2020) conducted a study on the utilization of recycled glass and aluminum packaging materials for the synthesis of zeolitic materials using an alkaline hydrothermal treatment (Noviello et al., 2020). The research was conducted within the framework of the circular economy. In their study, Sayehi, Delahay, and Tounsi (2022) employed an alkaline fusion process followed by hydrothermal treatment to synthesize Na-P1 and Na-FAU zeolites using waste glass derived from fluorescent tubes and aluminum scraps (Sayehi et al., 2022). The process of converting cathode-ray-tube funnel glass and liquid crystal display (LCD) panel glass into zeolites LTA, FAU, and Na-P1 is also conducted. To expedite the process of crystallization, microwave radiation was utilized during the initial phase of hydrothermal activation to induce the zeolitization process. In their study, Majdinasab et al. (2019) were able to successfully synthesize zeolite HS and Na-P1 using microwave radiation, utilizing a residue known as waste glass cullet (Majdinasab et al., 2019).

This chapter provides an overview of different approaches used for the synthesis of zeolite. Also, the utilization of various solid wastes for zeolite preparation, such as lithium sludge, alum sludge, glass waste, opal waste rock, cupola slag, porcelain waste, coal fly ash, and bauxite residue, among others, have been elaborately discussed, and summarized. Further, the challenges and future recommendations on zeolite synthesis from different solid wastes have also been included toward the end of the chapter.

## 7.2  Overview of zeolite synthesis approaches

The conventional approach to synthesizing zeolite LTA was utilizing alkaline solutions comprising sodium silicate and sodium aluminate. The aluminosilicate gel was created by combining the components, followed by hydrothermal treatment at temperatures of 100°C to produce zeolite LTA crystals. Significantly, the introduction of zeolite into water resulted in an elevation in the solution's pH within the range of $9-11$. The occurrence of this scenario cannot be attributed to the presence of residual alkali inside the zeolite structure. In contrast, it has been observed that zeolites characterized by a low Si/Al ratio undergo partial hydrolysis when exposed to a solution. The addition of sodium chloride to the zeolite/water mixture resulted in the reversal of the hydrolytic equilibrium. The solution's pH decreased to a range of $6.0-6.5$ as a result (compared to a pH range of $10.0-10.5$ when the solution included no salt) (Jafari, Mohammadi, & Kazemimoghadam, 2014).

The key distinction observed in the experimental procedure was the gradual addition of sodium aluminate to the sodium silicate solution. The process of heating sodium silicate and sodium silicate solutions was carried out to achieve a gel temperature of around 65°C. The aforementioned methodology seemingly facilitated the subsequent hydrothermal process at a temperature of 95°C to be efficiently executed within a very short duration. The success of the indicated strategy was contingent upon the viscosity of the gel (Sayehi et al., 2022). The water glass was combined with LTA zeolite powder and subsequently subjected to rolling on a granulating plate. Following this, the bead was subjected to activation through heating in a muffle furnace at a temperature of 400°C (Yu, Kwon, & Na, 2021).

Several writers have provided descriptions of the utilization of clay as a primary ingredient in the preparation of zeolite LTA. Kaolin, with the chemical formula $Si_2Al_2O_5(OH)_4$, is widely recognized as a prominent clay mineral that has undergone conversion into zeolites. The conventional process involves an initial stage in which kaolin is subjected to thermal treatment at around 550°C, resulting in the conversion of kaolin to the more reactive metakaolin. This metakaolin is then reacted with alkali to produce zeolite. The reaction of kaolin sourced from Iraq was documented by Ugal, Hassan, and Ali (2010) at a temperature of 550°C for a duration of 1.5 hours, which was subsequently followed by reflux at 90°C with a sodium hydroxide solution at a ratio of 1:5 for a period of 4 hours (Ugal et al., 2010). The formation of zeolite LTA occurred. The particle size of zeolite LTA generated using the metakaolin approach was determined by Chaudhuri, Dey, and Pal (2002) to fall within the range of 3−7 μm, which was found to be less than the desired 10 μm size often recommended for the purpose of detergent builder utilization (Chaudhuri et al., 2002). Nevertheless, the zeolite LTA material obtained from clay exhibited impurities, including minor amounts of potassium, magnesium, calcium, titanium, and iron. Chandrasekhar (1996) has recommended the use of elevated temperatures during the metakaolin production stage (Chandrasekhar, 1996). The recommendation for subjecting clay from Kerala to a temperature of 900°C for a duration of 1 hour was made considering parameters such as brightness and crystallinity. Various techniques have been explored to mitigate the presence of iron species to enhance the brightness of zeolite LTA. Yamane and Nakazawa (1986) effectively eliminated iron impurities as well as other elements, including aluminum, calcium, and magnesium, by the utilization of a chemical reaction between clay and sulfuric acid (Yamane & Nakazawa, 1986). The clay framework underwent decomposition as a result of the acid, leading to the formation of a gel constituted of silicic acid. Subsequently, this gel was subjected to a reaction with sodium aluminate to produce zeolite LT.

The presence of microporosity within the channels of zeolite LTA can give rise to challenges related to the migration of ions toward the exchange sites. As a result, innovative methods have been devised to enhance mass transfer in these materials by introducing mesoporosity. Chen et al. (2011) employed a three-dimensionally organized mesoporous carbon (3DOm) material in the synthesis of mesoporous zeolite LTA (Chen et al., 2011). The study utilized selected-area electron diffraction to present evidence of the formation of single crystals. These crystals were hypothesized to have developed within the neighboring cages of the 3DOm material. In an alternative approach, Hasan, Singh, Li, Zhao, and Webley (2012) employed cetyltrimethylammonium bromide (CTAB) micelles during the synthesis process of zeolite LTA, which was conducted over a duration of 2 weeks under ambient temperature conditions (Hasan et al., 2012). The zeolite LTA material, which includes mesopores, exhibited a notably higher sorption capacity for ethylene in comparison to standard zeolite LTA. Specifically, the sorption amounts were 4.32 mmol/g and 3.61 mmol/g, respectively. Ni, Zheng, Wang, Zhang, and Zhao (2014) conducted a study that verified the enhanced sorption capacity of hierarchical zeolite LTA spherical beads in comparison to zeolite LTA with regular cubic particles (Ni et al., 2014). The researchers found that the former exhibited roughly double the sorption capacity for the fire suppressant 2-bromo-3,3,3-trifluoropropene (BTP). In this study, the synthesis of hierarchical zeolite LTA was conducted using calcined clay as the precursor material. The key factor in the synthesis process was the inclusion of ethanol in the reaction mixture, which was subsequently heated to a temperature of 95°C to facilitate the formation of the zeolite. The authors Cho, Cho, de Ménorval, and Ryoo (2009) additionally documented the utilization of an organosilane surfactant that was introduced into the synthesis mixture to stimulate the formation of mesoporosity (Cho et al., 2009). It is worth noting that the diffusion of xenon in the mesoporous/microporous materials developed exhibited a significantly higher rate compared to the investigation solely focused on pure microporous materials.

One strategy that has been employed to facilitate the separation of powdered components contained in a solution involves the integration of magnetic nanoparticles. In the study conducted by Liu et al. (2013) magnetite ($Fe_3O_4$) was introduced into the aluminosilicate gel before the concluding hydrothermal synthesis phase (Liu et al., 2013). Various iron oxide loadings were assessed, revealing a rise in magnetic susceptibility that facilitated efficient separation from aqueous solution. Additionally, the sorption capacity of the zeolite for metal ions remained largely unaffected. In their study, Faghihian, Moayed, Firooz, and Iravani (2013) made modifications to the methodology by introducing iron chlorides into the zeolite mixture subsequent to the hydrothermal synthesis process (Faghihian et al., 2013). The aforementioned methodology resulted in the formation of nanomagnetite crystals, which subsequently enable efficient separation through the utilization of magnetic forces in a solution. One notable advancement involves the production of magnetic zeolite substances by the utilization of red mud waste obtained from the alumina sector. Xie et al. (2018) presented a methodology that yielded magnetic zeolite LTA, along with many other valuable materials, through the utilization of red mud (Xie et al., 2018). The fundamental approach involved the amalgamation of red mud and sodium hydroxide, followed by the crystallization of the solubilized salts into zeolite. Subsequently, the solid constituents were subjected to acid washing to remove dissolved iron, which was subsequently mixed with zeolite LTA. The strengths, weaknesses, opportunities, and threats (SWOT) assessment of the synthesis approaches for the preparation of zeolite LTA is shown in Table 7.1.

## 7.3 Zeolite synthesis from various solid wastes

### 7.3.1 Glass waste

Waste glass can be obtained from various sources, including beverage bottles, windscreens, and TV panels. The recycling of glass waste can be efficiently achieved using a technique that involves subjecting the glass to high temperatures of approximately 1500°C. As a result, there is a notable increase in energy use, accompanied by the emission of greenhouse gases. Furthermore, it is worth

**TABLE 7.1** SWOT analysis of alternative/novel synthesis approaches for zeolite LTA production.

|  | Strengths | Weaknesses |
|---|---|---|
|  | 1. Established commercial manufacture based upon well-known batch hydrothermal processes | 1. Time-consuming synthesis<br>2. Multistage process, which adds to the cost |
| **Opportunities** | S/O | W/O |
| 1. New technologies available for material processing | 1. Create a continuous process for zeolite LTA synthesis using microfluidics | 1. Evaluate rapid zeolite synthesis approaches using ultrasound and microwave energy<br>2. Develop "one-pot synthesis" strategies |
| **Threats** | S/T | W/T |
| 1. Economically unattractive synthesis conditions<br>2. Performance inadequate for application | 1. Avoid the use of organic templating methods for zeolite LTA synthesis<br>2. Pursue binderless zeolite LTA materials | 1. Study zeolite LTA synthesis via reverse microemulsions to enhance product crystallinity |

*Source*: From Collins, F., Rozhkovskaya, A., Outram, J. G., & Millar, G. J. (2020). *Microporous and Mesoporous Materials, 291*, 109667. https://doi.org/10.1016/j.micromeso.2019.109667.

noting that the rates of glass recycling exhibit significant variation across different countries (Ibrahim & Meawad, 2018; Matamoros-Veloza, Rendón-Angeles, Yanagisawa, Mejia-Martínez, & Parga, 2015). In Australia, the rates of glass recycling have experienced a decline, reaching a value of 42% (Heriyanto, Pahlevani, & Sahajwalla, 2018). In contrast, European countries exhibit higher levels of glass recycling, with an average rate of 73%. Terzano et al. (2015) utilized glass bottle waste obtained from a municipal solid waste facility in conjunction with aluminum garbage collected from soft drink cans (Terzano et al., 2015). The production of dissolved silicates involved subjecting a finely ground mixture of aluminum and glass to a sodium hydroxide solution. Following this, the specimens were subjected to thermal treatment in a controlled environment at a temperature of 60°C for a duration of 1 week, resulting in the formation of zeolites with a crystalline structure. The utilization of this particular method yielded a maximum zeolite LTA content of merely 30%. Tsujiguchi et al. (2014) initially conducted a process of reducing the concentration of boron, calcium, and magnesium in the glass sample by subjecting it to heating with nitric acid (Tsujiguchi et al., 2014). In contrast, Majdinasab et al. (2019) proposed the utilization of microwaves as a means to expedite the process of zeolite crystallization (Majdinasab et al., 2019). However, the degree of phase purity observed in zeolite LTA did not exceed 32%. In a study conducted by Lee, discarded LCD glass was utilized for the synthesis of zeolite LTA. The glass waste underwent hydrothermal treatment at a temperature of 100°C for a duration of 12 hours, resulting in the formation of zeolite LTA. The zeolite LTA exhibited a Si/Al ratio of 2. The presence of zeolite P was observed when the Si/Al molar ratio was increased to 2.8. In their study, Kim et al. (2015) employed a specific methodology to synthesize zeolite using waste material derived from windshield glass (Kim et al., 2015). The cullets of windshield glass trash were subjected to a process of fine grinding and milling, followed by an acid treatment conducted at a temperature of 70°C for a duration of 4 days. Following that, sodium aluminate was introduced, along with a solution of sodium hydroxide. The hydrothermal treatment was successfully conducted at a temperature of 90°C for a duration of 24 hours. It is worth noting that in cases when the glass was not subjected to milling, the resulting product was sodalite rather than zeolite LTA. The observed behavior can be ascribed to the enhanced efficiency of the acid in eliminating contaminants, such as calcium, magnesium, and sodium, through its interaction with the smaller particles.

## 7.3.2 Alum sludge

The process of coagulation/flocculation is widely employed in traditional drinking water treatment facilities due to its effectiveness in eliminating turbidity, organic substances, inorganic compounds, and other pollutants present in surface water sources, as shown in Fig. 7.1. Aluminum-based coagulants, such as aluminum sulfate (commonly known as alum) with the chemical

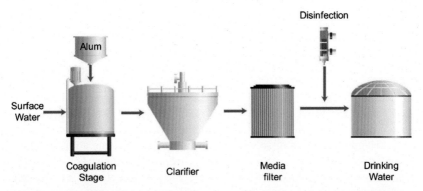

**FIGURE 7.1** Conventional drinking water treatment process. *From Collins, F., Rozhkovskaya, A., Outram, J. G., & Millar, G. J. (2020).* Microporous and Mesoporous Materials, 291, 109667. *https://doi.org/10.1016/j.micromeso.2019.109667.*

formula $Al_2(SO_4)_3 \cdot 14-18H_2O$, are extensively employed in various applications owing to their notable cost-effectiveness and high performance. Upon introduction to water, alum undergoes hydrolysis, resulting in the formation of aluminum hydroxide flocs. These flocs play a crucial role in facilitating the process of water clearing. Consequently, the production of alum sludge can be significant. As an illustration, Malaysia has an annual production of approximately 2 million tons. As a reference, the volumetric proportion of sludge corresponds to approximately 1%–3% of the volume of water subjected to treatment, and the solid content can reach up to 23% when the sludge is subjected to centrifugation for the purpose of dewatering. In recent years, a number of extensive evaluations have been published with the aim of assessing alternate methods for managing alum sludge (Babatunde & Zhao, 2007; Kaish, Breesem, & Abood, 2018; Keeley, Jarvis, & Judd, 2014; Nishat Ashraf, Rajapakse, Dawes, & Millar, 2018). The utilization of alum sludge as a sorbent has been extensively documented in the literature for addressing many wastewater issues, including but not limited to phosphate, arsenic, fluoride, colors, turbidity, and chromium. The investigation of zeolite materials has been explored due to the composition of alum sludge, which mostly consists of aluminum and silicon. The creation of zeolite X from alum sludge was described by Espejel-Ayala and Ramírez Z (2011) utilizing a two-stage synthesis process (Espejel-Ayala & Ramírez Z, 2011). The alum sludge underwent a fusion with sodium hydroxide at high temperatures, followed by hydrothermal synthesis. In the most favorable scenario, the resulting exchange capacity was measured to be 2.11 mill equivalents per gram. Espejel-Ayala, Schouwenaars, Durán-Moreno, and Ramírez-Zamora (2014) subsequently demonstrated the feasibility of synthesizing zeolite LTA using alum waste as a precursor material (Espejel-Ayala et al., 2014). The chosen approach involved the initial combination of alum sludge and sodium hydroxide at a temperature of 550°C for a duration of 2 hours. Subsequently, the adjustment of reaction parameters was conducted during the process of alkaline hydrothermal synthesis.

### 7.3.3  Coal fly ash

The process of coal combustion in power plants leads to the generation of significant quantities of residual fly ash. According to recent data, the annual production of fly ash amounts to a minimum of 800 million tons, with China, India, and the United States being the primary contributors to this output. Fly ash has the potential to serve as a significant source of silicon due to its high concentration of this element. Additionally, fly ash contains smaller proportions of aluminum, iron, calcium, magnesium, and potassium (Hosseini Asl et al., 2019; Kazemian, Naghdali, Ghaffari Kashani, & Farhadi, 2010). The silicon content present in fly ash was extracted by Behin, Bukhari, Dehnavi, Kazemian, and Rohani (2014) by a fusion process using sodium hydroxide at a temperature of 60°C for a duration of 12 hours (Behin et al., 2014). Following that, the Si/Al ratio was modified to 1 with the introduction of a sodium aluminate solution. Subsequently, the gel obtained was subjected to microwave irradiation to facilitate the proliferation of zeolite LTA crystals. It is worth noting that the synthesis procedure did not involve the use of pure water; rather, wastewater derived from plasma electrolytic oxidation was utilized. In an alternative approach, Cardoso, Horn, Ferret, Azevedo, and Pires (2015) employed a typical two-step fusion and hydrothermal treatment methodology to synthesize zeolite LTA using fly ash as the precursor material (Cardoso et al., 2015). Despite the fact that the purity of zeolite LTA was only 82% in this particular instance, the material had physical characteristics that were comparable to those of commercially available zeolite LTA. The investigations that have been mentioned demonstrate that the composition of coal fly ash exhibits significant variability. As an example, it can be observed that bituminous coal generally contains a silica, alumina, and iron oxide concentration over 70%, but lignite coals often exhibit a range of 50%−70% for this particular value. Hence, it may be necessary to customize the procedures for preparing zeolite LTA based on the quality of the fly ash.

### 7.3.4  Lithium sludge

Lithium possesses significant value as a metal, mostly driven by the escalating demand for its utilization in various applications, notably batteries employed in electronic devices such as mobile phones, digital cameras, laptops, and tablets. These aforementioned devices collectively account for approximately 35% of the overall consumption of lithium. Lithium is commonly encountered in several geological formations, including lithium-rich pegmatite, salt water, and salt lakes or brines (Kesler et al., 2012; Martin, Rentsch, Höck, & Bertau, 2017; Nishihama, Onishi, & Yoshizuka, 2011). Grosjean and colleagues provided a comprehensive summary of the lithium inventory present in brine and mineral resources (Grosjean, Miranda, Perrin, & Poggi, 2012). The estimation conducted by these authors suggests that brines have the potential to provide around 26.9 million tons of lithium, while mineral deposits contain approximately 16.7 million tons of lithium. The extraction of lithium from rocks containing lithium deposits leads to the generation of a byproduct known as lithium slag. The conventional process for extracting lithium from spodumene ore involves subjecting the ore to elevated temperatures followed by treatment with sulfuric acid (Tan et al., 2018). Typically, the production of lithium slag is tenfold the amount of lithium carbonate produced. Lithium slag is mostly composed of a combination of silicon and aluminum, along with calcium, iron, magnesium, manganese, titanium, sulfur, and phosphorus in diverse crystalline forms. Therefore efforts have been undertaken to transform lithium slag into materials that have practical use. In a study conducted by Lin, Zhuang, Cui, Wang, and Yao (2015), the authors examined the transformation of lithium slag in alkaline environments under high-pressure circumstances (Lin et al., 2015). Their findings revealed the formation of composite materials consisting of zeolite LTA and zeolite X. The synthesis conditions, including the inclusion of extra aluminate, influenced the ratio of zeolite LTA/X. Consequently, the observed range for the proportion of zeolite LTA was between 45% and 82%.

### 7.3.5    Aluminum scrap

Aluminum is a pivotal metallic element that holds significant importance in several industries, such as aerospace, automotive, food and beverage, construction, and electrical/electronic sectors. Hence, a substantial quantity of aluminum scrap is readily accessible from various sources, including but not limited to soft drink cans. Consequently, there is a pressing need to effectively recycle and/or repurpose this aluminum waste through novel approaches. In their study, Tounsi, Mseddi, and Djemel (2009) employed a dissolution process to extract aluminum scrap obtained from a turner manufacturing facility (Tounsi et al., 2009). This scrap was next subjected to a reaction with sodium hydroxide, followed by the utilization of sodium silicate derived from Tunisian sand. In contrast, Selim, EL-Mekkawi, Aboelenin, Sayed Ahmed, and Mohamed (2017) employed a commercially accessible sodium silicate compound to undergo a reaction with sodium aluminate derived from aluminum waste material (Selim et al., 2017). In both cases, the successful synthesis of zeolite LTA with a high degree of purity and crystallinity was accomplished.

### 7.3.6  Bauxite residue

The alumina industry utilizes bauxite ore as a primary raw material for the production of alumina by the well-established Bayer process. In a similar vein, a substantial quantity of rubbish known as bauxite residue or red mud is generated, exhibiting a very alkaline nature. According to current estimations, the annual production of bauxite residue amounts to approximately 150 million tons, while the existing accumulation of this residue is estimated to be over 2.7 billion tons. The waste primarily consists of iron species; however, aluminum and silicon are the subsequent most abundant components found. The process of bauxite flotation has the potential to enhance the alumina-to-silica ratio, albeit at the expense of generating bauxite tailings. The composition of these tailings mostly consists of alumina, silica, and iron oxides, predominantly on an oxide basis. In addition, lesser amounts of titanium, potassium oxide, calcium oxide, and sodium oxide are also observed. In line with past methodologies employed in cases where the raw material exhibited impurities, an acid digestion process was implemented to eliminate the presence of iron species. The tailings that underwent acid treatment were next subjected to fusion with sodium hydroxide at high temperatures within a muffle furnace, resulting in the formation of sodium aluminate and sodium silicate (Ma, Wang, Guo, Zhang, & Liu, 2014; Pepper, Couperthwaite, & Millar, 2018; Ujaczki et al., 2018; Verma, Suri, & Kant, 2017). An increased amount of sodium silicate was required to modify the Si/Al ratio. Subsequently, the fused materials were subjected to hydrothermal treatment in the presence of a sodium hydroxide solution, which was maintained at a temperature of 100°C for a duration of up to 16 hours. The synthesized LTA zeolite exhibited a high level of purity and excellent crystallinity. The ion exchange process involving calcium was conducted, and the results indicated that the exchange capacity of the zeolite was determined to be 296 mg $CaCO_3$ per gram of zeolite.

### 7.3.7  Crushed stone powder

Crushed stone powder is a byproduct generated from the process of crushing aggregates or sawing stone, commonly found in industrial settings. The composition of this material mostly consists of silicon, along with smaller quantities of aluminum, potassium, calcium, and iron species. The mineral phases that were detected in the study encompassed quartz, albite, muscovite, kaolinite, calcite, and orthoclase (Song, Lee, & Kim, 2014). Kuroki, Hashishin, Morikawa, Yamashita, and Matsuda (2019) conducted a reaction between crushed stone powder and aluminum ash obtained from the aluminum refining industry (Kuroki et al., 2019). Consequently, both zeolite LTA and zeolite X were synthesized. The experimental procedure consisted of three sequential stages. In the first stage, quartz was subjected to a hydrothermal reaction at a temperature of 150°C, using a 2 M sodium hydroxide solution for a duration of 24 hours. This resulted in the dissolution of the quartz. The second stage involved subjecting the solid product obtained from the first stage to a heat treatment at 80°C, using a 2 M hydrochloric acid solution for a duration of 24 hours. This step aimed to further modify the solid product. Finally, in the third stage, the dried filtrate obtained from the second stage was subjected to hydrothermal treatment at 80°C for 24 hours, leading to the formation of zeolite. It was determined that the inclusion of calcium species in the crushed stone hindered the development of zeolite LTA, resulting in the production of zeolite P $(Ca[Al_8Si_8O_{32}].16H_2O)$ instead. The calcium species were effectively eliminated from the crushed stone powder by subjecting it to a 1 M HCl solution for a duration of 1 minute. This process facilitated the formation of zeolite LTA during the subsequent hydrothermal stage, resulting in a yield of 79 wt.%. It was imperative to exercise caution during the drying process, as a prolonged duration was found to promote the formation of zeolite X.

### 7.3.8   Cupola slag

Cupola furnaces continue to be often employed in foundry operations for the purpose of producing liquid iron. The waste product that is produced is commonly referred to as slag, which mostly consists of silicon but also contains aluminum, as well as notable quantities of calcium and magnesium (Aristizábal, Pérez, Katz, & Bauer, 2014; Ladomerský, Janotka, Hroncová, & Najdená, 2016). Anuwattana and Khummongkol (Anuwattana & Khummongkol, 2009) utilized a demanding yet readily accessible waste material to synthesize zeolite LTA. The issue of insufficient aluminum content in the slag was initially resolved through the utilization of leftover aluminum sludge obtained from a plating facility. Significantly, it was necessary to subject the slag to treatment with hydrochloric and sulfuric acid to eliminate calcium and iron impurities. Furthermore, the process of calcination at high temperatures was necessary to eliminate organic compounds. The modified slag that was obtained was combined with aluminum sludge and sodium hydroxide and then subjected to fusion at a temperature of 700°C for a duration of 1 hour. The ultimate hydrothermal synthesis procedure involved the introduction of a 3 M sodium hydroxide solution to the fused substance, followed by heating at a temperature of 90°C for a duration of several hours. In spite of the meticulous methodology employed, the degree of crystallinity exhibited by the end product was quite modest, with the maximum proportion of zeolite LTA detected being 64%. In addition, the formation of other minerals, like as sodalite, may diminish the suitability of zeolite LTA for some applications.

### 7.3.9    Porcelain waste

According to Mallapur and Kennedy Oubagaranadin (2017), it has been noted that the ceramics sector generates a significant amount of waste porcelain, which holds the potential for conversion into zeolite materials (Mallapur & Kennedy Oubagaranadin, 2017). China is the leading producer in the field of ceramics manufacturing, with Guangdong province accounting for a significant portion of global production capacity, specifically 30%. In their study, Wajima and Ikegami (2007) conducted an analysis of porcelain waste powder, revealing its composition in terms of oxides (Wajima & Ikegami, 2007). The results indicated that the powder consisted of the following oxides: $SiO_2$ (69.8%), $Al_2O_3$ (18.5%), $K_2O$ (6.2%), CaO (3%), as well as trace amounts of iron, zinc, magnesium, and sodium. The waste porcelain underwent a two-stage treatment at a temperature of 80C to be converted into zeolite LTA, zeolite P, or zeolite X. The extraction of silicon from the waste material was conducted using a 4-M sodium hydroxide solution with a solid-to-liquid ratio of 1:4 and a reaction period of 12 hours. A maximum concentration of 26,000 mg/L of silicon was successfully extracted, whereas the remaining solid residue consisted of zeolite P and hydroxysodalite. In the subsequent phase, the aluminate solution was introduced to the silicon solution obtained from the initial stage. Zeolite LTA was synthesized by adjusting the Si/Al ratio to 1. The researchers, Nezamzadeh-Ejhieh and Banan (2011), employed the methods outlined by Wajima and Ikegami (2007) to synthesize zeolite LTA using discarded porcelain as a precursor material (Nezamzadeh-Ejhieh & Banan, 2011). In this particular instance, the researchers introduced CdS into the zeolite matrix and subsequently employed this composite material to effectively facilitate the photocatalytic degradation of crystal violet dye.

### 7.3.10    Opal waste rock

Opal waste rock (OWR) is commonly encountered in palygorskite mines, where a substantial annual production of over 100,000 tons is observed in a single area within China, for instance. The composition of this material consists of a minimum of 83% silica, accompanied by minor quantities of magnesium, iron, calcium, titanium, and potassium species. The synthesis and ammonium exchange capability of zeolite LTA derived from OWR were assessed by Wu et al. (2018). In the first step, the OWR underwent a crushing process to decrease the size of its particles, hence enhancing the effectiveness of impurity elimination through treatment with hydrochloric acid at a temperature of 70°C for a duration of 4 hours. The OWR sample that had undergone leaching was subsequently combined with a suitable amount of sodium aluminate, sodium hydroxide, and water. The experimental results indicated that the most favorable temperature for the synthesis of zeolite

LTA was determined to be 85°C. It was noted that a reduction in temperature led to the presence of cristobalite. The promotion of sodalite production was seen when the water-to-sodium ratio in the synthesis mixture was reduced. Therefore it was necessary for the ratio of $H_2O$ to $Na_2O$ to be more than 20.

## 7.4 Challenges and future recommendations

The utilization of diverse solid waste sources for the synthesis of zeolites holds significant potential in terms of waste management and resource recuperation. Nevertheless, it presents several hurdles and obstacles. Several significant obstacles are associated with the manufacturing of zeolites from solid waste.

1. Variability of feedstock: The variability of feedstock poses a considerable challenge in the development of a standardized zeolite production process due to the substantial variations in composition observed in solid waste materials. The quality and yield of zeolites can be influenced by the variable quantities of pollutants, impurities, and mineral compositions found in different waste streams.

2. Preprocessing and pretreatment: Preprocessing and pretreatment are often necessary for the effective management of solid waste products, as they involve the removal of pollutants, metals, and other undesirable components. The manufacture of zeolite can see an escalation in both cost and energy consumption as a result of this phenomenon.

3. Resource availability: The accessibility of specific solid waste materials suitable for the manufacturing of zeolites may be constrained in some geographical areas. Obtaining a reliable and cost-effective supply of feedstock can be a significant obstacle, especially for enterprises operating on a smaller scale.

4. Regulatory and environmental concerns: Regulatory and environmental considerations are of utmost importance when it comes to the manufacturing of zeolite from solid waste. Adherence to environmental standards is imperative, as any discharge of hazardous by-products or emissions might pose significant risks. The implementation of appropriate waste management practices and the reduction of environmental consequences are of utmost importance.

5. Product quality and purity: Adherence to specified standards and applications is vital in ensuring the quality and purity of zeolite products. The presence of contaminants or deviations in zeolite qualities can impose constraints on its use and diminish its commercial worth.

6. Economic viability: The economic viability of zeolite production from solid waste is crucial to ensure cost-effectiveness and make the overall process financially sustainable. The feasibility of the process can be

influenced by factors such as the initial investment, operational costs, and market demand for zeolites.

7. Market competition: Market competitiveness in the zeolite industry is a significant factor, as the production of zeolite goods of superior quality at a competitive price is of utmost importance. The presence of competition from established zeolite sources, as well as developing technologies, poses a significant issue.

The utilization of mixed solid waste for the production of zeolite is a novel strategy in the fields of waste management and resource recovery. To further the progress in this particular discipline, it is advisable to take into account the subsequent recommendations and emerging trends.

1. Waste characterization and selection: Emphasize the identification and categorization of different types of waste materials, as well as the determination of appropriate methods for their management and disposal. Establish databases or rules that can be utilized to facilitate the process of waste selection for the synthesis of zeolites.
2. Innovative waste-to-zeolite pathways: Explore novel methods for transforming diverse waste products into zeolites. This may entail conducting research on novel synthesis processes or modifying established methodologies to handle diverse waste sources.
3. Multiwaste utilization: Explore the feasibility of integrating various waste streams to generate zeolites. The combination of various waste materials can lead to enhanced characteristics of zeolites and more efficient exploitation of resources.
4. Integration with circular economy principles: Incorporate circular economy principles which emphasize waste reduction, reuse, and recycling, to establish a closed-loop system for the manufacture of zeolite from solid waste.
5. Green synthesis methods: Priority should be given to green synthesis methods, specifically to the advancement of environmentally conscious and sustainable processes for the production of zeolites from waste materials. To mitigate the environmental impact, it is imperative to adopt strategies that aim to minimize energy use, limit chemical usage, and mitigate emissions.
6. Modular and decentralized production: The viability of implementing small-scale, modular zeolite manufacturing systems near waste sources is to be established, with the aim of minimizing transportation expenses and facilitating local resource recovery.
7. Economic viability and business models: The economic feasibility of producing zeolite from solid waste should be examined, along with an exploration of various business models. This should include an analysis of partnerships that may be formed between waste management businesses, zeolite manufacturers, and end-users.

8. Pilot projects and case studies: The implementation of pilot projects and case studies is recommended to showcase the viability and advantages of producing zeolite from diverse solid waste materials. Case studies serve as valuable tools for highlighting successful implementations and elucidating their environmental and economic benefits.

## 7.5 Conclusion

The synthesis of zeolites encompasses a wide range of approaches, including the utilization of sodium aluminate, sodium silicate, and clay as raw materials, as well as the incorporation of mesoporosity and the formulation of binderless zeolites. These diverse methods collectively contribute to a diverse array of possibilities in the field of material science and its various applications. The utilization of sodium aluminate, sodium silicate, and clay as principal constituents offers a sustainable and economically feasible approach to the production of zeolites. The combination of these unprocessed substances, under the influence of meticulous adjustment of synthesis settings, facilitates the generation of zeolitic structures possessing customized characteristics, rendering them appropriate for diverse applications. The introduction of mesoporosity into the structures of zeolites enhances their functional flexibility. The intentional incorporation of mesopores serves to augment the accessibility to active sites, accelerate diffusion kinetics, and broaden the scope of potential applications, particularly in the fields of catalysis and adsorption. The deliberate manipulation of mesoporosity in zeolites enables the tailoring of their properties to fulfill specific demands, hence enhancing their versatility in various industrial applications. The development of binderless zeolites signifies a significant advancement in tackling the practical obstacles linked to conventional zeolite catalysts and adsorbents. The higher stability and purity of binderless zeolites are attributed to their ability to eliminate the use of binders, which can introduce impurities and potentially impact performance. The increasing demand for adaptable and eco-friendly materials has led to the development of novel strategies that are driving the advancement of zeolite-based technologies, resulting in improved performance and a wider range of applications.

Moreover, the process of synthesizing zeolites using a wide range of solid waste materials, such as glass waste, alum sludge, lithium sludge, bauxite residue, OWR, cupola slag, and porcelain waste, signifies a pioneering and environmentally-friendly methodology within the field of materials research. This comprehensive investigation not only tackles the urgent matter of waste management but also converts these abandoned materials into valuable resources with substantial practical uses. The incorporation of various waste streams in the synthesis of zeolites is consistent with the concepts of a circular economy, which advocates for the transformation of waste into valuable resources. Every waste item contributes a distinct chemical composition to the synthesis process, hence impacting the characteristics of the zeolites produced. The presence of diverse zeolite features not only enhances resource efficiency but also provides opportunities for customization to align with unique industrial requirements. The ecological principles behind the process of zeolite synthesis from solid waste extend beyond the mere conversion of waste materials. Through the repurposing of these materials, the strategy effectively reduces the environmental impact linked to conventional disposal methods while also making a valuable contribution to the advancement of sustainable materials for a range of applications such as catalysis, adsorption, and environmental remediation.

## AI disclosure

During the preparation of this work, the author(s) used Quillbot to improve the English and for smooth writing. After using this tool/service, the author (s) reviewed and edited the content as needed and take(s) full responsibility for the content of the publication.

## References

Anuwattana, R., & Khummongkol, P. (2009). Conventional hydrothermal synthesis of Na-A zeolite from cupola slag and aluminum sludge. *Journal of Hazardous Materials*, *166*(1), 227−232. Available from https://doi.org/10.1016/j.jhazmat.2008.11.020, https://www.sciencedirect.com/science/article/pii/S0304389408016786.

Anweshan., Das, P. P., Dhara, S., Purkait, M. K., Husen, A., & Salahuddin Siddiqi, K. (2023). *Chapter 12 - Nanosensors in food science and technology micro and nano technologies* (pp. 247−272). Elsevier. Available from https://www.sciencedirect.com/science/article/pii/B978032399546700015X, 10.1016/B978-0-323-99546-7.00015-X.

Aristizábal, R. E., Pérez, P. A., Katz, S., & Bauer, M. E. (2014). Studies of a quenched cupola. *International Journal of Metalcasting*, *8*(3), 13−22. Available from https://doi.org/10.1007/BF03355586, https://doi.org/10.1007/BF03355586.

Babatunde, A. O., & Zhao, Y. Q. (2007). Constructive approaches toward water treatment works sludge management: An international review of beneficial reuses. *Critical Reviews in Environmental Science and Technology*, *37*(2), 129−164. Available from https://doi.org/10.1080/10643380600776239, https://doi.org/10.1080/10643380600776239.

Behin, J., Bukhari, S. S., Dehnavi, V., Kazemian, H., & Rohani, S. (2014). Using coal fly ash and wastewater for microwave synthesis of LTA zeolite. *Chemical Engineering & Technology*, *37*(9), 1532−1540. Available from https://doi.org/10.1002/ceat.201400225, https://doi.org/10.1002/ceat.201400225.

Cardoso, A. M., Horn, M. B., Ferret, L. S., Azevedo, C. M. N., & Pires, M. (2015). Integrated synthesis of zeolites 4A and Na−P1 using coal fly ash for application in the formulation of detergents and swine wastewater treatment. *Journal of Hazardous Materials*, *287*, 69−77. Available from https://doi.org/10.1016/j.jhazmat.2015.01.042, https://www.sciencedirect.com/science/article/pii/S0304389415000448.

Chakraborty, S., Das, P. P., Mondal, P., Sillanpää, M., Khadir, A., & Gurung, K. (2023). *34 - Recent advances in membrane technology for the recovery and reuse of valuable resources* (pp. 695−719). Elsevier. Available from https://www.sciencedirect.com/science/article/pii/ B9780323953276000282, 10.1016/B978-0-32395327-6.00028-2.

Chakraborty, S., Gautam, S. P., Das, P. P., & Hazarika, M. K. (2019). Instant controlled pressure drop (DIC) treatment for improving process performance and milled rice quality. *Journal of The Institution of Engineers (India): Series A, 100*(4), 683−695. Available from https://doi. org/10.1007/s40030-019-00403-w, https://doi.org/10.1007/s40030-019-00403-w.

Chandrasekhar, S. (1996). Influence of metakaolinization temperature on the formation of zeolite 4A from kaolin. *Clay Minerals, 31*(2), 253−261. Available from https://doi.org/10.1180/ claymin.1996.031.2.11, https://www.cambridge.org/core/article/influence-of-metakaoliniza-tion-temperature-on-the-formation-of-zeolite-4a-from-kaolin/ 867167452BB18A2160A429B88986A65B.

Changmai, M., Das, P. P., Mondal, P., Pasawan, M., Sinha, A., Biswas, P., ... Purkait, M. K. (2022). Hybrid electrocoagulation−microfiltration technique for treatment of nanofiltration rejected steel industry effluent. *International Journal of Environmental Analytical Chemistry, 102*(1), 62−83. Available from https://doi.org/10.1080/03067319.2020.1715381, https://doi.org/10.1080/03067319.2020.1715381.

Chaudhuri, A. R., Dey, G. K., & Pal, T. K. (2002). Synthesis and characterization of detergent-grade zeolite from indian clay. *Chemical Engineering & Technology, 25*(1), 91−95. Available from https://doi.org/10.1002/1521-4125.

Cheng, Y., Xu, L., & Liu, C. (2021). NaP1 zeolite synthesized via effective extraction of Si and Al from red mud for methylene blue adsorption. *Advanced Powder Technology, 32*(10), 3904−3914. Available from https://doi.org/10.1016/j.apt.2021.08.036, https://www.science-direct.com/science/article/pii/S0921883121004106.

Chen, H., Wydra, J., Zhang, X., Lee, P.-S., Wang, Z., Fan, W., ... Tsapatsis, M. (2011). Hydrothermal synthesis of zeolites with three-dimensionally ordered mesoporous-imprinted structure. *Journal of the American Chemical Society, 133*(32), 12390−12393. Available from https://doi.org/10.1021/ja2046815, https://doi.org/10.1021/ja2046815.

Cho, K., Cho, H. S., de Ménorval, L.-C., & Ryoo, R. (2009). Generation of mesoporosity in LTA zeolites by organosilane surfactant for rapid molecular transport in catalytic applica-tion. *Chemistry of Materials: A Publication of the American Chemical Society, 21*(23), 5664−5673. Available from https://doi.org/10.1021/cm902861y, https://doi.org/10.1021/ cm902861y.

Cundy, C. S., & Cox, P. A. (2005). The hydrothermal synthesis of zeolites: Precursors, inter-mediates and reaction mechanism. *Microporous and Mesoporous Materials, 82*(1), 1−78. Available from https://doi.org/10.1016/j.micromeso.2005.02.016, https://www.sciencedirect. com/science/article/pii/S1387181105000934.

Das, P. P., Anweshan., Mondal, P., Sinha, A., Biswas, P., Sarkar, S., ... Purkait, M. K. (2021). Integrated ozonation assisted electrocoagulation process for the removal of cyanide from steel industry wastewater. *Chemosphere, 263*128370. Available from https://doi.org/10.1016/j.chemo-sphere.2020.128370, https://www.sciencedirect.com/science/article/pii/S0045653520325650.

Das, P. P., Anweshan., & Purkait, M. K. (2021). Treatment of cold rolling mill (CRM) effluent of steel industry. *Separation and Purification Technology, 274*119083. Available from https://doi.org/10.1016/j.seppur.2021.119083, https://www.sciencedirect.com/science/article/ pii/S1383586621007930.

Das, P. P., Mondal, P., Anweshan., Sinha, A., Biswas, P., Sarkar, S., ... Purkait, M. K. (2021). Treatment of steel plant generated biological oxidation treated (BOT) wastewater by hybrid pro-cess. *Separation and Purification Technology, 258*118013. Available from https://doi.org/10.1016/ j.seppur.2020.118013, https://www.sciencedirect.com/science/article/pii/S1383586620324862.

Das, P. P., Deepti., & Purkait, M. K. (2023). *Industrial wastewater to biohydrogen production via potential bio-refinery route* (pp. 159−179). Cham: Springer International Publishing. Available from https://doi.org/10.1007/978-3-03120822-5_8, 10.1007/978-3-03120822-5_8.

Das, P. P., Dhara, S., & Purkait, M. K. (2023). *Hybrid electrocoagulation and ozonation techniques for industrial wastewater treatment* (pp. 107−128). Singapore: Springer Nature Singapore. Available from https://doi.org/10.1007/978-981-99-2560-5_6, 10.1007/978-981-99-2560-5_6.

Das, P. P., Duarah, P., Purkait, M. K., & Jafari, S. M. (2023). *5 - Fundamentals of food roasting process. Unit operations and processing equipment in the food industry* (pp. 103−130). Woodhead Publishing. Available from https://www.sciencedirect.com/science/article/pii/ B9780128186183000057, 10.1016/B978-0-12-818618-3.00005-7.

Das, P. P., Mondal, P., Sillanpää, M., Khadir, A., & Gurung, K. (2023). *14 - Membrane-assisted potable water reuses applications: benefits and drawbacks* (pp. 289−309). Elsevier. Available from https://www.sciencedirect.com/science/article/pii/B9780323993449000141, 10.1016/B978-0-323-99344-9.00014-1.

Das, P. P., Samanta, N. S., Dhara, S., Purkait, M. K., & Shah, M. P. (2023). *Chapter 13 - Biofuel production from algal biomass* (pp. 167−179). Elsevier. Available from https://www.sciencedirect.com/science/article/pii/B9780128243183000096, 10.1016/B978-0-12-824318-3.00009-6.

Das, P. P., Sontakke, A. D., Purkait, M. K., & Shah, M. P. (2023a). *Chapter 12 - Rice straw for biofuel production* (pp. 153−166). Elsevier. Available from https://www.sciencedirect.com/ science/article/pii/B9780128243183000345, 10.1016/B978-0-12-824318-3.00034-5.

Das, P. P., Sontakke, A. D., Purkait, M. K., & Shah, M. P. (2023b). *Chapter 2 - Electrocoagulation process for wastewater treatment: applications, challenges, and prospects* (pp. 23−48). Elsevier. Available from https://www.sciencedirect.com/science/article/ pii/B9780323956840000154, 10.1016/B978-0-323-95684-0.00015-4.

Das, P. P., Sontakke, A. D., Samanta, N. S., & Purkait, M. K. (2023). *Emerging contaminants in wastewater: Eco-toxicity and Sustainability sssessment* (pp. 63−87). Singapore: Springer Nature Singapore. Available from https://doi.org/10.1007/978-981-99-2489-9_4, 10.1007/ 978-981-99-2489-9_4.

Das, P. P., Mondal, P., & Purkait, M. K. (2022). *Recent advances in synthesis of iron nanoparticles via green route and their application in biofuel production* (pp. 79−104). Singapore: Springer Nature Singapore. Available from https://doi.org/10.1007/978-981-16-9356-4_4, 10.1007/978-981-16-9356-4_4.

Das, P. P., Sharma, M., & Purkait, M. K. (2022). Recent progress on electrocoagulation process for wastewater treatment: A review. *Separation and Purification Technology, 292*121058. Available from https://doi.org/10.1016/j.seppur.2022.121058, https://www.sciencedirect. com/science/article/pii/S1383586622006153.

Das, P. P., Dhara, S., & Purkait, M. K. (2024). Ozone-based oxidation processes for the removal of pharmaceutical products from wastewater. In Maulin P. Shah, & Pooja Ghosh (Eds.), *Development in wastewater treatment research and processes* (pp. 287−308). Elsevier. Available from https://doi.org/10.1016/B978-0-443-19207-4.00003-3.

Dhara, S., Das, P. P., Uppaluri, R., Purkait, M. K., & Shah, M. P. (2023). *Chapter 3 - Phosphorus recovery from municipal wastewater treatment plants* (pp. 49−72). Elsevier. Available from https://www.sciencedirect.com/science/article/pii/B9780323956840000142, 10.1016/B978-0-323-95684-0.00014-2.

Dhara, S., Das, P. P., Uppaluri, R., Purkait, M. K., Sillanpää, M., Khadir, A., . . . Gurung, K. (2023). *12 - Biological approach for energy self-sufficiency of municipal wastewater treatment plants* (pp. 235−260). Elsevier. Available from https://www.sciencedirect.com/science/ article/pii/B9780323993487000060, 10.1016/B978-0-323-99348-7.00006-0.

Dhara, S., Samanta, N. S., Uppaluri, R., & Purkait, M. K. (2023). High-purity alkaline lignin extraction from *Saccharum ravannae* and optimization of lignin recovery through response surface methodology. *International Journal of Biological Macromolecules, 234*123594. Available from https://doi.org/10.1016/j.ijbiomac.2023.123594, https://www.sciencedirect. com/science/article/pii/S0141813023004877.

Dhara, S., Shekhar Samanta, N., Das, P. P., Uppaluri, R. V. S., & Purkait, M. K. (2023). Ravenna grass-extracted alkaline lignin-based polysulfone mixed matrix membrane (MMM) for aqueous Cr(VI) removal. *ACS Applied Polymer Materials, 5*(8), 6399−6411. Available from https://doi.org/10.1021/acsapm.3c00999, https://doi.org/10.1021/acsapm.3c00999.

Duarah, P., Das, P. P., & Purkait, M. K. (2023). *Technological advancement in the synthesis and application of nanocatalysts.* Singapore: Springer Nature Singapore. Available from https:// doi.org/10.1007/978-981-99-3292-4_10, 10.1007/978-981-99-3292-4_10.

Espejel-Ayala, F., & Ramírez Z, R. M. (2011). Production process of zeolite X using alum sludge issued from drinking water treatment plants. *MRS Online Proceedings Library, 1380*(1), 7. Available from https://doi.org/10.1557/opl.2012.403, https://doi.org/10.1557/opl.2012.403.

Espejel-Ayala, F., Schouwenaars, R., Durán-Moreno, A., & Ramírez-Zamora, R. M. (2014). Use of drinking water sludge in the production process of zeolites. *Research on Chemical Intermediates, 40*(8), 2919−2928. Available from https://doi.org/10.1007/s11164-013-1138-8, https://doi.org/10.1007/s11164-013-1138-8.

Faghihian, H., Moayed, M., Firooz, A., & Iravani, M. (2013). Synthesis of a novel magnetic zeolite nanocomposite for removal of Cs + and Sr2 + from aqueous solution: Kinetic, equilibrium, and thermodynamic studies. *Journal of Colloid and Interface Science, 393*, 445−451. Available from https://doi.org/10.1016/j.jcis.2012.11.010, https://www.sciencedirect.com/science/article/pii/S0021979712012970.

Grosjean, C., Miranda, P. H., Perrin, M., & Poggi, P. (2012). Assessment of world lithium resources and consequences of their geographic distribution on the expected development of the electric vehicle industry. *Renewable and Sustainable Energy Reviews, 16*(3), 1735−1744. Available from https://doi.org/10.1016/j.rser.2011.11.023, https://www.sciencedirect.com/science/article/pii/S1364032111005594.

Hasan, F., Singh, R., Li, G., Zhao, D., & Webley, P. A. (2012). Direct synthesis of hierarchical LTA zeolite via a low crystallization and growth rate technique in presence of cetyltrimethylammonium bromide. *Journal of Colloid and Interface Science, 382*(1), 1−12. Available from https://doi.org/10.1016/j.jcis.2012.05.027, https://www.sciencedirect.com/science/article/pii/S002197971200570X.

Heriyanto., Pahlevani, F., & Sahajwalla, V. (2018). From waste glass to building materials − An innovative sustainable solution for waste glass. *Journal of Cleaner Production, 191*, 192−206. Available from https://doi.org/10.1016/j.jclepro.2018.04.214, https://www.sciencedirect.com/science/article/pii/S0959652618312502.

Hosseini Asl, S. M., Javadian, H., Khavarpour, M., Belviso, C., Taghavi, M., & Maghsudi, M. (2019). Porous adsorbents derived from coal fly ash as cost-effective and environmentally-friendly sources of aluminosilicate for sequestration of aqueous and gaseous pollutants: A review. *Journal of Cleaner Production, 208*, 1131−1147. Available from https://doi.org/10.1016/j.jclepro.2018.10.186, https://www.sciencedirect.com/science/article/pii/S0959652618332025.

Ibrahim, S., & Meawad, A. (2018). Assessment of waste packaging glass bottles as supplementary cementitious materials. *Construction and Building Materials, 182*, 451−458. Available from https://doi.org/10.1016/j.conbuildmat.2018.06.119, https://www.sciencedirect.com/science/article/pii/S0950061818315162.

Jafari, M., Mohammadi, T., & Kazemimoghadam, M. (2014). Synthesis and characterization of ultrafine sub-micron Na-LTA zeolite particles prepared via hydrothermal template-free method. *Ceramics International, 40*(8, Part A), 12075−12080. Available from https://doi.org/10.1016/j.ceramint.2014.04.047, https://www.sciencedirect.com/science/article/pii/S0272884214005756.

Kaish, A. B. M. A., Breesem, K. M., & Abood, M. M. (2018). Influence of pre-treated alum sludge on properties of high-strength self-compacting concrete. *Journal of Cleaner Production, 202*, 1085−1096. Available from https://doi.org/10.1016/j.jclepro.2018.08.156, https://www.sciencedirect.com/science/article/pii/S0959652618324892.

Kazemian, H., Naghdali, Z., Ghaffari Kashani, T., & Farhadi, F. (2010). Conversion of high silicon fly ash to Na-P1 zeolite: Alkaline fusion followed by hydrothermal crystallization. *Advanced Powder Technology, 21*(3), 279−283. Available from https://doi.org/10.1016/j.apt.2009.12.005, https://www.sciencedirect.com/science/article/pii/S0921883109002271.

Keeley, J., Jarvis, P., & Judd, S. J. (2014). Coagulant recovery from water treatment residuals: A review of applicable technologies. *Critical Reviews in Environmental Science and Technology, 44*(24), 2675−2719. Available from https://doi.org/10.1080/10643389.2013.829766, https://doi.org/10.1080/10643389.2013.829766.

Kesler, S. E., Gruber, P. W., Medina, P. A., Keoleian, G. A., Everson, M. P., & Wallington, T. J. (2012). Global lithium resources: Relative importance of pegmatite, brine and other deposits. *Ore Geology Reviews, 48*, 55−69. Available from https://doi.org/10.1016/j.oregeorev.2012.05.006, https://www.sciencedirect.com/science/article/pii/S0169136812001539.

Khaleque, A., Alam, M. M., Hoque, M., Mondal, S., Haider, J. B., Xu, B., ... Moni, M. A. (2020). Zeolite synthesis from low-cost materials and environmental applications: A review. *Environmental Advances*, 2100019. Available from https://doi.org/10.1016/j.envadv.2020.100019, https://www.sciencedirect.com/science/article/pii/S2666765720300193.

Kim, J.-C., Choi, M., Song, H. J., Park, J. E., Yoon, J.-H., Park, K.-S., ... Kim, D.-W. (2015). Synthesis of uniform-sized zeolite from windshield waste. *Materials Chemistry and Physics, 166*, 20−25. Available from https://doi.org/10.1016/j.matchemphys.2015.09.028, https://www.sciencedirect.com/science/article/pii/S0254058415303473.

Kuroki, S., Hashishin, T., Morikawa, T., Yamashita, K., & Matsuda, M. (2019). Selective synthesis of zeolites A and X from two industrial wastes: Crushed stone powder and aluminum ash. *Journal of Environmental Management, 231*, 749−756. Available from https://doi.org/10.1016/j.jenvman.2018.10.082, https://www.sciencedirect.com/science/article/pii/S0301479718312167.

Ladomerský, J., Janotka, I., Hroncová, E., & Najdená, I. (2016). One-year properties of concrete with partial substitution of natural aggregate by cupola foundry slag. *Journal of Cleaner Production, 131*, 739−746. Available from https://doi.org/10.1016/j.jclepro.2016.04.101, https://www.sciencedirect.com/science/article/pii/S0959652616303778.

Lin, G., Zhuang, Q., Cui, Q., Wang, H., & Yao, H. (2015). Synthesis and adsorption property of zeolite FAU/LTA from lithium slag with utilization of mother liquid. *Chinese Journal of Chemical Engineering, 23*(11), 1768−1773. Available from https://doi.org/10.1016/j.cjche.2015.10.001, https://www.sciencedirect.com/science/article/pii/S1004954115003560.

Liu, H., Peng, S., Shu, L., Chen, T., Bao, T., & Frost, R. L. (2013). Magnetic zeolite NaA: Synthesis, characterization based on metakaolin and its application for the removal of Cu2 + , Pb2 + . *Chemosphere, 91*(11), 1539−1546. Available from https://doi.org/10.1016/j.chemosphere.2012.12.038, https://www.sciencedirect.com/science/article/pii/S0045653512015391.

Majdinasab, A. R., Manna, P. K., Wroczynskyj, Y., van Lierop, J., Cicek, N., Tranmer, G. K., ... Yuan, Q. (2019). Cost-effective zeolite synthesis from waste glass cullet using energy efficient microwave radiation. *Materials Chemistry and Physics, 221*, 272−287. Available from https://doi.org/10.1016/j.matchemphys.2018.09.057, https://www.sciencedirect.com/science/article/pii/S0254058418308083.

Mallapur, V. P., & Kennedy Oubagaranadin, J. U. (2017). A brief review on the synthesis of zeolites from hazardous wastes. *Transactions of the Indian Ceramic Society, 76*(1), 1−13. Available from https://doi.org/10.1080/0371750X.2016.1231086, https://doi.org/10.1080/0371750X.2016.1231086.

Martin, G., Rentsch, L., Höck, M., & Bertau, M. (2017). Lithium market research − Global supply, future demand and price development. *Energy Storage Materials, 6*, 171−179. Available from https://doi.org/10.1016/j.ensm.2016.11.004, https://www.sciencedirect.com/science/article/pii/S2405829716302392.

Matamoros-Veloza, Z., Rendón-Angeles, J. C., Yanagisawa, K., Mejia-Martínez, E. E., & Parga, J. R. (2015). Low temperature preparation of porous materials from TV panel glass compacted via hydrothermal hot pressing. *Ceramics International, 41*(10, Part A), 12700−12709. Available from https://doi.org/10.1016/j.ceramint.2015.06.102, https://www.sciencedirect.com/science/article/pii/S0272884215012183.

Ma, D., Wang, Z., Guo, M., Zhang, M., & Liu, J. (2014). Feasible conversion of solid waste bauxite tailings into highly crystalline 4A zeolite with valuable application. *Waste Management, 34*(11), 2365−2372. Available from https://doi.org/10.1016/j.wasman.2014.07.012, https://www.sciencedirect.com/science/article/pii/S0956053X1400316X.

Millar, G. J., Couperthwaite, S. J., & Alyuz, K. (2016). Behaviour of natural zeolites used for the treatment of simulated and actual coal seam gas water. *Journal of Environmental Chemical Engineering, 4*(2), 1918−1928. Available from https://doi.org/10.1016/j.jece.2016.03.014, https://www.sciencedirect.com/science/article/pii/S2213343716300938.

Millar, G. J., Winnett, A., Thompson, T., & Couperthwaite, S. J. (2016). Equilibrium studies of ammonium exchange with Australian natural zeolites. *Journal of Water Process Engineering, 9*, 47−57. Available from https://doi.org/10.1016/j.jwpe.2015.11.008, https://www.sciencedirect.com/science/article/pii/S2214714415300660.

Mondal, P., Samanta, N. S., Kumar, A., & Kumar Purkait, M. (2020). Recovery of H2SO4 from wastewater in the presence of NaCl and KHCO$_3$ through pH responsive polysulfone membrane: Optimization approach. *Polymer Testing, 86*106463. Available from https://doi.org/10.1016/j.polymertesting.2020.106463, https://www.sciencedirect.com/science/article/pii/S0142941820301239.

Mondal, P., Samanta, N. S., Meghnani, V., & Purkait, M. K. (2019). Selective glucose permeability in presence of various salts through tunable pore size of pH responsive PVDF-co-HFP membrane. *Separation and Purification Technology, 221*, 249−260. Available from https://doi.org/10.1016/j.seppur.2019.04.001, https://www.sciencedirect.com/science/article/pii/S1383586618346252.

Montalvo, S., Guerrero, L., Borja, R., Sánchez, E., Milán, Z., Cortés, I., ... Angeles de la la Rubia, M. (2012). Application of natural zeolites in anaerobic digestion processes: A review. *Applied Clay Science, 58*, 125−133. Available from https://doi.org/10.1016/j.clay.2012.01.013, https://www.sciencedirect.com/science/article/pii/S016913171200035X.

Nezamzadeh-Ejhieh, A., & Banan, Z. (2011). A comparison between the efficiency of CdS nanoparticles/zeolite A and CdO/zeolite A as catalysts in photodecolorization of crystal violet. *Desalination, 279*(1), 146−151. Available from https://doi.org/10.1016/j.desal.2011.06.006, https://www.sciencedirect.com/science/article/pii/S0011916411005194.

Nishat Ashraf, S., Rajapakse, J., Dawes, L. A., & Millar, G. J. (2018). Coagulants for removal of turbidity and dissolved species from coal seam gas associated water. *Journal of Water*

*Process Engineering*, *26*, 187−199. Available from https://doi.org/10.1016/j. jwpe.2018.10.017, https://www.sciencedirect.com/science/article/pii/S2214714418305762.

Nishihama, S., Onishi, K., & Yoshizuka, K. (2011). Selective recovery process of lithium from seawater using integrated ion exchange methods. *Solvent Extraction and Ion Exchange*, *29* (3), 421−431. Available from https://doi.org/10.1080/07366299.2011.573435, https://doi. org/10.1080/07366299.2011.573435.

Ni, X., Zheng, Z., Wang, X., Zhang, S., & Zhao, M. (2014). Fabrication of hierarchical zeolite 4A microspheres with improved adsorption capacity to bromofluoropropene and their fire suppression performance. *Journal of Alloys and Compounds*, *592*, 135−139. Available from https://doi.org/10.1016/j.jallcom.2013.12.025, https://www.sciencedirect.com/science/article/ pii/S0925838813030016.

Noviello, M., Gattullo, C. E., Allegretta, I., Terzano, R., Gambacorta, G., & Paradiso, V. M. (2020). Synthetic zeolite materials from recycled glass and aluminium food packaging as potential oenological adjuvant. *Food Packaging and Shelf Life*, *26*100572. Available from https://doi.org/10.1016/j.fpsl.2020.100572, https://www.sciencedirect.com/science/article/pii/ S2214289420305901.

Pepper, R. A., Couperthwaite, S. J., & Millar, G. J. (2018). Value adding red mud waste: Impact of red mud composition upon fluoride removal performance of synthesised akaganeite sorbents. *Journal of Environmental Chemical Engineering*, *6*(2), 2063−2074. Available from https://doi.org/10.1016/j.jece.2018.02.048, https://www.sciencedirect.com/science/article/pii/ S2213343718301209.

Přech, J. (2018). Catalytic performance of advanced titanosilicate selective oxidation catalysts − A review. *Catalysis Reviews*, *60*(1), 71−131. Available from https://doi.org/10.1080/ 01614940.2017.1389111, https://doi.org/10.1080/01614940.2017.1389111.

Samanta, N. S., Banerjee, S., Mondal, P., Anweshan., Bora, U., & Purkait, M. K. (2021). Preparation and characterization of zeolite from waste Linz-Donawitz (LD) process slag of steel industry for removal of Fe3 + from drinking water. *Advanced Powder Technology*, *32* (9), 3372−3387. Available from https://doi.org/10.1016/j.apt.2021.07.023, https://www. sciencedirect.com/science/article/pii/S0921883121003605.

Samanta, N., Das, P. P., & Mondal, P. (2022). Physico-chemical and adsorption study of hydrothermally treated zeolite A and FAU-type zeolite X prepared from LD (Linz−Donawitz) slag of the steel industry. *International Journal of Environmental Analytical Chemistry*, *13*, 1−23. Available from https://doi.org/10.1080/03067319.2022.2079082.

Samanta, N. S., Das, P. P., Mondal, P., Changmai, M., & Purkait, M. K. (2022). Critical review on the synthesis and advancement of industrial and biomass waste-based zeolites and their applications in gas adsorption and biomedical studies. *Journal of the Indian Chemical Society*, *99*(11)100761. Available from https://doi.org/10.1016/j.jics.2022.100761, https:// www.sciencedirect.com/science/article/pii/S001945222200423X.

Samanta, N. S., Das, P. P., Sharma, M., Purkait, M. K., Sillanpää, M., Khadir, A., ... Gurung, K. (2023). *12* - Recycle of water treatment plant sludge and its utilization for wastewater treatment (pp. 239−264). Elsevier. Available from https://www.sciencedirect.com/science/ article/pii/B9780323993449000104, 10.1016/B978-0-323-99344-9.00010-4.

Samanta, N. S., Mondal, P., & Purkait, M. K. (2023). *Nanofiltration technique for the treatment of industrial wastewater* (pp. 165−190). Singapore: Springer Nature Singapore. Available from https://doi.org/10.1007/978-981-99-3292-4_9, 10.1007/978-981-99-3292-4_9.

Sayehi, M., Delahay, G., & Tounsi, H. (2022). Synthesis and characterization of ecofriendly materials zeolite from waste glass and aluminum scraps using the hydrothermal technique. *Journal of*

*Environmental Chemical Engineering, 10*(6)108561. Available from https://doi.org/10.1016/j. jece.2022.108561, https://www.sciencedirect.com/science/article/pii/S2213343722014348.

Selim, M. M., EL-Mekkawi, D. M., Aboelenin, R. M. M., Sayed Ahmed, S. A., & Mohamed, G. M. (2017). Preparation and characterization of Na-A zeolite from aluminum scrub and commercial sodium silicate for the removal of Cd2 + from water. *Journal of the Association of Arab Universities for Basic and Applied Sciences. null, 24*(1), 19−25. Available from https://doi.org/10.1016/j.jaubas.2017.05.002, https://doi.org/10.1016/j. jaubas.2017.05.002.

Sharma, M., Das, P. P., Chakraborty, A., Purkait, M. K., Sillanpää, M., Khadir, A., . . . Gurung, K. (2023). *29 - Extraction of clean energy from industrial wastewater using bioelectrochemical process* (pp. 601−620). Elsevier. Available from https://www.sciencedirect.com/science/article/pii/B9780323953276000038, 10.1016/B978-0-323-95327-6.00003-8.

Sharma, M., Das, P. P., Chakraborty, A., & Purkait, M. K. (2022). Clean energy from salinity gradients using pressure retarded osmosis and reverse electrodialysis: A review. *Sustainable Energy Technologies and Assessments, 49*101687. Available from https://doi.org/10.1016/j. seta.2021.101687, https://www.sciencedirect.com/science/article/pii/S2213138821007013.

Sharma, M., Das, P. P., Kumar, S., & Purkait, M. K. (2023). *Polyurethane foams as packing and insulating materials, . Polyurethanes: Preparation, properties, and applications, Vol. 3: Emerging applications* (1454). American Chemical Society. Available from 10.1021/bk-2023-1454.ch004.

Sharma, M., Das, P. P., Purkait, M. K., Husen, A., & Siddiqi, K. S. (2023). *Chapter 16 - Energy storage properties of nanomaterials. Micro and nano technologies* (pp. 337−350). Elsevier. Available from https://www.sciencedirect.com/science/article/pii/B9780323995467000057, 10.1016/B978-0-323-99546-7.00005-7.

Sharma, M., Das, P. P., Sood, T., Chakraborty, A., & Purkait, M. K. (2021). Ameliorated polyvinylidene fluoride based proton exchange membrane impregnated with graphene oxide, and cellulose acetate obtained from sugarcane bagasse for application in microbial fuel cell. *Journal of Environmental Chemical Engineering, 9*(6), 2213−3437. Available from https://doi.org/10.1016/j.jece.2021.106681, https://www.sciencedirect.com/science/article/pii/S2213343721016584.

Sharma, M., Das, P. P., Sood, T., Chakraborty, A., & Purkait, M. K. (2022). Reduced graphene oxide incorporated polyvinylidene fluoride/cellulose acetate proton exchange membrane for energy extraction using microbial fuel cells. *Journal of Electroanalytical Chemistry, 907*115890. Available from https://doi.org/10.1016/j.jelechem.2021.115890, https://www. sciencedirect.com/science/article/pii/S1572665721009164.

Sharma, M., Samanta, N. S., Chakraborty, A., Purkait, M. K., Sillanpää, M., Khadir, A., . . . Gurung, K. (2023). *30 - Simultaneous treatment of industrial wastewater and resource recovery using microbial fuel cell* (pp. 621−637). Elsevier. Available from https://www.sciencedirect.com/science/article/pii/B9780323953276000026, 10.1016/B978-0-323-95327-6.00002-6.

Shekhar Samanta, N., Das, P. P., Dhara, S., & Purkait, M. K. (2023). An overview of precious metal recovery from steel industry slag: Recovery strategy and utilization. *Industrial & Engineering Chemistry Research, 62*(23), 9006−9031. Available from https://doi.org/ 10.1021/acs.iecr.3c00604, https://doi.org/10.1021/acs.iecr.3c00604.

Song, T.-H., Lee, S.-H., & Kim, B. (2014). Recycling of crushed stone powder as a partial replacement for silica powder in extruded cement panels. *Construction and Building Materials, 52*, 105−115. Available from https://doi.org/10.1016/j.conbuildmat.2013.10.060, https://www.sciencedirect.com/science/article/pii/S0950061813009811.

Sontakke, A. D., Das, P. P., Mondal, P., & Purkait, M. K. (2022). Thin-film composite nanofiltration hollow fiber membranes toward textile industry effluent treatment and environmental
remediation applications: review. *Emergent Materials*, *5*(5), 1409−1427. Available from
https://doi.org/10.1007/s42247-021-00261-y, https://doi.org/10.1007/s42247-021-00261-y.

Sontakke, A. D., Deepti., Samanta, N. S., Purkait, M. K., Husen, A., & Salahuddin Siddiqi, K.
(2023). *Chapter 2 - Smart nanomaterials in the medical industry. Micro and nano technologies* (pp. 23−50). Elsevier. Available from https://www.sciencedirect.com/science/article/pii/
B9780323995467000252, 10.1016/B978-0-32399546-7.00025-2.

Tan, H., Zhang, X., He, X., Guo, Y., Deng, X., Su, Y., . . . Wang, Y. (2018). Utilization of lithium
slag by wet-grinding process to improve the early strength of sulphoaluminate cement paste.
*Journal of Cleaner Production*, *205*, 536−551. Available from https://doi.org/10.1016/j.jcle-
pro.2018.09.027, https://www.sciencedirect.com/science/article/pii/S0959652618327392.

Terzano, R., D'Alessandro, C., Spagnuolo, M., Romagnoli, M., & Medici, L. (2015). Facile zeolite synthesis from municipal glass and aluminum solid wastes. *CLEAN − Soil, Air, Water*,
*43*(1), 133−140. Available from https://doi.org/10.1002/clen.201400091, https://doi.org/
10.1002/clen.201400091.

Tounsi, H., Mseddi, S., & Djemel, S. (2009). Preparation and characterization of Na-LTA zeolite
from Tunisian sand and aluminum scrap. *Proceedings of the JMSM 2008 Conference*, *2*(3),
1065−1074. Available from https://doi.org/10.1016/j.phpro.2009.11.064, https://www.scien-
cedirect.com/science/article/pii/S1875389209001473.

Tsujiguchi, M., Kobashi, T., Oki, M., Utsumi, Y., Kakimori, N., & Nakahira, A. (2014). Synthesis
and characterization of zeolite A from crushed particles of aluminoborosilicate glass used in
LCD panels. *Journal of Asian Ceramic Societies. null*, *2*(1), 27−32. Available from https://doi.
org/10.1016/j.jascer.2013.12.005, https://doi.org/10.1016/j.jascer.2013.12.005.

Ugal, J. R., Hassan, K. H., & Ali, I. H. (2010). Preparation of type 4A zeolite from Iraqi kaolin:
Characterization and properties measurements. *Journal of the Association of Arab Universities
for Basic and Applied Sciences*, *9*(1), 2−5. Available from https://doi.org/10.1016/j.jau-
bas.2010.12.002, https://www.sciencedirect.com/science/article/pii/S1815385210000039.

Ujaczki, É., Feigl, V., Molnár, M., Cusack, P., Curtin, T., Courtney, R., . . . Lenz, M. (2018).
Re-using bauxite residues: benefits beyond (critical raw) material recovery. *Journal of
Chemical Technology & Biotechnology*, *93*(9), 2498−2510. Available from https://doi.org/
10.1002/jctb.5687, https://doi.org/10.1002/jctb.5687.

Verma, A. S., Suri, N. M., & Kant, S. (2017). Applications of bauxite residue: A mini-review.
*Waste Management & Research: the Journal of the International Solid Wastes and Public
Cleansing Association, ISWA*, *35*(10), 999−1012. Available from https://doi.org/10.1177/
0734242X17720290, https://doi.org/10.1177/0734242X17720290.

Wajima, T., & Ikegami, Y. (2007). Synthesis of zeolitic materials from waste porcelain at low
temperature via a two-step alkali conversion. *Ceramics International*, *33*(7), 1269−1274.
Available from https://doi.org/10.1016/j.ceramint.2006.05.020, https://www.sciencedirect.
com/science/article/pii/S0272884206001969.

Wen, J., Dong, H., & Zeng, G. (2018). Application of zeolite in removing salinity/sodicity from
wastewater: A review of mechanisms, challenges and opportunities. *Journal of Cleaner
Production*, *197*, 1435−1446. Available from https://doi.org/10.1016/j.jclepro.2018.06.270,
https://www.sciencedirect.com/science/article/pii/S0959652618319292.

Wu, Z., Xie, J., Liu, H., Chen, T., Cheng, P., Wang, C., . . . Kong, D. (2018). Preparation, characterization, and performance of 4A zeolite based on opal waste rock for removal of ammonium ion. *Adsorption Science & Technology*, *36*(9−10), 1700−1715. Available from https://
doi.org/10.1177/0263617418803012, https://doi.org/10.1177/0263617418803012.

Xie, W.-M., Zhou, F.-P., Bi, X.-L., Chen, D.-D., Li, J., Sun, S.-Y., ... Chen, X.-Q. (2018). Accelerated crystallization of magnetic 4A-zeolite synthesized from red mud for application in removal of mixed heavy metal ions. *Journal of Hazardous Materials, 358,* 441−449. Available from https://doi.org/10.1016/j.jhazmat.2018.07.007, https://www.sciencedirect.com/science/article/pii/S0304389418305193.

Yamane, I., & Nakazawa, T. (1986). Development of zeolite for non-phosphated detergents in Japan. Available from https://doi.org/10.1351/pac198658101397.

Yu, S., Kwon, S., & Na, K. (2021). Synthesis of LTA zeolites with controlled crystal sizes by variation of synthetic parameters: Effect of Na + concentration, aging time, and hydrothermal conditions. *Journal of Sol-Gel Science and Technology, 98*(2), 411−421. Available from https://doi.org/10.1007/s10971-018-4850-4, https://doi.org/10.1007/s10971-018-4850-4.

# Chapter 8

# Various aspects in the application of waste-based zeolite

## 8.1 Introduction

Zeolites, known for their easily adjustable physicochemical properties, have gained significant attention as a research hotspot in various industrial applications. They can be naturally occurring in sediments or synthesized from high-silica and high-aluminum raw materials. However, the performance of natural zeolites is hampered by complex formation conditions, unstable chemical composition, and the presence of impurities, limiting their industrial use. Consequently, there is a pressing need in production practice to replace natural zeolites with synthetic ones. In recent years, researchers have explored alternative sources of silica-aluminum for zeolite syntheses, such as solid wastes like coal fly ash, coal gangue, rice husk, sugarcane bagasse, and coal gasification slag (Mondal, Samanta, Kumar, & Purkait, 2020; Samanta et al., 2021; Samanta, Anweshan, Mondal, Bora, & Purkait, 2023; Samanta, Das, Mondal, Bora, & Purkait, 2022; Samanta, Das, Mondal, Changmai, & Purkait, 2022; Shekhar Samanta, Das, Dhara, & Purkait, 2023). These materials, characterized by low cost and abundant reserves, are considered favorable raw materials for synthesizing zeolites.

Over the past decade, the combination of national governments advocating sustainable green development and the rapid advancements in the field of microporous materials has spurred numerous noteworthy research papers and reviews on zeolites. Since 2002, literature reviews on zeolite synthesis from diverse wastes, such as fly ash (Querol et al., 2002) and biomass ash (Munawar et al., 2021), have been consistently published. These reviews delve into synthesis methods and applications of zeolites, revealing similarities. Currently, significant strides have been made in research related to synthesizing zeolites using various industrial and agricultural solid wastes, with the range of solid waste sources expanding from fly ash to gangue, biomass, gasification slag, municipal solid wastes, and more. Furthermore, synthetic zeolites have undergone modifications, extending their utility beyond pollutant adsorption in wastewater (Das, Deepti, & Purkait, 2023; Das, Dhara, &

Waste-based Zeolite. DOI: https://doi.org/10.1016/B978-0-443-22316-7.00008-5
**197**

Purkait, 2023a, 2023b; Das, Samanta, Dhara, & Purkait, 2023; Dhara, Das, Uppaluri, & Purkait, 2023) to applications in atmospheric purification, agriculture (Das, Dhara, & Purkait, 2024; Das, Mondal, & Purkait, 2022; Das, Sharma, & Purkait, 2023; Das, Sontakke, & Purkait, 2023; Das, Sontakke, Samanta, & Purkait, 2023), industrial catalysis, and other fields (Dhara, Das, Uppaluri, & Purkait, 2023; Sharma & Das, 2023; Sharma, Das, & Purkait, 2023).

## 8.2 Waste-based zeolites for environmental remediation

### 8.2.1 Hazardous gas removal

Apart from the water pollution issue (Bharti, Das, & Purkait, 2023; Chakraborty, Gautam, Das, & Hazarika, 2019; Changmai & Das, 2022; Das, Anweshan, & Mondal, 2021; Das, Anweshan, & Purkait, 2021; Das & Mondal, 2021; Das, Sharma, & Purkait, 2022; Dhara, Samanta, & Das, 2023; Sontakke & Das, 2021), air pollution poses a serious threat to human health. Common air pollutants include volatile organic compounds (VOCs) and carbon dioxide ($CO_2$). Currently, adsorption is the predominant method for gas capture. Solid adsorbents employed for carbon dioxide capture encompass carbonaceous materials like activated carbon, carbon nanotubes, and graphene (Naushad, Ahamad, Al-Maswari, Abdullah Alqadami, & Alshehri, 2017), as well as noncarbonaceous materials such as mesoporous silica, metal-organic frameworks, resins, and zeolite (Li, Wang, Guo, Zhu, & Xu, 2021). Zeolite molecular sieves exhibit a substantial volume of crystal cavities and an empty mineral skeleton structure, providing a large specific surface area with numerous uniform micropores, all possessing polarity. Among these adsorbents, zeolite plays a pivotal role in carbon dioxide capture due to its molecular sieve nature, adjustable physicochemical properties, and high selectivity for carbon dioxide. Considering cost-effectiveness, noteworthy strides have been made in enhancing the adsorption of VOCs and $CO_2$ using zeolite materials synthesized from solid waste, such as fly ash.

### 8.2.1.1 Removal of $CO_2$

Amidst rising concerns about global climate change and the urgent need for $CO_2$ emission reduction in the atmosphere, researchers have delved into the adsorption capabilities of zeolite A and a combination of zeolites A and X, focusing on $CO_2$, a significant contributor to climate change (Liu, Singh, Xiao, Webley, & Zhai, 2011). The study revealed that the mixture of zeolites exhibited superior efficiency in capturing $CO_2$ compared to individual zeolite X synthesized at 90°C. Another investigation by Sarmah et al. explored $CO_2$ capture on coal fly ash (CFA)-based zeolite, amended with diethylamine/N,N-dimethylaniline and monoethanolamine/N,N-dimethylaniline. The results showed respective adsorption capacities of 5.99 and 6.89 mmol/g, demonstrating that modified zeolite had a more pronounced impact on $CO_2$ adsorption than commercially available zeolites, suggesting an economically viable option (Sarmah, Baruah, & Khare, 2013).

In a separate study, Wang et al. synthesized zeolite X adsorbents through the hydrothermal method, utilizing rice hull ash (RHA) as a potential Si source for zeolite synthesis. Employing the ion-exchange process, they embedded lanthanum (La) and cerium (Ce) in zeolite X for $CO_2$ adsorption. The introduction of La and Ce ions, with three positive charges replacing the one positive charge of Na or Li ions, maintained the zeolite's overall charge balance. The findings indicated that La-zeolite X exhibited a notable $CO_2$ adsorption capacity of 4.451 mmol/g compared to Ce-zeolite X. This difference was attributed to the higher ionic density and substitution induced by La modification, highlighting its superior $CO_2$ capture capabilities (Wang, Jia, Chen, Fang, & Du, 2020).

### 8.2.1.2 Removal of sulfur compounds

Sulfur-based compounds, particularly $SO_2$ and $SO_3$, are significant contributors to pollutants, with a considerable portion of acid rain comprising these harmful sulfur oxides. Zeolites present a noteworthy solution for the removal of such sulfur-based compounds. The effectiveness of CFA-based zeolite in adsorbing $SO_2$ is highly contingent on factors such as the type of zeolite, drying rate, and zeolite content, demonstrating varying efficiency compared to zeolite X, Y, and A varieties (Liu & Zhang, 2011). In contrast, thermally activated Fly ash-derived zeolite (FAZ) was found to be significantly more efficient than coal fly ash-based zeolite. Thermal activation increased the adsorption capacity of the latter by at least 2−3 times (Suchecki, Wałek, & Banasik, 2004).

To further understand the gas adsorption performance, a comparative study involving commercial zeolite 3A and SM-27 was conducted. The investigation encompassed activated and nonactivated commercial and synthesized zeolites, with activation achieved through a 10-hour treatment at an elevated temperature of 400°C. Adsorption data over nontreated zeolites indicated $SO_2$ expulsion efficiency ranging from 0.5 to 20 mg $SO_2$/g of zeolite. In contrast, activated adsorbents exhibited removal capacities almost three times higher than commercial zeolite 3A and nearly two times higher than zeolite SM-27. The order of zeolite activity on $SO_2$ sorption was observed as 3A < SM-27 < FAZ. The study suggested that zeolite FAZ possessed hydrophobic characteristics.

The authors noted that thermal activation of zeolites at lower temperatures eliminated adsorbed water molecules, while at elevated temperatures, it also removed structural water. However, the structural changes induced by higher temperatures might increase the interior surface areas of zeolites, influencing their adsorption capabilities. In contrast, zeolite 3A exhibited sustainability up to 700°C, showing no significant variations in $SO_2$ uptake between thermally activated and nonactive types. The study authors proposed that greater exposure and prolonged contact time between sorbent and adsorbate could enhance removal efficiency. They suggested the use of a fluidized bed reduction experiment for industrial scaling up, as opposed to utilizing a fixed bed reactor. Additionally, Izquierdo et al. highlighted that the presence of $H_2O$ in flue gas could affect the structure and performance of CFA-based CaY zeolite, resulting in structural changes in the zeolite framework and reduced activity during the adsorption process (Izquierdo, Juan, & Rubio, 2013).

From the preceding section, it was observed that the efficacy of the above systems is yet to be determined for the industrialization of waste-based zeolites toward toxic gas removal from flue gas. To determine feasibility, energy consumption, and a cost-benefit analysis are also necessary. Apart from the zeolite activation at elevated temperatures, research should be focused toward advanced techniques for zeolite fabrication to overcome these challenges.

### 8.2.1.3   Removal of N compounds

Nitrogen-based compounds stand as a significant pollution source, prompting studies on the efficacy of P-type zeolite derived from CFA in conjunction with copper-modified zeolite for nitric oxide (NO) reduction. At a temperature of 500K, the copper-modified zeolite exhibited a superior performance efficiency of 47%, surpassing untreated zeolite by nearly 10% (López-Salinas et al., 1997). A comparative analysis was conducted on commercial 13-X zeolite and ion-exchange Ag-X, Fe-X, and Ni-X zeolites prepared from CFA for NOx removal catalysis (Karthikeyan & Saravanan, 2013). Results indicated that CFA-based zeolites demonstrated enhanced catalytic activity compared to commercially available 13-X zeolites. Additionally, no zeolite deactivation occurred after a 50-hour reaction duration, suggesting significant NOx reduction, particularly with Fe and Ag-incorporated zeolite catalysts.

Another investigation focused on fly ash-based zeolites exchanged with copper ($Cu^{2+}$) and iron ($Fe^{3+}$), revealing a dependence on water presence for NO reduction at temperatures below 200°C. The study highlighted varying temperatures of active substances ($Fe^{3+}$ and $Cu^{2+}$) in zeolite LY-Cu and ion-exchanged zeolites, influencing NO reduction differently (Izquierdo, Juan, Rubio, & Gómez-Giménez, 2016). The entire equation may be written as follows: $4NO + 4NH_3 + O_2 \leftrightarrow 4N_2 + 6H_2O$, where water is a product according to the reaction scheme. The disturbance in equilibrium due to water vapor in the airflow enhanced $NO + O_2$ concentration, leading to increased $Cu^+$ oxidation and a higher number of active sites at low temperatures. However, water vapor did not affect NO reduction at a steady state for all ion-exchanged zeolites. The investigation found that zeolite LY-Cu exhibited a higher surface area than LY-Fe zeolite, contributing to better reduction efficiency.

CFA-based modified zeolite Y (zeolite LY-Cu) demonstrated noteworthy NO reduction from flue gas below 200°C. Further exploration involving commercial zeolites and their ion-exchanged counterparts is warranted to assess feasibility and economic sustainability. Researchers may consider these findings as a foundation for future studies, particularly if cost-effective utilization is sought with cheaper silica supplies and environmental advantages taken into account.

### 8.2.1.4 Elimination of volatile organic compounds

Zeolites play a crucial role in effectively removing harmful VOCs, major contributors to environmental issues like smog. In the hydrothermal synthesis of zeolite X using CFA as a silica source, a comparison study was conducted with commercial activated carbon for VOC removal. Despite activated carbon having a higher surface area ($1020.8$ m$^2$/g) compared to X-type zeolite ($990.3$ m$^2$/g), the synthesized zeolite demonstrated a higher adsorption capacity (Zhang, Chen, Wei, & Zu, 2012). The VOCs studied included isopropanol, cyclohexane, and benzene, with the adsorption capacity ranking in the order of isopropanol > benzene > cyclohexane. The molecules' polarity strongly influenced the adsorbent's ability, with isopropanol exhibiting higher polarity than cyclohexane and benzene. Additionally, the unsaturation of molecules played a vital role, with greater unsaturation leading to higher adsorption capability. Activated carbon, due to covalent bond unsaturation during synthesis, demonstrated significant adsorption characteristics influenced by its polarity (Prajapati, Bhaduri, Joshi, Srivastava, & Verma, 2016).

In contrast, zeolite X, with its unique pore size and honeycomb-like framework possessing a negative charge, entrapped adsorbate molecules through electrostatic attraction, extending its adsorption capacity (Kim & Ahn, 2012). Another investigation explored the preparation of different zeolite shapes, including composite hydroxyapatite-zeolite (HAP-ZE), Y-type zeolite, and zeolite P1, using commercial steel slag (SS) as a precursor. VOCs (toluene and acetaldehyde) adsorption assessment revealed better performance for FAU-type zeolite (zeolite Y). The significant sorption capacity of slag-based zeolite type-Y and HAP-ZE was attributed to high porosity created by different nucleation times (Kuwahara, Ohmichi, Kamegawa, Mori, & Yamashita, 2009).

Nucleation time-induced variations in zeolite formation, along with distinct properties like surface charge, pore size, and shape, influenced the sorption performance of the final products. The synthetic approach offered advantages such as low aging temperature, shortened preparation stages (without acid hydrolysis), and cost-effectiveness, aligning with the requirements for alternative pathways in zeolitic mineral synthesis. It is presumed that the aging synthetic approach could find application in converting other complex multicomponent materials, benefiting resource efficiency and providing unique solutions to waste management and environmental issues.

### 8.2.1.5 Elimination of Hg vapor

Zeolites and their modified forms synthesized from CFA have proven to be highly effective in mercury removal from flue gas. Specifically, zeolite X/Ag has demonstrated superior efficiency in adsorbing mercury compared to untreated zeolite, attributed to the heightened surface area and exceptional porosity resulting from the Ag treatment. Additionally, factors such as contact time, contact temperature, and flue gas flow rate were identified as crucial parameters influencing adsorption efficiency (Wdowin et al., 2014).

Adsorption performance was further explored using the supercritical hydrothermal process, employing a solid/liquid ratio of 1.5, NaOH concentration of 1 mol/L, and a temperature of 400°C (Wang et al., 2015). The variation in NaOH molar concentration influenced the formation of different zeolite types, with an 8 mol/L NaOH concentration inducing the formation of Cancrinite and Sodalite zeolite crystals. Sodalite exhibited a maximum Hg removal efficiency of 75%, surpassing the 60% achieved by Cancrinite type zeolite in batch adsorption studies conducted at 100°C for 480 min. A reduction in adsorption time to 300 min led to a dramatic decline in Hg elimination efficiency from 75% to 20%. The study demonstrated that both prepared adsorbents exhibited significant Hg removal capability from simulated coal gas, outperforming as-fabricated zeolite. This suggests that during supercritical hydrothermal treatment, Fe or other metals can be activated, displaying substantial catalytic activity.

Using the same synthesis strategy of supercritical hydrothermal treatment, zeolite adsorbent was manufactured from CFA and Fe-enriched stainless steel for successive removal of elemental mercury (Hg°) from simulated flue gas (Ma et al., 2019). Unmodified and HCl-modified zeolites were employed in the Hg removal experiment. Untreated magnetic zeolites exhibited a 30% Hg removal effectiveness after 240 minutes of sorption at 200°C, whereas Hg removal efficiency reached approximately 100% over modified zeolites under the same conditions. The varying HCl concentration during zeolite modification showed that as HCl concentration increased, the removal efficiency also improved, likely due to the formation of more active sites on the zeolite surface (Rungnim, Promarak, Hannongbua, Kungwan, & Namuangruk, 2016).

The synthesis of zeolites from CFA and stainless steel (SS) waste, along with their utilization for mercury removal, exemplifies the concept of waste valorization. Upon a comprehensive review of the research, it becomes evident that zeolites boasting favorable properties can also be crafted from silicon and aluminum-containing biomass wastes, rendering them viable candidates for investigations into gas removal. Additionally, the scarcity of research on biomass ash-based zeolites for gas adsorption opens up opportunities for researchers to explore this particular gap in the field. Table 8.1 provides an overview of hazardous gas removal using various zeolites derived from diverse sources.

The above research works present the manifold applications of CFA-based zeolites and their composites concerning the adsorption of hazardous gaseous pollutants. However, there has been limited exploration into the preparation of zeolites from biomass for the removal of toxic gaseous compounds such as $CO_2$, $SO_2$, $SO_3$, and VOCs. Thus further research is imperative to develop suitable methods for zeolite synthesis and modification from bio-ash, contributing to the removal of harmful gases from industrial effluents and potentially aiding in environmental mitigation.

### 8.2.2 Wastewater treatment application

The rapid expansion of agriculture and industry has resulted in profound water pollution, with common pollutants in wastewater including heavy metals, ammonia nitrogen, phosphate, and dye. These contaminants are crucial monitoring indicators for industrial wastewater discharge, presenting a pressing ecological challenge. In the context of wastewater treatment costs and waste recycling, the utilization of synthetic zeolites from industrial solid waste proves to be highly advantageous. However, comparing the adsorption properties and parameters of various molecular sieves is challenging due to differences in modification methods, adsorbent preparation, and adsorbent/solution ratios, as indicated by the diverse results in Table 8.2. Over the past decade, numerous scholars have dedicated their efforts to developing various zeolite-based composites, focusing primarily on the adsorption efficiency of molecular sieve materials for water source pollutants and exploring the associated adsorption mechanisms.

The primary mechanisms involved in the adsorption of molecular sieve-based materials encompass surface adsorption, ion exchange, electrostatic interaction, diffusion, complexation, hydrogen bonding, and others. For zeolite materials, the predominant mechanism in heavy metal adsorption is ion exchange. In the context of organic dye adsorption, electrostatic and hydrogen bonding interactions often coexist. As illustrated in Figs. 8.1 and 8.2, electrostatic interaction arises from a robust covalent bond between the Si—O— (deprotonated silanol) in the zeolite and the amine functional group present in the dye molecule.

Hydrogen bonding arises from the interaction between the silanol group (Si—OH) in the zeolite and either the amine functional group or carbonyl group (—C = O) present in the dye molecule (Ameh, Oyekola, & Petrik, 2022). Additionally, in Fig. 8.2, the cationic dye Rh-6 G, containing amine groups, can be adsorbed onto the negative charges of $[SiO_4]^{4-}$ and $[AlO_4]^{5-}$

**TABLE 8.1** Hazardous gas removal using various zeolites derived from different waste sources.

| Application | Sorbent type | Waste source | BET surface area ($m^2$ $g^{-1}$) | Temperature (°C) | Adsorption capacity or percent |
|---|---|---|---|---|---|
| Nitric oxide (NO) removal | Zeolite Y | CFA | – | 50–350 | – |
| Benzene | | | | | 1.49 μmol/g |
| Toluene | | | | | 10.52 μmol/g |
| o-Xylene | | | 94.49 | 25 | 26.22 μmol/g |
| BTX removal | Zeolite P1 | CFA | | | 25.96 μmol/g |
| m-Xylene | | | | | |
| p-xylene | | | | | 20.50 μmol/g |
| Benzene | | | | | 383.67 μmol/g |
| Toluene | | | | | 525.03 μmol/g |
| o-Xylene | | | 157.43 | 25 | 545.26 μmol/g |
| BTX removal | Zeolite X | CFA | | | 563.82 μmol/g |
| m-Xylene | | | | | |
| p-Xylene | | | | | 582.91 μmol/g |
| $CO_2$ adsorption | Zeolite X | CFA | 643 | 25 | 225 mg/g |
| $CO_2$ adsorption | MCM-41 | CFA | – | 75 | 13.31% |
| $CO_2$ adsorption | Zeolite X | CFA | 14.2 | 70 | 26 mg/g |
| $SO_2$ adsorption | Merlinoite perlialite | CFA | 102.4 | 25 | 46 mg/g |

| Application | Zeolite | Source | Value | Capacity |
| --- | --- | --- | --- | --- |
| $SO_2$ adsorption | Zeolite X | CFA | – | 1.68 mmol/g |
| $SO_2$ adsorption | Zeolite Y | CFA | 426 | 21.9 mg/g |
| $SO_2$ adsorption | Zeolite P | CFA | – | 38 mg/g |
| Benzene | | | 157.4 | 383.67 µg/g |
| Toluene | | | – | 525.03 µg/g |
| o-Xylene | Zeolite X | CFA | – | 545.26 µg/g |
| m-Xylene | | | – | 563.82 µg/g |
| p-Xylene | | | | 582.91 µg/g |
| Benzene | Zeolite P1 | CFA | 34.06 | 63.32% |
| Benzene | Zeolite X | CFA | 990.3 | 151 mg/g |
| Hg | Sodalite | CFA | 48.5 | 60 mg/g |
| Hg | Cancrinite | CFA and SS | 11.9 | 100% |
| Acetaldehyde | HAP-ZE | Steel slag | – | – |
| Toluene | | | – | – |

VOC removal

VOCs

*Source:* From Samanta, N. S., Das, P. P., Mondal, P., Changmai, M., & Purkait, M. K. (2022). Critical review on the synthesis and advancement of industrial and biomass waste-based zeolites and their applications in gas adsorption and biomedical studies. *Journal of the Indian Chemical Society, 99*(11). https://doi.org/10.1016/j.jics.2022.100761.

**TABLE 8.2** Application of zeolite from waste ash in wastewater treatment.

| Raw material | Zeolite types | Heavy metals and nutrients | Conc. of heavy metals and nutrients | Zeolite dosage | pH | Time | Adsorption capacity | Removal efficiency | Adsorption mechanism |
|---|---|---|---|---|---|---|---|---|---|
| Coal gangue | Hierarchical porous zeolites | $Cu^{2+}$ | 100 mL, 200 g/L | 0.10 g | – | 1 h | 167.00 mg/g | 83.50% | Physical adsorption |
| | | Rh-B | 100 mL, 10 mg/L | 0.10 g | – | 1 h | 9.50 mg/g | 95.10% | |
| | Zeolite-activated carbon composite | $Cu^{2+}$ | 100 mL, 240 mg/L | 0.20 g | – | 210 min | – | 92.80% | Physical adsorption |
| | | Rh-B | 250 mL, 10 mg/L | 0.08 g | – | 60 min | – | 94.20% | |
| | NaY | $Pb^{2+}$ | 50 mL, 10–200 mg/L | 0.25 g/L | 7.00 | 30 min | 431.60 ~ 482.10 mg/g | 100.00% | Ion exchange |
| SCBA | Zeolite A | $Pb^{2+}$ | 200 mL, 50 mg/L | 0.02 g | 5.00 | 5 h | 625.00 mg/g | 100.00% | Electrostatic attractions, chelate complexes |
| Rice husk | ZSM-5 | $Pb^{2+}$ | 150 rpm | – | – | 24 h | – | – | Ion exchange, complexation |

| CFA | | | | | | | | | |
|---|---|---|---|---|---|---|---|---|---|
| | Zeolite-geopolymer composites | Pb(II) | 50 mL, 200 mg/L | 0.05 g | 5.00 | 48 h | 446.73 mg/g | – | Electrostatic actions, ion exchange, interparticle diffusion |
| | Zeolite A | Sr$^{2+}$<br>Cs$^{+}$ | 10 mL, 1000 mg/L | 0.10 g | 7.00 | 24 h | 95.74 mg/g<br>54.12 mg/g | > 90.00%<br>> 50.00% | Ion exchange |
| | Zeolite X | Sr$^{2+}$<br>Cs$^{+}$ | | | | | 93.14 mg/g<br>53.14 mg/g | > 90.00%<br>> 50.00% | |
| | nZVI/Ni@FZA | Cr(VI) | 50 mL, 5–70 mg/L | 0.20 g/L | 3.00 | – | 48.31 mg/g | – | Reduction, adsorption, ion exchange |
| | | Cu(II) | 50 mL, 10–180 mg/L | | 5.00 | – | 147.06 mg/g | – | |
| | Zeolite A | Ni$^{2+}$ | 100 mL, 60–300 mg/L | 1.50–10.00 g/L | 7.00 | 3 h | 46.05 mg/g | 92.10% | Adsorption, ion exchange |
| | Na-X zeolite | As(V) | 50 mL, 22.83 mg/L | 0.10 g | 2.14 | – | 27.79 mg/g | 83.00% | Electrostatic attractions |
| | Na-P1 | Sr$^{2+}$<br>Cs$^{+}$ | 10 mL, 1000 mg/L | 0.10 g | 7.00 | –<br>– | 92.48 mg/g<br>39.30 mg/g | – | Ion exchange |
| | Na-P1 | Pb$^{2+}$ | 50 mL, 100–300 mg/L | 1.60 g | 6.00 | 1 h | – | 99.00% | Ion exchange, chemisorption |
| | MOR zeolite | Pb$^{2+}$<br>Sr$^{2+}$ | 51.30 mg/L<br>74.94 mg/L | 0.10 g | – | – | –<br>– | 99.99%<br>60.00% | Ion exchange, electrostatic attractions |
| | Na–P/SOD | Rh-6 G | 30 mg/L | 1.02 g | – | 1 h | 282.35 mL/g | 71.00% | Electrostatic interaction |
| | Zeolite X | PO$_4{}^{3-}$ | 50 mL, 10–200 mg/L | 0.05 g | 5.00 | 24 h | 87.70 mg/g | 65.40% | Electrostatic attractions, |

*(Continued)*

**TABLE 8.2 (Continued)**

| Raw material | Zeolite types | Heavy metals and nutrients | Conc. of heavy metals and nutrients | Zeolite dosage | pH | Time | Adsorption capacity | Removal efficiency | Adsorption mechanism |
|---|---|---|---|---|---|---|---|---|---|
| | | | | | | | | | surface interaction, precipitation |
| | ZSM-22 | Rh-6 G | — | 0.05 g | 6.00 | 2 h | 195.30 mg/g | — | Electrostatic interaction, hydrogen bonding |
| | Zeolite/ hydrous iron oxide | Methylene blue | 40 mL, 250 mg/L | 0.20 g | 11.20 | 4 h | 46.35 mg/g | 94.71% | — |
| | Zeolite/ hydrous zirconia | | | | | | 45.09 mg/g | 91.17% | |
| | ZFA/Fe$_2$O$_3$ | Phosphate | 40 mL, 5 mg/ L | 0.20 g | <7 | 24 h | 18.20 mg/L | 100.00% | — |
| | | Ammonium | 40 mL, 25 mg/L | | 7.00—9.50 | | — | 80.00% | |

*Source:* From Cao, C., Xuan, W., Yan, S., & Wang, Q. (2023). Zeolites synthesized from industrial and agricultural solid waste and their applications: A review. *Journal of Environmental Chemical Engineering, 11*(5), 110898. https://doi.org/10.1016/j.jece.2023.110898.

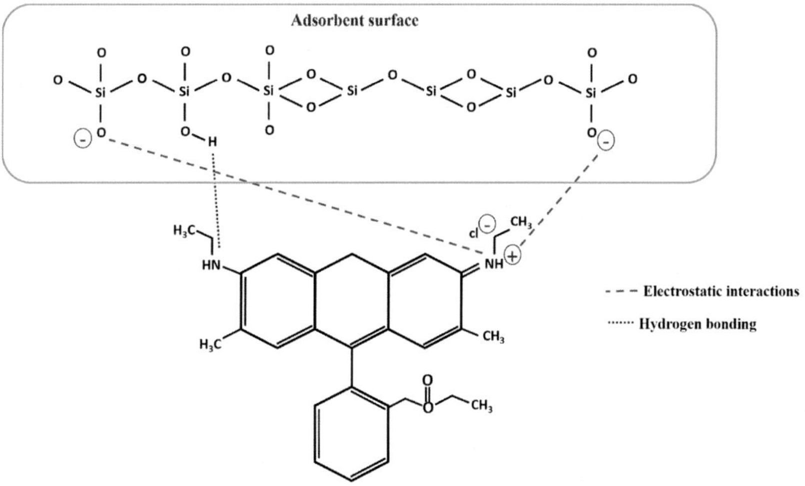

**FIGURE 8.1** **Electrostatic interaction and hydrogen bonding in ZSM-22 adsorbed cationic dye Rh-6G.** Schematic representation of electrostatic interaction and hydrogen bonding of organic dye onto modified zeolite. *From Gollakota, A. R. K., Volli, V., Munagapati, V. S., Wen, J. C., & Shu, C. M. (2020). Synthesis of novel ZSM-22 zeolite from Taiwanese coal fly ash for the selective separation of Rhodamine 6G.* Journal of Materials Research and Technology, *9(6), 15381−15393. https://doi.org/10.1016/j.jmrt.2020.10.070.*

within the zeolite through electrostatic interactions. Another study conducted by Mokrzycki et al. utilized a $Cu(NO_3)_2 \cdot 3\,H_2O$ solution to modify fly ash X zeolite for the removal of phosphate ions from aqueous solutions. The adsorption mechanism involved the interaction of phosphate ions with copper ions on the zeolite surface, precipitation of calcium phosphate, and electrostatic attraction (Mokrzycki et al., 2022).

In another investigation, Angaru et al. synthesized fly ash-based zeolite-loaded nano-zerovalent FeNi bimetallic composites (nZVI/Ni@FZA) for the removal of Cr(VI) and Cu(II). The removal process was primarily influenced by reduction, adsorption, and ion exchange, as depicted in Fig. 8.3. Cr(VI) and Cu(II) in the solution enter near the zeolite or adsorb directly on the active site, followed by oxidative reduction in solution by nZVI/Ni to Cr(III) and $Cu^\circ/Cu^+$, respectively. The reduced Cr precipitates as hydroxide or oxide, and a small fraction of Cu(II) and Cr(III) is removed directly through cation exchange in the zeolite (Angaru et al., 2021).

Moreover, parameters such as solution pH, adsorbent quantity, and adsorption duration play crucial roles in influencing the adsorption efficiency of molecular sieve-based materials. The pH of the aqueous solution emerges as a critical factor impacting the adsorption performance of the adsorbent. Its influence on adsorption is substantial as it affects both the surface charge of the adsorbent and the form of hazardous metal ions within the solution. Studies

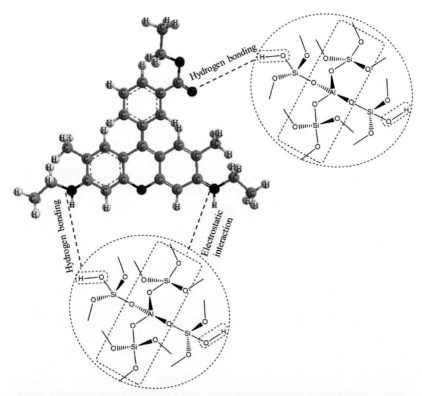

**FIGURE 8.2  Schematic of the columnar adsorption mechanism of Rh-6G on Na-P/SOD zeolite.** Column-based adsorption of organic dye Rh-6G on modified zeolite. *From Ameh, A. E., Oyekola, O. O., & Petrik, L. F. (2022). Column adsorption of Rhodamine 6G over Na−P/SOD zeolite synthesised from aluminosilicate secondary waste.* Journal of Cleaner Production, 338. *https://doi.org/10.1016/j.jclepro.2022.130571.*

have demonstrated variations in the adsorption efficiency of zeolite for different heavy metals at the same pH value, suggesting a correlation between adsorption efficiency and the pH value of the aqueous system, along with the zero-charge point value (Visa, 2016). Additionally, investigations have indicated that when the solution pH is below 7, electrostatic interactions between the organic dye Rhodamine 6G and modified ZSM-22 zeolite predominantly occur, thereby enhancing the dye adsorption process (Gollakota, Volli, Munagapati, Wen, & Shu, 2020). This underscores the pivotal role of pH in modulating the dissociation of functional groups at the active site of the adsorbent. In a study by Xie et al., a novel zeolite/Fe$_2$O$_3$ composite was prepared from fly ash, revealing that the removal of phosphate increased with decreasing pH, whereas the removal of ammonium exhibited an opposite trend. Hence, neutral pH conditions proved favorable for the simultaneous uptake of ammonium and phosphate (Xie, Wang, Wu, & Kong, 2014).

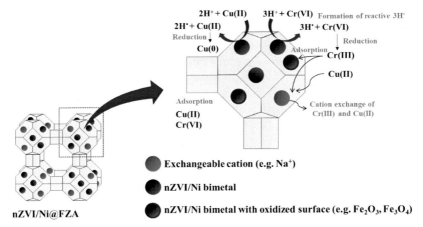

**FIGURE 8.3 Removal mechanism of Cr(VI) and Cu(II) by nZVI/Ni@FZA bimetallic composite.** Chromium and copper removal mechanism by zerovalent iron-modified zeolites. *From Angaru, G. K. R., Choi, Y. L., Lingamdinne, L. P., Choi, J. S., Kim, D. S., Koduru, J. R., Yang, J. K., & Chang, Y. Y. (2021). Facile synthesis of economical feasible fly ash—based zeolite—supported nano zerovalent iron and nickel bimetallic composite for the potential removal of heavy metals from industrial effluents.* Chemosphere, 267. *https://doi.org/10.1016/j.chemosphere.2020.128889.*

The quantity of zeolite represents another pivotal parameter governing the adsorption process. A significant escalation in zeolite dosage can lead to the occlusion of active pores within the adsorbent, an augmentation in the number of active sites, and the saturation of these active sites. Consequently, this results in a decline in the adsorption capacity of the adsorbent. Ma et al. demonstrated that with an increase in the adsorbent amount, the removal of $Pb^{2+}$ exhibited a gradual rise, surpassing 99% due to the numerous adsorption sites in the wastewater. However, with further increments in the adsorbent amount, certain adsorption sites on the adsorbent became less effective due to the limited amount of $Pb^{2+}$ in the solution, consequently leading to a reduction in adsorption capacity (Ma et al., 2022).

The adsorption time serves as a crucial benchmark for assessing the applicability of zeolite materials. Previous research indicates that the zeolite adsorption rate gradually decelerates with an increase in contact time, ultimately reaching equilibrium. This phenomenon is attributed to the gradual occupation of adsorption sites on the zeolite surface as the adsorption time prolongs, thereby constraining the transfer of adsorbed material from the solution to the synthetic zeolite surface (Hossini Asl, Masomi, & Tajbakhsh, 2020).

The adsorption temperature plays a significant role in influencing the physical and chemical mechanisms of adsorption. An elevation in temperature correlates with an increase in adsorption, signifying the heat-absorbing nature of the reaction. This is ascribed to enhanced ion mobility and the stability of the active site. Ma et al. synthesized zeolite materials with rapid adsorption rates for $Pb^{2+}$ in aqueous solutions, favoring the adsorption of $Pb^{2+}$ on the adsorbent at high temperatures. This process is primarily considered a heat adsorption process, predominantly through chemisorption (Z. Ma et al., 2022). Conversely, it has been observed that the removal efficiency of zeolites for benzene, toluene, and m-xylene (BTX) diminishes with increasing temperature, confirming that the adsorption process on molecular sieves is spontaneous and thermodynamically favorable at lower temperatures but transitions to a nonspontaneous mode at higher temperatures (Hossini Asl et al., 2020).

### 8.2.2.1 Removal of heavy metals

The imbalance in charge resulting from the inconsistent valence of Si and Al in the zeolite skeleton is typically compensated by alkali metals to neutralize it. However, alkali metals exhibit weak binding to the zeolite skeleton and are easily exchanged with other cations. The cation exchange capacity of zeolites is generally contingent on the extent of loading they carry, with zeolites possessing higher aluminum content displaying increased ion-exchange capacity. Consequently, numerous researchers have explored the removal of hazardous metal ions from solutions based on this zeolite property.

Due to the limited adsorption performance of zeolites synthesized from waste ash, some researchers have enhanced coal-based zeolites by incorporating active ingredients to create composites with superior adsorption capabilities. Li et al. synthesized zeolite-calcium silicate hydrate composites from fly ash, demonstrating excellent adsorption properties for various heavy metals. The adsorption mechanism relies on the exchange of $Na^+$ and $Ca^{2+}$ with hazardous metal ions during the adsorption process in the composites (Li et al., 2022). Additionally, a fly ash-based zeolite-loaded nano-zerovalent iron and nickel bimetallic composite exhibited a larger specific surface area and higher removal efficiency of Cr(VI) and Cu(II) compared to the unmodified fly ash-based zeolite, where Ni played a crucial role in enhancing the reduction ability of this composite (Angaru et al., 2021). As illustrated in Fig. 8.4, copper ions migrate from the adsorbent solution to the adsorbent surface in the initial stage of the adsorption process. Subsequently, the adsorbent enters the internal space through the internal pore structure, and the interaction between the adsorbent and adsorbate, such as van der Waals forces, leads to adsorption at the adsorption sites (Li et al., 2020).

Traditionally, studies on metal ion adsorption using zeolites focused on single heavy metals. However, wastewater typically contains a mixture of pollutants and metal ions. Research in this field has evolved, especially concerning acid mine water, which has a complex composition with anions such as fluorine, chloride, and nitrate, cations like sodium, ammonium, magnesium, and potassium, and hazardous metal ions such as arsenic, copper, iron, manganese,

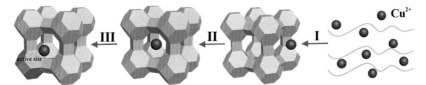

FIGURE 8.4  **Schematic diagram of zeolite adsorption of Cu2 + kinetic models.** Kinetic model representation of adsorption of copper onto zeolite. *From Li, H., Li, M., Zheng, F., Wang, J., Chen, L., Hu, P., Zhen, Q., Bashir, S., & Liu, J. L. (2021). Efficient removal of water pollutants by hierarchical porous zeolite-activated carbon prepared from coal gangue and bamboo.* Journal of Cleaner Production, 325, *129322. https://doi.org/10.1016/j.jclepro.2021.129322.*

nickel, and zinc. Cardoso et al. compared the efficacy of commercial and synthetic Na-P1 zeolites for treating acid mine water, finding that synthetic Na-P1 zeolites were more effective for removing fluoride, chloride, nitrate, and magnesium. They were comparable to commercial zeolites for sulfate and heavy metals removal, particularly for manganese and zinc, achieving up to 99.8% and 81%, respectively (Cardoso, Paprocki, Ferret, Azevedo, & Pires, 2015). The selectivity of faujasite for metal removal in acid mine water treatment was observed in the decreasing order of Fe > As > Pb > Zn > Cu > Ni > Cr (Ríos, Williams, & Roberts, 2008).

### 8.2.2.2   Removal of organic pollutants

The swift progress in industries like textiles, pharmaceuticals, agriculture, and others has led to an increased release of organic pollutants, including organic dyes, pharmaceuticals, phenols, derivatives, and pesticides. This surge in discharge has resulted in detrimental effects on both the environment and human health. In response to this concern, there has been a growing interest in the use of zeolites for adsorbing organic pollutants in wastewater in recent years. Li et al. produced composite materials combining zeolite and activated carbon derived from coal gangue, demonstrating high efficacy in adsorbing $Cu^{2+}$ and Rh-B. This efficiency is attributed to the uniform micropores on the zeolite's surface, providing an ideal environment for the adsorption of hazardous metal ions. Simultaneously, the multistage pore structure of activated carbon accommodates macromolecular organic matter, as illustrated in Fig. 8.5 (Li et al., 2020).

In a study by Lin et al., zeolite/hydrous metal oxide hybrids were synthesized from fly ash for methylene blue adsorption. The induction of negative charge by metal oxides at higher pH levels enhanced dye adsorption, and heat treatment coupled with pH adjustment enabled the regeneration of zeolite/hydrous metal oxides (Lin, Lin, Li, Wu, & Kong, 2016). Zhang et al. created NaP zeolite using various commercial silicon sources and $NaAlO_2$ extracted from CFA, showcasing excellent adsorption performance for Rhodamine B with a removal rate of 98.26%. Adsorption mechanisms

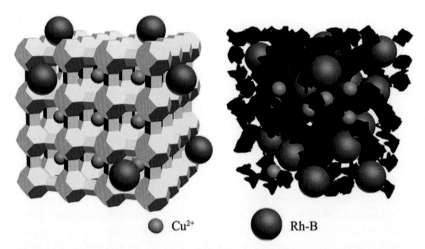

**FIGURE 8.5** **Schematic diagram of adsorption of Cu2 + and Rh-B by (I) zeolite and (II) activated carbon.** Adsorption mechanism of copper and Rh-B onto zeolite and activated carbon. *From Li, H., Zheng, F., Wang, J., Zhou, J., Huang, X., Chen, L., Hu, P., Gao, J., Zhen, Q., Bashir, S., & Liu, J. L. (2020). Facile preparation of zeolite-activated carbon composite from coal gangue with enhanced adsorption performance.* Chemical Engineering Journal, 390, *124513. https://doi.org/10.1016/j.cej.2020.124513.*

involved pore filling, electrostatic attraction, and hydrogen bonding, emphasizing physisorption (Zhang et al., 2021). Xie et al. synthesized Na-P1 from fly ash and modified it with a surfactant, significantly enhancing its adsorption capacity for both ionizable phenolic compounds and nonionizable organic compounds (Xie, Meng, Wu, Zhang, & Kong, 2012). Although limited research has focused on synthetic zeolites from solid wastes for adsorbing drugs or pesticides, the majority has centered on natural zeolites or zeolites directly synthesized from chemical reagents. Furthermore, a series of ammonium phosphotungstate/coal gangue (NH$_4$-PW/CG) adsorbents were synthesized, demonstrating up to 82% adsorption efficiency for ciprofloxacin (CIP) antibiotics within 10 minutes (Zhang, Zhao, Liu, Zhang, & Du, 2023).

### 8.2.2.3 Water softening

The broad application of zeolite in the removal of inorganic contaminants from wastewater or hard water sources is attributed to its high ion-exchange capacity and selectivity. Modified and synthetic zeolites have proven effective in removing Mg$^{2+}$ and Ca$^{2+}$ ions. The removal of ammonium by zeolites has been extensively investigated. Liu reported the solvent-free synthesis of zeolite P1 from fluidized bed fly ash, demonstrating a maximum ammonium adsorption capacity of 22.9 mg/g, surpassing many natural zeolites. Onyango et al. explored fluoride removal using Al-treated low-silica synthetic zeolites, showcasing enhanced adsorption capacity for F$^-$, PO$_4^-$, and $NO_3^-$ after modification. The study suggested fluoride adsorption onto modified zeolites occurred via ion exchange or chemisorption mechanisms, as indicated by the Dubinin-Radushkevitch model, and was influenced by initial pH and desorption.

Investigated regarding the synthesis of zeolite A and its defluoridation ability, comparing natural and synthetic zeolites was also performed by researchers. Cubic crystal zeolite A, with a cation exchange capacity of 4.6 meq $Ca^{2+}$/g, was successfully prepared from natural kaolin. Moreover, zeolites subjected to $NH_4^+$ and subsequent $Ca^{2+}$ ion-exchange treatment exhibited significantly improved fluoride removal capabilities from water. Hermassi et al. synthesized zeolitic material from fly ash and its modified form for studying phosphate removal from wastewater. The Ca modification increased the phosphate adsorption capacity on fly ash-prepared zeolitic material by fourfold. In a batch-wise experiment conducted by Schick et al., a surfactant-modified zeolite demonstrated rapid nitrate uptake (0.5−1 hour) and removed 80% of nitrate within an initial concentration range of 0.08−2.42 mmol/L. Importantly, the presence of competing anions ($SO_4^{2-}$, $HCO_3^-$, and $Cl^-$) did not affect the maximum nitrate uptake but slowed down the exchange kinetics.

### 8.2.3   Zeolites as a potential source of biodiesel production

Exploring a promising alternative to homogeneous catalysts, researchers investigated the use of chemically modified natural materials and rocks to develop low-cost, readily available, and environmentally friendly heterogeneous catalysts for transesterification reactions. Zeolite, with its microporous structure, high surface area, stability, mechanical strength, and cation exchange capacity, finds extensive applications in biodiesel synthesis. The controlled functionalization of natural and synthetic zeolite surfaces with acidic or basic groups has been shown to enhance biodiesel yield. Numerous studies have demonstrated that alkali metal ion activation of natural zeolite improves catalytic properties by increasing basicity, a crucial factor in the transesterification process. Faujasite zeolite (NaX), noted for its high surface area and abundance of basic sites attributed to its aluminum content, has been recognized as beneficial for biodiesel synthesis. The catalytic performance of nano-zeolite catalysts in biodiesel synthesis is detailed in Table 8.3.

developed a composite catalyst by trapping alkali in zeolite and utilized it as a base catalyst for biodiesel production from waste cooking oil (WCO). Four variations of alkali-modified clinoptilolite (K, Na, Ca, and Mg) extracted from green tea were prepared. Scanning electron microscope (SEM) images in Fig. 8.6 displayed distinct morphological characteristics for the studied samples. Additionally, diverse elemental compositions were revealed through energy dispersive X-ray (EDX) investigations in Fig. 8.7.

**TABLE 8.3** Catalytic performance of nano-zeolite catalysts in biodiesel synthesis.

| Biodiesel feedstock (oil) | Nanocatalyst | Catalyst preparation conditions | Particle size of catalyst (nm) | Surface area ($m^2/g$) | MTOMR | Reaction conditions | | | Product yield (%) |
|---|---|---|---|---|---|---|---|---|---|
| | | | | | | Catalyst (wt.%) | Temp (°C) | Time (h) | |
| Waste cooking oil | Cerium-doped MCM-41 | Hydrothermal method, calcined at 600°C for 3 h | 17.3 | 1200 | 9:1 | 5 | 70 | 6 | 94.3 |
| Waste cooking oil | K/ clinoptilolite | Clinoptilolite dispersed in water containing alkali-bearing salts, added green tea-based reducing reagent, centrifuged and dried at 70°C for 12 h | 17.6 | 263 | 16:1 | 4 | 70 | 2 | 93.6 |
| Waste cooking oil | Na/ clinoptilolite | Clinoptilolite dispersed in water containing alkali-bearing salts, added green tea-based reducing reagent, centrifuged and dried at 70°C for 12 h | 17.3 | 312.7 | 16:1 | 4 | 70 | 2 | 95.2 |

| | | | | | | | | | |
|---|---|---|---|---|---|---|---|---|---|
| Waste cooking oil | Ca/clinoptilolite | Clinoptilolite dispersed in water containing alkali-bearing salts, added green tea-based reducing reagent, centrifuged and dried at 70°C for 12 h | 15.4 | 252.4 | 16:1 | 4 | 70 | 3 | 96.4 |
| Waste cooking oil | Mg/clinoptilolite | Clinoptilolite dispersed in water containing alkali-bearing salts, added green tea-based reducing reagent, centrifuge and dried at 70°C for 12 h | 19.6 | 342.5 | 16:1 | 4 | 70 | 2.5 | 98.7 |
| Sunflower oil | CaO nanoparticles/NaX zeolite | NaX zeolite was impregnated with a mixture of CaO nanoparticles/ethanol, evaporated, dried, and calcined at 550°C for 3 h | – | – | 6:1 | 16 | 60 | 6 | 93.5 |
| Soybean oil | KNa/ZIF-8 (zeolite imidazolate framework [ZIF-8] doped with K) | Sol—gel method, dried at 120°C | – | 1195 | 10:1 | 0.013 | 100 | 3.5 | 98 |

(Continued)

**TABLE 8.3** (Continued)

| Biodiesel feedstock (oil) | Nanocatalyst | Catalyst preparation conditions | Particle size of catalyst (nm) | Surface area (m²/g) | Reaction conditions | | | | |
| | | | | | MTOMR | Catalyst (wt.%) | Temp (°C) | Time (h) | Product yield (%) |
|---|---|---|---|---|---|---|---|---|---|
| Waste cooking oil | Sono-enhanced CaO-dispersed over Zr-doped MCM-41 nanocatalyst | Hydrothermal method and ultrasound-assisted impregnation synthesis method, calcined at 650°C for 6 h | 15.9 | 350 | 9:1 | 5 | 70 | 6 | 88.5 |

*Source:* From Islam, S., Basumatary, B., Rokhum, S. L., Mochahari, P. K., & Basumatary, S. (2022). Advancement in the utilization of nanomaterials as efficient and recyclable solid catalyst for biodiesel synthesis. *Cleaner Chemical Engineering, 3*, 100043. https://doi.org/10.1016/j.clce.2022.100043.

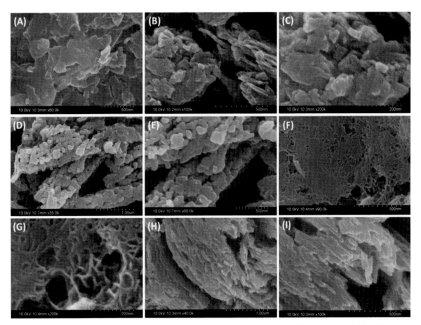

**FIGURE 8.6   SEM images of raw clinoptilolite (A, B), K-modified clinoptilolite (C), Na-modified clinoptilolite (D and E), Mg-modified clinoptilolite (G and H), and Ca-modified clinoptilolite (H and I).** Morphological analysis of raw clinoptilolite and various others modified clinoptilolite. *From Abukhadra, M. R., Basyouny, M. G., El-Sherbeeny, A. M., & El-Meligy, M. A. (2020). The effect of different green alkali modification processes on the clinoptilolite surface as adsorbent for ammonium ions; characterization and application.* Microporous and Mesoporous Materials, 300, *110145. https://doi.org/10.1016/j.micromeso.2020.110145.*

The average pore sizes of clinoptilolite, K/clinoptilolite, Na/clinoptilolite, Ca/clinoptilolite, and Mg/clinoptilolite nanoparticles were determined as 18.3, 17.6, 17.3, 15.4, and 19.6 nm, with corresponding surface areas of 258, 263, 312.7, 252.4, and 342.5 $m^2$/g, respectively. Biodiesel yields obtained with the modified catalyst were 93.6%, 95.2%, 96.4%, and 98.7% for K/clinoptilolite, Na/clinoptilolite, Ca/clinoptilolite, and Mg/clinoptilolite, respectively, under the operating reaction conditions of 70°C temperature, 16:1 methanol-to-oil molar ratio, and 4 wt.% catalyst loading. The reaction times were 120, 120, 180, and 150 minutes for K/clinoptilolite, Na/clinoptilolite, Ca/clinoptilolite, and Mg/clinoptilolite catalysts, respectively. Furthermore, they reported the catalyst's reusability for up to five cycles, and after the fifth cycle, a decline in catalytic activity was observed due to the coating of by-products on the active sites of the catalysts.

Luz Martinez et al. showcased the synthesis of CaO nanoparticles/NaX zeolite for sunflower oil *trans*-esterification. Achieving a 93.5% biodiesel yield, they utilized the catalyst with a 16 wt.% loading, a 6:1 methanol-to-oil

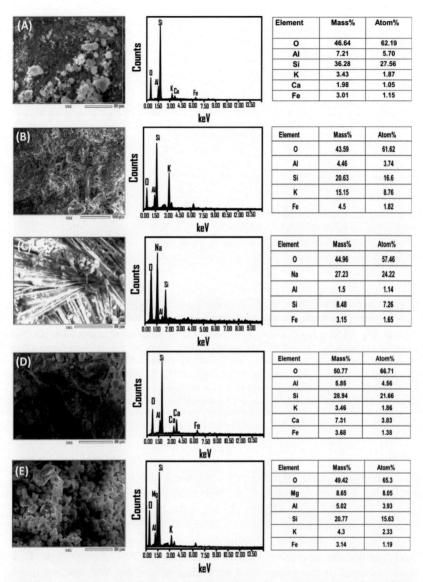

**FIGURE 8.7    EDX investigation of raw clinoptilolite (A), K/clinoptilolite (B), Na/clinoptilolite (C), Ca/clinoptilolite (D), and Mg/clinoptilolite (E).** Elemental analysis of various elements present in raw clinoptilolite and other modified clinoptilolite. *From Abukhadra, M. R., Basyouny, M. G., El-Sherbeeny, A. M., & El-Meligy, M. A. (2020). The effect of different green alkali modification processes on the clinoptilolite surface as adsorbent for ammonium ions; characterization and application.* Microporous and Mesoporous Materials, 300, *110145. https:// doi.org/10.1016/j.micromeso.2020.110145.*

molar ratio (MTOMR), and operated at 60°C for 6 hours. In another study, the sol−gel method was employed to prepare KNa/ZIF-8 (zeolite imidazolate framework, ZIF-8 doped with K) as a catalyst for soybean oil biodiesel production. The catalyst exhibited a surface area, pore volume, and pore diameter of 1195 m²/g, 0.527 cm³/g, and 1.21 nm, respectively. A biodiesel yield of 98% was attained under the operating reaction conditions of 0.0125 wt.% catalyst loading, 10:1 MTOMR, and 100°C in a 3.5-hour reaction time.

## 8.3    Limitations and challenges

Zeolite-based adsorbents and catalysts have found extensive applications in specific domains. However, the drawback of insufficient mass transfer for reactants, intermediates, and products results in reversible catalyst deactivation due to pore blockage, posing challenges for large-scale regeneration in industrial settings. Hollow MFI-type zeolites have been identified as a potential solution to enhance internal mass transport and catalytic efficiency in various Brønsted or Lewis acid-catalyzed reactions. Nonetheless, addressing these issues in the context of industrial waste ash-based zeolites and enhancing product selectivity and catalyst stability remains a formidable task.

The current limitations of detection methods and techniques necessitate further exploration for a comprehensive understanding, simulation, and summarization of the crystallization process. Identifying the structural species formed during the transformation of precursor sols or gels into crystalline nanoparticles poses a significant challenge. The advent of new observational methods at the molecular and atomic levels, coupled with the application of simulation techniques, introduces both challenges and opportunities for ongoing research in this field.

## 8.4    Conclusion and future perspective

Different type of synthetic zeolites contributes to their utilization in various fields. However, challenges in feedstock selection, synthesis approaches, and performance evaluation and application assessment need to be overcome in the future. The future scope of research work could be focused on the below ideas:

- Exploring environmentally friendly pathways characterized by minimal energy consumption and reduced or zero waste generation for synthesizing zeolites from economical raw materials is imperative.
- A comprehensive database encompassing precursor selection, synthesis methods, zeolite properties, and application performance must be systematically developed. Establishing standardized evaluation methods for products is crucial, not only for advancing data science studies but also for promoting large-scale commercialization.

- When selecting low-cost raw materials and developing synthesis methods, the focus should be on the target application rather than solely aiming to enhance overall performance.
- When designing physical adsorbents for environmental applications, practical engineering issues should also be considered. This includes taking into account factors such as the hierarchical pores in structured zeolite adsorbents.
- Conducting a techno-economic analysis of synthetic zeolites and related technologies is essential to facilitate their deployment in areas of urgent need.

## References

Ameh, A. E., Oyekola, O. O., & Petrik, L. F. (2022). Column adsorption of Rhodamine 6G over Na−P/SOD zeolite synthesised from aluminosilicate secondary waste. *Journal of Cleaner Production, 338*. Available from https://doi.org/10.1016/j.jclepro.2022.130571, https://www.journals.elsevier.com/journal-of-cleaner-production.

Angaru, G. K. R., Choi, Y. L., Lingamdinne, L. P., Choi, J. S., Kim, D. S., Koduru, J. R., . . . Chang, Y. Y. (2021). Facile synthesis of economical feasible fly ash-based zeolite-supported nano zerovalent iron and nickel bimetallic composite for the potential removal of heavy metals from industrial effluents. *Chemosphere, 267*. Available from https://doi.org/10.1016/j.chemosphere.2020.128889, http://www.elsevier.com/locate/chemosphere.

Bharti, M., Das, Pranjal P., & Purkait, Mihir K. (2023). A review on the treatment of water and wastewater by electrocoagulation process: Advances and emerging applications. *Journal of Environmental Chemical Engineering, 11*, 111558. Available from https://doi.org/10.1016/j.jece.2023.111558.

Cardoso, A. M., Paprocki, A., Ferret, L. S., Azevedo, C. M. N., & Pires, M. (2015). Synthesis of zeolite Na-P1 under mild conditions using Brazilian coal fly ash and its application in wastewater treatment. *Fuel, 139*, 59−67. Available from https://doi.org/10.1016/j.fuel.2014.08.016, http://www.journals.elsevier.com/fuel/.

Chakraborty, S., Gautam, S. P., Das, Pranjal P., & Hazarika, Manuj K. (2019). Instant controlled pressure drop (DIC) treatment for improving process performance and milled rice quality. *Journal of The Institution of Engineers (India): Series A, 100*, 683−695. Available from https://doi.org/10.1007/s40030-019-00403-w.

Changmai, M., Das, P. P., Mondal, P., Pasawan, M., Sinha, A., Biswas, P., Sarkar, S., & Purkait, M. K. (2022). Hybrid electrocoagulation−microfiltration technique for treatment of nanofiltration rejected steel industry effluent. *International Journal of Environmental Analytical Chemistry, 102*, 62−83. Available from https://doi.org/10.1080/03067319.2020.1715381.

Das, Pranjal P., Anweshan., Mondal, P., Sinha, A., Biswas, P., Sarkar, S., & Purkait, M. K. (2021a). Integrated ozonation assisted electrocoagulation process for the removal of cyanide from steel industry wastewater. *Chemosphere, 263*, 128370. Available from https://doi.org/10.1016/j.chemosphere.2020.128370.

Das, Pranjal P., Anweshan., & Purkait, Mihir K. (2021b). Treatment of cold rolling mill (CRM) effluent of steel industry. *Separation and Purification Technology, 274*, 119083. Available from https://doi.org/10.1016/j.seppur.2021.119083.

Das, Pranjal P., Deepti., & Purkait, Mihir K. (2023a). Industrial wastewater to biohydrogen production via potential bio-refinery route. In Maulin P. Shah (Ed.), *Biorefinery for water and wastewater treatment* (pp. 159–179). Springer. Available from http://doi.org/10.1007/978-3-031-20822-5_8.

Das, Pranjal P., Dhara, S., & Purkait, Mihir K. (2023b). Hybrid electrocoagulation and ozonation techniques for industrial wastewater treatment. In Maulin P. Shah (Ed.), *Sustainable industrial wastewater treatment and pollution control* (pp. 107–128). Springer. Available from http://doi.org/10.1007/978-981-99-2560-5_6.

Das, Pranjal P., Dhara, S., & Purkait, Mihir K. (2023c). The anaerobic ammonium oxidation process: Inhibition, challenges and opportunities. In Maulin P. Shah (Ed.), *Ammonia oxidizing bacteria: Applications in industrial wastewater treatment* (pp. 56–82). Royal Society of Chemistry. Available from http://doi.org/10.1039/9781837671960.

Das, Pranjal P., Dhara, S., & Purkait, Mihir K. (2024). Ozone-based oxidation processes for the removal of pharmaceutical products from wastewater. In Maulin P. Shah, & Pooja Ghosh (Eds.), *Development in wastewater treatment research and processes* (pp. 287–308). Elsevier. Available from http://doi.org/10.1016/B978-0-443-19207-4.00003-3.

Das, Pranjal P., Mondal, P., Anweshan., Sinha, A., Biswas, P., Sarkar, S., & Purkait, M. K. (2021c). Treatment of steel plant generated biological oxidation treated (BOT) wastewater by hybrid process. *Separation and Purification Technology*, *258*, 118013. Available from https://doi.org/10.1016/j.seppur.2020.118013.

Das, Pranjal P., Mondal, P., & Purkait, Mihir K. (2022a). Recent advances in synthesis of iron nanoparticles via green route and their application in biofuel production. In M. Srivastava, M. A. Malik, & P. K. Mishra (Eds.), *Green nano solution for bioenergy production enhancement* (pp. 79–104). Springer. Available from http://doi.org/10.1007/978-981-16-9356-4_4.

Das, Pranjal P., Sharma, M., & Purkait, Mihir K. (2022b). Recent progress on electrocoagulation process for wastewater treatment: A review. *Separation and Purification Technology*, *292*, 121058. Available from https://doi.org/10.1016/j.seppur.2022.121058.

Dhara, S., Das, Pranjal P., Uppaluri, R., & Purkait, Mihir K. (2023a). Biological approach for energy self-sufficiency of municipal wastewater treatment plants. In M. Sillanpaa, A. Khadir, & K. Gurung (Eds.), *Resource recovery in municipal waste waters* (pp. 235–260). Elsevier. Available from http://doi.org/10.1016/B978-0-323-99348-7.00006-0.

Dhara, S., Das, Pranjal P., Uppaluri, R., & Purkait, Mihir K. (2023b). Phosphorus recovery from municipal wastewater treatment plants. In Maulin P. Shah (Ed.), *Development in wastewater treatment research and processes* (pp. 49–72). Elsevier. Available from http://doi.org/10.1016/B978-0-323-95684-0.00014-2.

Dhara, S., Samanta, N. S., Das, Pranjal P., Uppaluri, R. V. S., & Purkait, M. K. (2023c). Ravenna grass-extracted alkaline lignin-based polysulfone Mixed Matrix Membrane (MMM) for aqueous Cr(VI) removal. *Applied Polymer Materials*, *5*, 6399–6411. Available from https://doi.org/10.1021/acsapm.3c00999.

Das, Pranjal P., Samanta, N. S., Dhara, S., & Purkait, Mihir K. (2023d). Biofuel production from algal biomass. In Maulin P. Shah (Ed.), *Green approach to alternative fuel for a sustainable future* (pp. 167–179). Elsevier. Available from http://doi.org/10.1016/B978-0-12-824318-3.00009-6.

Das, Pranjal P., Sharma, M., & Purkait, Mihir K. (2023e). Ammonia oxidizing bacteria in wastewater treatment. In Maulin P. Shah (Ed.), *Ammonia oxidizing bacteria: Applications in industrial wastewater treatment* (pp. 83–102). Royal Society of Chemistry. Available from http://doi.org/10.1039/BK9781837671960-00083.

Das, Pranjal P., Sontakke, A. D., & Purkait, Mihir K. (2023f). Electrocoagulation process for wastewater treatment: Applications, challenges, and prospects. In Maulin P. Shah (Ed.), *Development in wastewater treatment research and processes* (pp. 23–48). Elsevier. Available from http://doi.org/10.1016/B978-0-323-95684-0.00015-4.

Das, Pranjal P., Sontakke, A. D., Samanta, N. S., & Purkait, Mihir K. (2023g). Emerging contaminants in wastewater: Eco-toxicity and sustainability assessment. In Maulin P. Shah (Ed.), *Industrial wastewater reuse* (pp. 63−87). Springer. Available from http://doi.org/10.1007/978-981-99-2489-9_4.

Gollakota, A. R. K., Volli, V., Munagapati, V. S., Wen, J. C., & Shu, C. M. (2020). Synthesis of novel ZSM-22 zeolite from Taiwanese coal fly ash for the selective separation of Rhodamine 6G. *Journal of Materials Research and Technology, 9*(6), 15381−15393. Available from https://doi.org/10.1016/j.jmrt.2020.10.070, http://www.elsevier.com/journals/journal-of-materials-research-and-technology/2238-7854.

Hossini Asl, S. M., Masomi, M., & Tajbakhsh, M. (2020). Hybrid adaptive neuro-fuzzy inference systems for forecasting benzene, toluene & m-xylene removal from aqueous solutions by HZSM-5 nano-zeolite synthesized from coal fly ash. *Journal of Cleaner Production, 258*120688. Available from https://doi.org/10.1016/j.jclepro.2020.120688.

Izquierdo, M., Juan, R., & Rubio, B. (2013). $SO_2$ adsorption on type y zeolite synthesized from coal fly ash. *Revista Investigaciones Aplicadas, 7*, 11−19.

Izquierdo, M. T., Juan, R., Rubio, B., & Gómez-Giménez, C. (2016). No removal in the selective catalitic reduction process over Cu and Fe exchanged type Y zeolites synthesized from coal fly ash. *Energy Sources, Part A: Recovery, Utilization and Environmental Effects, 38*(9), 1183−1188. Available from https://doi.org/10.1080/15567036.2014.881932, http://www.tandf.co.uk/journals/titles/15567036.asp.

Karthikeyan, D., & Saravanan, C.G. (2013). Experimental analysis of flyash based, ion exchanged Fe, Cu and Ni-X zeolite catalytic converter for SI engine exhaust emission control. In *Proc. 9th Asia-Pacific Conf. Combust.*, ASPACC 2013 (pp. 657−660).

Kim, K. J., & Ahn, H. G. (2012). The effect of pore structure of zeolite on the adsorption of VOCs and their desorption properties by microwave heating. *Microporous and Mesoporous Materials, 152*, 78−83. Available from https://doi.org/10.1016/j.micromeso.2011.11.051, 13871811.

Kuwahara, Y., Ohmichi, T., Kamegawa, T., Mori, K., & Yamashita, H. (2009). A novel synthetic route to hydroxyapatite-zeolite composite material from steel slag: Investigation of synthesis mechanism and evaluation of physicochemical properties. *Journal of Materials Chemistry, 19* (39), 7263−7272,. Available from https://doi.org/10.1039/b911177h, 13645501.

Li, G., Li, M., Zhang, X., Cao, P., Jiang, H., Luo, J., & Jiang, T. (2022). Hydrothermal synthesis of zeolites-calcium silicate hydrate composite from coal fly ash with co-activation of Ca $(OH)_2$-NaOH for aqueous heavy metals removal. *International Journal of Mining Science and Technology, 32*(3), 563−573. Available from https://doi.org/10.1016/j.ijmst.2022.03.001, http://www.elsevier.com/wps/find/journaldescription.cws_home/727915/description#description.

Li, H., Zheng, F., Wang, J., Zhou, J., Huang, X., Chen, L., ... Liu, J. L. (2020). Facile preparation of zeolite-activated carbon composite from coal gangue with enhanced adsorption performance. *Chemical Engineering Journal, 390*124513. Available from https://doi.org/10.1016/j.cej.2020.124513.

Li, X., Wang, J., Guo, Y., Zhu, T., & Xu, W. (2021). Adsorption and desorption characteristics of hydrophobic hierarchical zeolites for the removal of volatile organic compounds. *Chemical Engineering Journal, 411*128558. Available from https://doi.org/10.1016/j.cej.2021.128558.

Lin, L., Lin, Y., Li, C., Wu, D., & Kong, H. (2016). Synthesis of zeolite/hydrous metal oxide composites from coal fly ash as efficient adsorbents for removal of methylene blue from water. *International Journal of Mineral Processing, 148*, 32−40. Available from https://doi.org/10.1016/j.minpro.2016.01.010, http://www.elsevier.com/inca/publications/store/5/0/3/3/3/8/.

Liu, L., Singh, R., Xiao, P., Webley, P. A., & Zhai, Y. (2011). Zeolite synthesis from waste fly ash and its application in $CO_2$ capture from flue gas streams. *Adsorption*, *17*(5), 795−800. Available from https://doi.org/10.1007/s10450-011-9332-8.

Liu, Y. X., & Zhang, J. (2011). Photochemical oxidation removal of NO and $SO_2$ from simulated flue gas of coal-fired power plants by wet scrubbing using $UV/H_2O_2$ advanced oxidation process. *Industrial and Engineering Chemistry Research*, *50*(7), 3836−3841. Available from https://doi.org/10.1021/ie1020377.

López-Salinas, E., Salas, P., Schifter, I., Morán, M., Castillo, S., & Mogica, E. (1997). Reduction of NO by CO using a zeolite catalyst obtained from fly ash. *Studies in Surface Science and Catalysis*, *105*, 1565−1570. Available from https://doi.org/10.1016/s0167-2991(97)80800-2, https://www.sciencedirect.com/bookseries/studies-in-surface-science-and-catalysis.

Ma, L., Han, L., Chen, S., Hu, J., Chang, L., Bao, W., & Wang, J. (2019). Rapid synthesis of magnetic zeolite materials from fly ash and iron-containing wastes using supercritical water for elemental mercury removal from flue gas. *Fuel Processing Technology*, *189*, 39−48. Available from https://doi.org/10.1016/j.fuproc.2019.02.021, https://www.journals.elsevier.com/fuel-processing-technology.

Ma, Z., Zhang, X., Lu, G., Guo, Y., Song, H., & Cheng, F. (2022). Hydrothermal synthesis of zeolitic material from circulating fluidized bed combustion fly ash for the highly efficient removal of lead from aqueous solution. *Chinese Journal of Chemical Engineering*, *47*, 193−205. Available from https://doi.org/10.1016/j.cjche.2021.05.043, https://www.journals.elsevier.com/chinese-journal-of-chemical-engineering.

Mokrzycki, J., Fedyna, M., Marzec, M., Szerement, J., Panek, R., Klimek, A., ... Mierzwa-Hersztek, M. (2022). Copper ion-exchanged zeolite X from fly ash as an efficient adsorbent of phosphate ions from aqueous solutions. *Journal of Environmental Chemical Engineering*, *10*(6)108567. Available from https://doi.org/10.1016/j.jece.2022.108567.

Mondal, P., Samanta, N. S., Kumar, A., & Purkait, M. K. (2020). Recovery of $H_2SO_4$ from wastewater in the presence of NaCl and $KHCO_3$ through pH responsive polysulfone membrane: Optimization approach. *Polymer Testing*, *86*. Available from https://doi.org/10.1016/j.polymertesting.2020.106463, https://www.journals.elsevier.com/polymer-testing.

Munawar, M. A., Khoja, A. H., Naqvi, S. R., Mehran, M. T., Hassan, M., Liaquat, R., & Dawood, U. F. (2021). Challenges and opportunities in biomass ash management and its utilization in novel applications. *Renewable and Sustainable Energy Reviews*, *150*. Available from https://doi.org/10.1016/j.rser.2021.111451, https://www.journals.elsevier.com/renewable-and-sustainable-energy-reviews.

Naushad, M., Ahamad, T., Al-Maswari, B. M., Abdullah Alqadami, A., & Alshehri, S. M. (2017). Nickel ferrite bearing nitrogen-doped mesoporous carbon as efficient adsorbent for the removal of highly toxic metal ion from aqueous medium. *Chemical Engineering Journal*, *330*, 1351−1360. Available from https://doi.org/10.1016/j.cej.2017.08.079, http://www.elsevier.com/inca/publications/store/6/0/1/2/7/3/index.htt.

Prajapati, Y. N., Bhaduri, B., Joshi, H. C., Srivastava, A., & Verma, N. (2016). Aqueous phase adsorption of different sized molecules on activated carbon fibers: Effect of textural properties. *Chemosphere*, *155*, 62−69. Available from https://doi.org/10.1016/j.chemosphere.2016.04.040, http://www.elsevier.com/locate/chemosphere.

Querol, X., Moreno, N., Umaa, J. C., Alastuey, A., Hernández, E., López-Soler, A., & Plana, F. (2002). Synthesis of zeolites from coal fly ash: an overview. *International Journal of Coal Geology*, *50*(1−4), 413−423. Available from https://doi.org/10.1016/S0166-5162(02)00124-6, 01665162.

Rungnim, C., Promarak, V., Hannongbua, S., Kungwan, N., & Namuangruk, S. (2016). Complete reaction mechanisms of mercury oxidation on halogenated activated carbon. *Journal of Hazardous Materials, 310,* 253–260. Available from https://doi.org/10.1016/j.jhazmat.2016.02.033, http://www.elsevier.com/locate/jhazmat.

Ríos, C. A., Williams, C. D., & Roberts, C. L. (2008). Removal of heavy metals from acid mine drainage (AMD) using coal fly ash, natural clinker and synthetic zeolites. *Journal of Hazardous Materials, 156*(1–3), 23–35. Available from https://doi.org/10.1016/j.jhazmat.2007.11.123.

Samanta, N. S., Banerjee, S., Mondal, P., Anweshan., Bora, U., & Purkait, M. K. (2021). Preparation and characterization of zeolite from waste Linz-Donawitz (LD) process slag of steel industry for removal of Fe3 + from drinking water. *Advanced Powder Technology, 32* (9), 3372–3387. Available from https://doi.org/10.1016/j.apt.2021.07.023, http://www.elsevier.com.

Samanta, N. S., Das, P. P., Mondal, P., Bora, U., & Purkait, M. K. (2022). Physico-chemical and adsorption study of hydrothermally treated zeolite A and FAU-type zeolite X prepared from LD (Linz–Donawitz) slag of the steel industry. *International Journal of Environmental Analytical Chemistry.* Available from https://doi.org/10.1080/03067319.2022.2079082, http://www.tandf.co.uk/journals/titles/03067319.asp.

Samanta, N. S., Das, P. P., Mondal, P., Changmai, M., & Purkait, M. K. (2022). Critical review on the synthesis and advancement of industrial and biomass waste-based zeolites and their applications in gas adsorption and biomedical studies. *Journal of the Indian Chemical Society, 99*(11). Available from https://doi.org/10.1016/j.jics.2022.100761, https://www.sciencedirect.com/journal/journal-of-the-indian-chemical-society.

Samanta, N. S., Anweshan., Mondal, P., Bora, U., & Purkait, M. K. (2023). Synthesis of precipitated calcium carbonate from LD-slag using $CO_2$. *Materials Today Communications, 36.* Available from https://doi.org/10.1016/j.mtcomm.2023.106588, http://www.journals.elsevier.com/materials-today-communications/.

Sarmah, M., Baruah, B. P., & Khare, P. (2013). A comparison between $CO_2$ capturing capacities of fly ash based composites of MEA/DMA and DEA/DMA. *Fuel Processing Technology, 106,* 490–497. Available from https://doi.org/10.1016/j.fuproc.2012.09.017.

Sharma, M., Das, Pranjal P., Kumar, S., & Purkait, M. K. (2023a). Polyurethane foams as packing and insulating materials. In Ram K. Gupta (Ed.), *Polyurethanes: Preparation, properties, and applications* (pp. 83–99). American Chemical Society. Available from http://doi.org/10.1021/bk-2023-1454.ch004

Sharma, M., Das, Pranjal P., & Purkait, Mihir K. (2023b). Energy storage properties of nanomaterials. In A. Husen, & K. S. Siddiqi (Eds.), *Advances in smart nanomaterials and their applications* (pp. 337–350). Elsevier. Available from http://doi.org/10.1016/B978-0-323-99546-7.00005-7

Shekhar Samanta, N., Das, P. P., Dhara, S., & Purkait, M. K. (2023). An overview of precious metal recovery from steel industry slag: Recovery strategy and utilization. *Industrial and Engineering Chemistry Research, 62*(23), 9006–9031. Available from https://doi.org/10.1021/acs.iecr.3c00604, http://pubs.acs.org/journal/iecred.

Sontakke, A. D., Das, Pranjal P., Mondal, P., & Purkait, M. K. (2021). Thin-film composite nanofiltration hollow fiber membranes toward textile industry effluent treatment and environmental remediation applications: review. *Emergent Materials, 5,* 1409–1427. Available from https://doi.org/10.1007/s42247-021-00261-y.

Suchecki, T. T., Wałek, T., & Banasik, M. (2004). Fly ash zeolites as sulfur dioxide adsorbents. *Polish Journal of Environmental Studies, 13*(6), 723–727.

Visa, M. (2016). Synthesis and characterization of new zeolite materials obtained from fly ash for heavy metals removal in advanced wastewater treatment. *Powder Technology*, *294*, 338−347. Available from https://doi.org/10.1016/j.powtec.2016.02.019, http://www.elsevier.com/locate/powtec.

Wang, J., Li, D., Ju, F., Han, L., Chang, L., & Bao, W. (2015). Supercritical hydrothermal synthesis of zeolites from coal fly ash for mercury removal from coal derived gas. *Fuel Processing Technology*, *136*, 96−105. Available from https://doi.org/10.1016/j.fuproc.2014.10.020.

Wang, Y., Jia, H., Chen, P., Fang, X., & Du, T. (2020). Synthesis of La and Ce modified X zeolite from rice husk ash for carbon dioxide capture. *Journal of Materials Research and Technology*, *9*(3), 4368−4378. Available from https://doi.org/10.1016/j.jmrt.2020.02.061, http://www.elsevier.com/journals/journal-of-materials-research-and-technology/2238-7854.

Wdowin, M., Wiatros-Motyka, M. M., Panek, R., Stevens, L. A., Franus, W., & Snape, C. E. (2014). Experimental study of mercury removal from exhaust gases. *Fuel*, *128*, 451−457. Available from https://doi.org/10.1016/j.fuel.2014.03.041.

Xie, J., Meng, W., Wu, D., Zhang, Z., & Kong, H. (2012). Removal of organic pollutants by surfactant modified zeolite: Comparison between ionizable phenolic compounds and non-ionizable organic compounds. *Journal of Hazardous Materials*, *231−232*, 57−63. Available from https://doi.org/10.1016/j.jhazmat.2012.06.035.

Xie, J., Wang, Z., Wu, D., & Kong, H. (2014). Synthesis and properties of zeolite/hydrated iron oxide composite from coal fly ash as efficient adsorbent to simultaneously retain cationic and anionic pollutants from water. *Fuel*, *116*, 71−76. Available from https://doi.org/10.1016/j.fuel.2013.07.126.

Zhang, B., Chen, Y., Wei, L., & Zu, Z. (2012). Preparation of molecular sieve X from coal fly ash for the adsorption of volatile organic compounds. *Microporous and Mesoporous Materials*, *156*, 36−39. Available from https://doi.org/10.1016/j.micromeso.2012.02.016.

Zhang, H., Zhao, R., Liu, Z., Zhang, X., & Du, C. (2023). Enhanced adsorption properties of polyoxometalates/coal gangue composite:The key role of kaolinite-rich coal gangue. *Applied Clay Science*, *231*, 106730. Available from https://doi.org/10.1016/j.clay.2022.106730.

Zhang, Y., Han, H., Wang, X., Zhang, M., Chen, Y., Zhai, C., . . . Zhang, C. (2021). Utilization of NaP zeolite synthesized with different silicon species and NaAlO2 from coal fly ash for the adsorption of Rhodamine B. *Journal of Hazardous Materials*, *415*125627. Available from https://doi.org/10.1016/j.jhazmat.2021.125627.

# Chapter 9

# Differences between commercial zeolite and waste-derived zeolite

## 9.1 Introduction

Zeolites are remarkable minerals that possess unique properties of adsorption and ion exchange (Samanta et al., 2021; Samanta, Anweshan, Mondal, Bora, & Purkait, 2023). In wastewater treatment, composite membranes play a significant role wherein zeolite is employed as a nanofiller (Bulasara, Thakuria, Uppaluri, & Purkait, 2011; Chakraborty, Das, & Mondal, 2023; Das & Mondal, 2023; Das, Mondal, & Purkait, 2022; Dhara, Shekhar Samanta, Das, Uppaluri, & Purkait, 2023; Mondal, Samanta, Meghnani, & Purkait, 2019; Mondal, Samanta, Kumar, & Purkait, 2020; Purkait, DasGupta, & De, 2005; Samanta, Mondal, & Purkait, 2023; Setiawan, Alvin Setiadi, & Mukti, 2023; Sharma, Das, & Purkait, 2023; Sontakke, Das, Mondal, & Purkait, 2022). Nowadays, the electrocoagulation process and microbial fuel cells made a remarkable approach for industrial as well as wastewater treatment (Changmai et al., 2022; Das et al., 2021; Das, Sharma, & Purkait, 2022; Das, Anweshan, & Purkait, 2021; Das, Mondal, et al., 2021; Sharma, Das, Sood, Chakraborty, & Purkait, 2021; Sharma, Das, Chakraborty, & Purkait, 2022; Sharma, Das, Sood, Chakraborty, & Purkait, 2022). However, zeolites are extensively used in various industrial and food science applications due to their versatility (Anweshan, Das, Dhara, & Purkait, 2023; Samanta et al., 2021; Samanta, Das, Mondal, Changmai, & Purkait, 2022; Shekhar Samanta, Das, Sharma, & Purkait, 2023; Shekhar Samanta, Das, Dhara, & Purkait, 2023). Two distinct forms of zeolites exist—commercial and lab-made, each with its own set of characteristics and applications (Dhara, Samanta, Uppaluri, & Purkait, 2023; Samanta, Das, Mondal, Changmai, et al., 2022). This duality prompts an interesting inquiry into the differences between these two forms of zeolites. By examining their synthesis, properties, and applications, this discussion aims to provide a constructive exploration of the intriguing disparities between commercial zeolites and their waste-derived counterparts.

Commercial zeolites, synthesized with precision and expertise in controlled laboratory processes, have emerged as a cornerstone of various industries,

Waste-based Zeolite. DOI: https://doi.org/10.1016/B978-0-443-22316-7.00009-7
**229**

including petrochemicals and pharmaceuticals, environmental remediation, agriculture, biofuel production, and healthcare (Das, Sontakke, & Purkait, 2023; Dhara, Das, Uppaluri, & Purkait, 2023; Kianfar, Hajimirzaee, mousavian, & Mehr, 2020; Sharma, Das, Chakraborty, & Purkait, 2023; Sowunmi et al., 2018). These crystals are engineered with meticulous attention to detail, ensuring uniformity in purity, size, and composition, making them an ideal choice for industries where performance metrics are critical (Kianfar et al., 2020). Their synthetic genesis offers the flexibility to tailor zeolite structures to suit a wide range of applications, delivering consistent and predictable performance across diverse industrial contexts (Asselman, Kirschhock, & Breynaert, 2023; Li, Corma, & Yu, 2015; Wu, Luan, & Xiao, 2022).

Waste-derived and/or lab-made zeolites are a promising field in sustainable materials science and environmental engineering (Das, Sontakke, Samanta, & Purkait, 2023). These zeolites are developed by converting waste materials like coal fly ash, mining residues, and industrial by-products into valuable resources (Sharma, Samanta, Chakraborty, & Purkait, 2023). This approach not only helps in reducing the environmental impact of waste disposal but also repurposes otherwise discarded materials into high-value, functional products. However, the complex and diverse nature of waste-derived zeolites poses some challenges in terms of composition, purity, and crystal structure (Samanta et al., 2021; Samanta, Das, Mondal, Changmai, et al., 2022). Thus, a thorough understanding of their properties is required to effectively utilize them and further improve their potential (Khan, Saeed, & Khan, 2019).

Zeolites, whether commercial-grade or waste-derived, have unique thermal properties that can be harnessed to optimize their performance across various industrial applications (Das, Deepti, & Purkait, 2023; Dhara, Das, Uppaluri, & Purkait, 2023; Pérez-Botella, Valencia, & Rey, 2022; Samanta, Anweshan, & Purkait, 2023). Commercial-grade zeolites, manufactured under controlled laboratory conditions, exhibit a high degree of thermal stability due to their well-defined structures and optimal thermal resistance (Manrique, Guzmán, Pérez-Pariente, Márquez-Álvarez, & Echavarría, 2016). This makes them valuable in high-temperature industrial processes that involve thermal treatments like catalysis (Das, Samanta, Dhara, & Purkait, 2023; Duarah, Das, & Purkait, 2023; Sontakke, Deepti, Samanta, & Purkait, 2023). Although waste-derived zeolites can vary in thermal stability due to their diverse range of raw materials and synthesis routes, they offer a unique opportunity to apply a nuanced approach to their use in high-temperature environments (Bharti, Das, & Purkait, 2023; Das, Dhara, & Purkait, 2023, 2024). By understanding the composition and impurity profile of the parent materials, structural irregularities and impurities can be minimized for improved thermal stability (Samanta, Das, Mondal, Bora, & Purkait, 2022).

In summary, both commercial-grade and waste-derived zeolites offer distinct thermal properties that can be leveraged to optimize their performance. With careful consideration and understanding of their unique characteristics, we can utilize zeolites to their full potential in a range of industrial applications.

This chapter basically focuses on the fascinating world of zeolites and explores their synthesis methodologies, structural characteristics, and functional applications in both commercial and waste-derived forms. Through a comparative analysis, we can uncover the unique advantages and limitations of each type and seek synergistic integration opportunities that can lead to innovative applications in sustainable technologies, environmental remediation, and beyond. This exploration not only enhances our comprehension of zeolite science but also emphasizes the significant role these remarkable materials play in shaping a sustainable and resource-conscious future.

## 9.2    Characterization study of pure vs synthetic zeolite

### 9.2.1    X-ray diffraction analysis

The X-ray diffraction (XRD) pattern of kaolin-based zeolite 13X and 4A and commercially available zeolite 4A and 13X samples have been shown in Fig. 9.1A and B, respectively (Sowunmi et al., 2018). From the figure, the number of intense peaks in synthetic zeolite was found higher than in commercial-grade zeolite samples. This can be attributed to the presence of other mineral phases in the metakaolin used as zeolite precursor material. Moreover, the high intense peak for both the synthetic zeolites arrived at $\sim 2\theta = 7°$ ascribed to the more crystalline nature as compared to less intense peak zeolite, which is readily available in the market (Fig. 9.1).

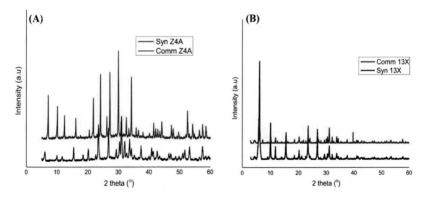

**FIGURE 9.1    Commercial and synthetic zeolite 4A and zeolite 13X.** XRD pattern of (A) commercial and synthetic 4A zeolite and (B) commercial and synthetic zeolite 13X. *From Sowunmi, A. R., Folayan, C. O., Anafi, F. O., Ajayi, O. A., Omisanya, N. O., Obada, D. O., & Dodoo-Arhin, D. (2018). Dataset on the comparison of synthesized and commercial zeolites for potential solar adsorption refrigerating system.* Data in Brief, 20, *90−95. https://doi.org/10.1016/j.dib.2018.07.040.*

## 9.2.2 FESEM analysis

Sowunmi et al. (2018) hydrothermally synthesized zeolite 13X and zeolite type-4A using metakaolin as silica and alumina feedstock. Fig. 9.2A–D illustrates the morphological comparison between synthetic and commercial zeolite type-4A (Fig. 9.2A and B) and synthetic and commercially available zeolite 13X (Fig. 9.2C and D). Regular and crystalline cubical and octahedral shape of commercial zeolite 4A was observed (Fig. 9.2B), on the other hand, irregular and brittle cubical-shaped type 4A-type zeolite morphology was found in the synthetic zeolite surface. Fig. 9.2C depicts the amorphous and noncrystalline nature of synthetic zeolite 13X; whereas, the well crystalline and regular shape of commercial zeolite 13X was found on the zeolite powder (Fig. 9.2D). The field emission scanning electron microscopy (FESEM) investigation showed that the produced zeolites are of a generally respectable grade.

In a different study, Nosrati, Olad, and Nofouzi (2015) prepared a $TiO_2$/Ag-exchanged-Na-A zeolite-like composite additive, which was used in the polyacrylic latex matrix to obtain a photocatalytic and hydrophilic coating.

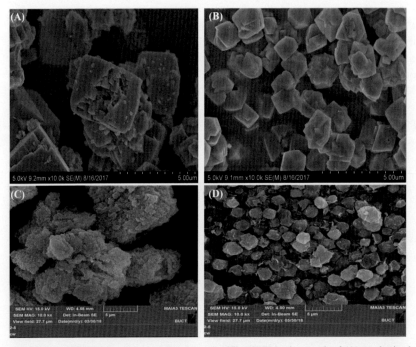

**FIGURE 9.2   FESEM images of zeolite 4A and 13X.** FESEM micrograph of (A) synthesized zeolite 4A, (B) commercial 4A zeolite, (C) synthesized zeolite 13X, and (D) readily available zeolite 13X. *From Sowunmi, A. R., Folayan, C. O., Anafi, F. O., Ajayi, O. A., Omisanya, N. O., Obada, D. O., & Dodoo-Arhin, D. (2018). Dataset on the comparison of synthesized and commercial zeolites for potential solar adsorption refrigerating system. Data in Brief, 20, 90–95. https://doi.org/10.1016/j.dib.2018.07.040.*

**FIGURE 9.3    Commercial zeolite A and slag-derived zeolite A.** (A) Commercial Na-A zeolite and (B) LD-slag-based Na-A zeolite. *From Samanta, N. S., Banerjee, S., Mondal, P., Anweshan, Bora, U., & Purkait, M. K. (2021). Preparation and characterization of zeolite from waste Linz-Donawitz (LD) process slag of steel industry for removal of Fe3 + from drinking water.* Advanced Powder Technology, 32(9), 3372–3387. https://doi.org/10.1016/j.apt.2021.07.023.

The $TiO_2$/Ag-exchanged-Na-A zeolite-based acrylic material has shown good UV absorption capability, which in turn enhances the degradation of organic pollutants and water stability. The commercial-grade Na-A zeolite was used in this work, which was procured from Azar Kimia Khatam Co., Iran. The morphological result of the Na-A zeolite has been shown in Fig. 9.3A. In another investigation, Samanta et al. (2021) synthesized cubical crystal structure A-type zeolite (Fig. 9.3B) from steel industry by-product LD-slag as silica and alumina feedstock.

### 9.2.3    BET analysis

The specific surface area of marketable and kaolin-based zeolite 4A is shown in Table 9.1. The Brunauer–Emmett–Teller (BET) studies of zeolite 4A, which was synthesized from kaolin, revealed pore sizes of 12.086 nm, pore volumes of 0.0065 $cm^3$/g, and surface areas of 22 $m^2$/g. In contrast, commercial zeolite 4A, which was used as a reference, had pore sizes of 58.143 nm, pore volumes of 0.2462 $cm^3$/g, and surface areas of 559.13 $m^2$/g. The total surface analysis of marketable and as-synthesized zeolite 13X mineral has been documented in Table 9.2. According to Table 9.2, the synthetic 13X zeolite has a specific surface area of 324.584 $m^2$/g, a volume of the pore of 0.135847 $cm^3$/g, and an average pore size of 10.5059 nm. On the contrary, for commercial zeolite type-13X, the specific surface area, pore size, and pore volume were found to be 310.0906 $m^2$/g, 7.2752 nm, and 0.135951 $cm^3$/g, respectively. From Table 9.1, it can be seen that the BET surface area of

**TABLE 9.1** BET surface analysis of commercially available and synthetic zeolite 4A.

| BET analysis | Commercial 4A | Synthetic 4A |
| --- | --- | --- |
| Pore size (nm) | 58.14 | 12.086 |
| Volume of pore ($cm^3/g$) | 0.2462 | 0.0065 |
| Specific surface area ($m^2/g$) | 559.13 | 22 |

Source: From Sowunmi, A. R., Folayan, C. O., Anafi, F. O., Ajayi, O. A., Omisanya, N. O., Obada, D. O., & Dodoo-Arhin, D. (2018). Dataset on the comparison of synthesized and commercial zeolites for potential solar adsorption refrigerating system. *Data in Brief, 20*, 90−95. https://doi.org/10.1016/j.dib.2018.07.040.

**TABLE 9.2** BET analysis of marketable and synthetic zeolite 13X.

| BET analysis | Commercial 13X | Synthetic 13X |
| --- | --- | --- |
| Specific surface area ($m^2/g$) | 310.0906 | 324.584 |
| Pore size (nm) | 7.2752 | 10.5059 |
| Volume of pore ($cm^3/g$) | 0.135951 | 0.135847 |

Source: From Sowunmi, A. R., Folayan, C. O., Anafi, F. O., Ajayi, O. A., Omisanya, N. O., Obada, D. O., & Dodoo-Arhin, D. (2018). Dataset on the comparison of synthesized and commercial zeolites for potential solar adsorption refrigerating system. *Data in Brief, 20*, 90−95. https://doi.org/10.1016/j.dib.2018.07.040.

synthesized zeolite type-4A was higher as compared to commercial-grade 4A zeolite. The reason could be after synthesis, commercial zeolites may be subjected to further treatments or changes that affect their surface characteristics.

## 9.2.4 X-ray fluorescence (XRF) analysis

Tables 9.3 and 9.4 show the elemental analysis of synthesized zeolite types 4A and 13X and commercially available 4A and 13X-type zeolite (Sowunmi et al., 2018). From the analysis report, it can be seen that the kankara kaolin used in their study comprised an abundant amount of silica and alumina which are feedstock for zeolite minerals. However, other minerals like MgO, $Na_2O$, and $TiO_2$ were also detected from the obtained kaolin-based and commercial-grade zeolite with varying concentrations (refer Tables 9.3 and 9.4). From both the tables, the silica/alumina ratio of synthetic zeolite 4A and 13X was found to be 2.440 and 3.110, respectively.

**TABLE 9.3** Elemental analysis of kaolin-based zeolite 4A, commercial-grade zeolite 4A, and reference zeolite type-4A.

| Constituents (wt.%) | Synthetic 4A zeolite | Commercial 4A zeolite | Reference 4A zeolite |
|---|---|---|---|
| MgO | 0.316 | 0.215 | 0.23 |
| $Na_2O$ | 13.77 | 12.601 | 12.60 |
| $Al_2O_3$ | 34.673 | 30.645 | 31.19 |
| $SiO_2$ | 49.719 | 55.695 | 35.90 |
| $K_2O$ | 0.187 | 0.239 | 0.400 |
| $Fe_2O_3$ | 0.664 | 0.059 | 1.13 |
| $TiO_2$ | 0.039 | 0.013 | 2.01 |
| Si/Al | 2.440 | 3.090 | 2.000 |

*Source*: From Sowunmi, A. R., Folayan, C. O., Anafi, F. O., Ajayi, O. A., Omisanya, N. O., Obada, D. O., & Dodoo-Arhin, D. (2018). Dataset on the comparison of synthesized and commercial zeolites for potential solar adsorption refrigerating system. *Data in Brief, 20*, 90–95. https://doi.org/10.1016/j.dib.2018.07.040.

**TABLE 9.4** Elemental analysis of kaolin-based zeolite 13X, commercial-grade 13X zeolite, and reference zeolite type-13X.

| Constituents (wt.%) | Synthetic zeolite 13X | Commercial 13X zeolite | Reference 13X zeolite |
|---|---|---|---|
| MgO | 0.502 | 0.8035 | ND |
| $Na_2O$ | 9.757 | 15.930 | 12.49 |
| $Al_2O_3$ | 30.180 | 28.580 | 30.17 |
| $SiO_2$ | 55.099 | 53.7975 | 49.28 |
| $K_2O$ | 0.400 | 0.0360 | ND |
| $Fe_2O_3$ | 0.836 | 2.080 | ND |
| $TiO_2$ | 0.069 | 0.3600 | ND |
| Si/Al | 3.110 | 3.200 | 2.770 |

*Source*: From Sowunmi, A. R., Folayan, C. O., Anafi, F. O., Ajayi, O. A., Omisanya, N. O., Obada, D. O., & Dodoo-Arhin, D. (2018). Dataset on the comparison of synthesized and commercial zeolites for potential solar adsorption refrigerating system. *Data in Brief, 20*, 90–95. https://doi.org/10.1016/j.dib.2018.07.040.

## 9.3 Applications of pure and waste-derived zeolite

### 9.3.1 Catalysis

In a catalytic application, Al-Ani, Mordvinova, Lebedev, Khodakov, and Zholobenko (2019) prepared cesium (Cs)-modified zeolite using several types of commercial-grade zeolite minerals like zeolite X, Y, maximum aluminum P (MAP), and Na-A zeolite via ion-exchange method. Cesium nitrate and cesium hydroxide were used as Cs sources. Ion exchange was performed using a 0.5 M solution including $KNO_3$ and KOH (10/1 v/v) under the same circumstances to manufacture K-containing zeolites. After exchanging zeolites, they were cleaned with deionized water and left to dry overnight at 80°C. The zeolites containing CsK were obtained using the same technique. By first transforming the Na-containing zeolites into their K-forms and then exchanging them with the Cs-containing solution, these catalysts were created. Before usage, the zeolites were heated to the target temperatures up to 450°C at a rate of 1°C per minute and kept there for two hours. This activated the zeolites in an airflow. They utilized the obtained zeolite catalysts for biodiesel production in a microwave batch reactor (Tangy, Pulidindi, & Gedanken, 2016). The biodiesel conversion was found to be 90% at a low oil-to-methanol ratio over 5 wt.% K-form of gismondine based on maximum aluminium potential (KMAP) catalyst at 160°C. After a 7-hour reaction period, the conversion was at almost 100% with a fair ratio of methyl ester and fatty acid over $K_2O$/Na-X zeolite (Muciño et al., 2016). Zeolite KMAP has better transesterification activity than KY, KA, and KX types combined with superior recyclability and resistance to deactivation over the course of three runs. In fact, zeolites MAP and A have the greatest Al concentration (silica/alumina ratio maintained at 1), while zeolite MAP, with a K/Al ratio of 0.90, has the highest degree of ion exchange of sodium for potassium cations. It should be noted that the activity of faujasite-type catalysts decreases when Cs cations are added to the Na forms. This decrease can be attributed to the larger size of the Cs cations, which causes a lower degree of ion exchange. Additionally, there is hardly any increase in the activity of CsKA and CsKMAP zeolites when compared to their potassium forms.

In a different study, Volli and Purkait (2015) prepared potassium ($K^+$)-modified zeolite Na-A and Na-X using power plant waste coal fly ash as zeolite feedstock. The zeolite substrate was prepared by a fusion-facilitated hydrothermal process. Before hydrothermal treatment, fly ash was converted to zeolite via alkaline fusion. The unburned carbon and other volatile components found in fly ash were eliminated using calcination, which was done for 2 hours at 900 ($\pm$ 10)°C. To enhance the fly ash sample's zeolite-forming activity, it was subjected to an additional treatment of 10% hydrochloric acid for 1 hour at 80°C. By carefully grinding and combining the ingredients in a 1:1−1:2.5 ratio of NaOH to fly ash, a uniform fusion combination of the two was created. To investigate the impact of the NaOH/fly ash ratio, fusion temperature, and duration on the degree of zeolite formation, the sample was heated for about 1−2 hours in the range of 450°C−600°C. To adjust the silica/alumina ratio and investigate its impact on zeolite formation, the resulting fused mass was cooled, ground, and thoroughly mixed with deionized water at a ratio of 1:10 (fly ash to water). At the same time, sodium aluminate (10−30 wt.%) was added. After that, the slurry was aged for 12−16 hours and given a 24-hour window to crystallize between 90°C and 120°C to investigate the impact of hydrothermal temperature and duration on zeolitization. Filtration was used to recover the solid crystalline product, which was then dried at 110°C and extensively cleaned until the filtrate pH was around 10. Using potassium acetate as a precursor, an ion-exchange technique was used to modify the produced zeolite. After being dried and heated to 60°C for 24 hours, the zeolite was mixed with powder in a 1.0 M potassium acetate solution at a ratio of 1:10. To get the ion-exchanged zeolite, the obtained mixture was periodically rinsed with distilled water, dried in the air at 110°C, and then calcined at 500°C for two hours. For the transesterification of mustard oil, the catalysts AZ-X, AZ-KX, UZ-KX, and UZ-X, were chosen due to their higher surface area in comparison to zeolite A. All tests were conducted in triplicate, and samples were taken at 1-hour intervals utilizing a 3-wt.% catalyst load, a 6:1 molar ratio of methanol to oil, and a 65°C reaction temperature. Following a 7-hour response, the conversion sequence was AZ-KX (72.2%) < UZ-KX (65.7%) < AZ-X (53.7%) < UZ-X (41.9%). Remarkably, when compared with other zeolites, acid-treated ion-exchanged zeolite showed higher conversion. This may be because ion-exchanged zeolites with components possessing a more electropositive nature ($K^+$) from parent zeolites have stronger bases (Babajide, Musyoka, Petrik, & Ameer, 2012; Xie & Huang, 2006).

### 9.3.2 Adsorption experiment

In a study, Belova (2019) employed the zeolite Yagodnisky deposit of the Kamchatka region to remove several heavy metals like $Co^{2+}$, $Fe^{2+}$, $Cu^{2+}$, and $Ni^{2+}$ from the contaminated water. The maximum adsorption capacity for the targeted metal was found to be 0.020 mg-equ/L for $Ni^{2+}$, 0.021 mg-equ/L for $Fe^{2+}$, 0.011 mg-equ/L for $Co^{2+}$, and 0.023 mg-equ/L for $Cu^{2+}$, respectively. The findings also reported that zeolite's sorption capacity rises as the concentration of metal ions rises. The following are the heavy metal ions in ascending order of sorption capacity: $Fe^{2+} > Ni^{2+} > Co^{2+} > Cu^{2+}$.

In another study, Samanta, Das, Mondal, Bora, et al. (2022) prepared Na-X and Na-A-type zeolite minerals using steel industry by-products such as silica and alumina sources. The adsorption study was focused on the methylene blue (MB) adsorption from the synthetic water with varying MB concentrations. The findings show the effective adsorption removal as well as the capacity of the as-synthesized zeolite materials. The MB adsorption capacity and removal efficacy were found to be 25.30 and 23.57 mg/g and 98.13% and 94.47%, respectively. The study also proposes that the LD-slag-derived zeolite minerals could be utilized for industrial wastewater treatment to extract heavy metals and water softening applications that may provide economic and environmental sustainability (Changmai, Banerjee, Nahar, & Purkait, 2018; Duarah et al., 2022; Taghizadeh et al., 2013).

### 9.3.3 Biomedical applications

Sandomierski, Zielińska, and Voelkel (2020) examined the use of zeolites to transport bisphosphonates, an antiosteoporotic medication. It was established that sodium risedronate, the medication that was used, has a greater affinity for calcium-exchanged Na-A zeolite than for Na-X-type zeolite. This was caused by the ionic exchange in Na-A zeolite being more efficient than in X-type zeolite. This finding prompted more investigation into the drug's structure and the kind of ion employed in the exchange mechanism. There were two types of bisphosphonates used: sodium risedronate and zoledronic acid. Risedronate exhibited the maximum level of adsorption on zinc-modified zeolite type-X, whereas zoledronic acid was shown to be most effective on magnesium-modified Na-X zeolite, according to the modification of zeolite with zinc, magnesium, and calcium ions. It was established that distinct adsorption outcomes might be seen based on the metal ion used. Zn-containing zeolites have been shown to act similarly to ZIF-8 when it comes to the release of medications that include nitrogen. This study demonstrated the versatility of zeolites as a therapeutic option for osteoporosis, mostly because of their accessibility, low cost, and ease of modification. The fact that large quantities of the medications are delivered in a short amount of time (the 100 hours analyzed) may be one of the research's shortcomings. It is well known that bisphosphonates taken orally have a limited bioavailability (Soares et al., 2016). Furthermore, the duration of this type of treatment necessitates the long-term administration of medication, which may intensify and multiply the unpleasant effects of the drug. Zeolites and biopolymers working together can also be useful in the creation of scaffolds that may be used to treat osteoporotic fractures. It was demonstrated that the chitosan-zeolite composite, as opposed to pure zeolite, enabled the release of significantly smaller dosages of risedronate over an extended period (Sandomierski, Adamska, Ratajczak, & Voelkel, 2022).

One possible use for the risedronate-ZIF-8 nanocomposite drug delivery system is the local therapy of bone defects (Cheng et al., 2020). To treat individuals with prospective or severe osteoporosis after an accident, surgical procedures are frequently needed to fix shattered bones. To provide customized bone grafts that support healing and appropriate osteogenesis for bone rebuilding, zeolite, and ZIF materials may be employed. Rats' application of chitosan-ZIF-8 hydrogel modified with catechol demonstrated much faster bone healing (Fig. 9.4) (Al-Baadani et al., 2022). Chen, Gao, Zhou, Wang, and Zhou (2022) examined the application of a biomaterial containing ZIF-8 as an alendronate carrier in another study. The drug's therapeutic efficacy may be enhanced by the ZIF-8 material's ability to deliver the medication steadily and under control, according to the research findings. Furthermore, this substance is noncytotoxic and, in hydrogel form, may find application in the surgical management of osteoporotic fractures. The combination of surface-decorated alendronate and encapsulated simvastatin with ZIF-8/gelatin was suggested by Wang et al. (2017). It has been demonstrated that simvastatin effectively treats osteoporosis (Tao et al., 2020), and the ZIF-8/gelatin surface enhanced with alendronate facilitates simple bone binding and more effective treatment. Other bioactive substances, other than bisphosphonates, might be employed in conjunction with one another to treat osteoporosis.

The aforementioned study presents the utilization of various zeolites in biomedical applications. It is seen that zeolite-composite-like materials possess

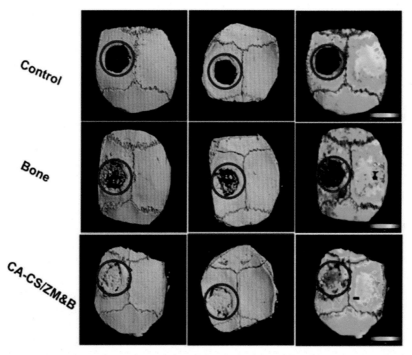

**FIGURE 9.4    Bone regeneration by zeolite ZIF-8.** Micro-CT demonstrating bone regeneration before and eight weeks following the implantation of the ZIF-8/bone powder material (ZM/B)-modified catechol/chitosan (CA/CS). *From Liu, Y., Zhu, Z., Pei, X., Zhang, X., Cheng, X., Hu, S., Gao, X., Wang, J., Chen, J., & Wan, Q. (2020). ZIF-8-modified multifunctional bone-adhesive hydrogels promoting angiogenesis and osteogenesis for bone regeneration.* ACS Applied Materials and Interfaces, 12(33), 36978—36995. https://doi.org/10.1021/acsami.0c12090.

a significant role in osteoporosis, drug delivery systems, bone healing, wound healing, and so on. Different commercially available zeolite and naturally occurring zeolites are being utilized for various medicinal purposes. The utilization of waste-derived zeolite in the biomedical sector has not been explored deeply, the reason could be the impurity of zeolite minerals as the silica and alumina wastes comprised other unwanted inorganic or hazardous constituents.

## 9.4   Summary

Zeolites are crystalline aluminosilicate minerals with distinct porosity architectures that are widely used in the biomedical industry, especially for adsorption procedures. Commercially accessible zeolites have been used extensively because of their remarkable selectivity and adsorption capabilities. They are generally produced by controlled hydrothermal procedures. They are useful in a variety of biological applications because of their characteristics. The comparative analysis of zeolite 13X and 4A-type zeolite suggests that the high BET surface area was found for synthetic Na-4A-type zeolite (324.584 m$^2$/g) rather than commercial zeolite type-4A. But, for zeolite type-13X the commercial zeolite shows a high surface area as compared to the synthesized zeolite. Moreover, a clear and well-defined zeolite crystal structure was seen for marketable zeolites namely Na-4A and 13X zeolite. Adsorption assessment reveals that commercial and waste-derived zeolite has considerable adsorptive affection toward the positively charged heavy metals and dye.

Commercial zeolites have been used in the biomedical environment for tissue engineering, wound healing, and drug delivery systems. Their large surface area and distinct pore diameters enable regulated medication release, enhancing therapeutic results and reducing adverse effects. Furthermore, because of their absorbent and antibacterial qualities, commercial zeolites have shown effectiveness in wound dressings, accelerating the healing process. They act as supports for cell proliferation and differentiation in tissue engineering, providing a framework for tissue regeneration. Conversely, synthetic zeolites are designed to possess particular characteristics that are necessary for specific biological uses. These specially-made zeolites have improved adsorption capacities for certain ions or biomolecules, which makes them indispensable for therapeutic and diagnostic applications. Synthetic zeolites, for instance, have been used to aid in detoxification procedures by eliminating heavy metals or pollutants from biological fluids. They also have a significant impact on hemodialysis processes, helping to effectively remove waste from the blood. Thus, based on the findings, it is anticipated that waste-derived zeolite will have a wide range of uses as long as its purity is comparable to that of commercially grade zeolite minerals.

# References

Al-Ani, A., Mordvinova, N. E., Lebedev, O. I., Khodakov, A. Y., & Zholobenko, V. (2019). Ion-exchanged zeolite P as a nanostructured catalyst for biodiesel production. *Energy Reports, 5*, 357−363. Available from http://www.journals.elsevier.com/energy-reports/, 10.1016/j.egyr.2019.03.003.

Al-Baadani, M. A., Xu, L., Hii Ru Yie, K., Sun, A., Gao, X., Cai, K., . . . Ma, P. (2022). In situ preparation of alendronate-loaded ZIF-8 nanoparticles on electrospun nanofibers for accelerating early osteogenesis in osteoporosis. *Materials and Design, 217*. Available from https://www. journals.elsevier.com/materials-and-design, https://doi.org/10.1016/j.matdes.2022.110596.

Anweshan, P. P., Das, S., Dhara, M. K., & Purkait. (2023). *Nanosensors in food science and technology. Advances in smart nanomaterials and their applications* (pp. 247−272). India: Elsevier. Available from https://www.sciencedirect.com/book/9780323995467, https://doi. org/10.1016/B978-0-323-99546-7.00015-X.

Asselman, K., Kirschhock, C., & Breynaert, E. (2023). Illuminating the black box: A perspective on zeolite crystallization in inorganic media. *Accounts of Chemical Research, 56*(18), 2391−2402. Available from https://doi.org/10.1021/acs.accounts.3c00269.

Babajide, O., Musyoka, N., Petrik, L., & Ameer, F. (2012). Novel zeolite Na-X synthesized from fly ash as a heterogeneous catalyst in biodiesel production. *Catalysis Today, 190*(1), 54−60. Available from https://doi.org/10.1016/j.cattod.2012.040.044.

Belova, T. P. (2019). Adsorption of heavy metal ions (Cu2 + , Ni2 + , Co2 + and Fe2 + ) from aqueous solutions by natural zeolite. *Heliyon, 5*(9), e02320. Available from https://doi.org/ 10.1016/j.heliyon.2019.e02320.

Bharti, M., Das, P. P., & Purkait, M. K. (2023). A review on the treatment of water and wastewater by electrocoagulation process: advances and emerging applications. *Journal of Environmental Chemical Engineering, 11*, 111558. Available from https://doi.org/10.1016/j. jece.2023.111558.

Bulasara, V. K., Thakuria, H., Uppaluri, R., & Purkait, M. K. (2011). Effect of process parameters on electroless plating and nickel-ceramic composite membrane characteristics. *Desalination, 268*(1−3), 195−203. Available from https://doi.org/10.1016/j.desal.2010. 100.025.

Chakraborty, S., Das, P. P., & Mondal, P. (2023). Elsevier BV *Recent advances in membrane technology for the recovery and reuse of valuable resources* (pp. 695−719). Elsevier BV. Available from https://doi.org/10.1016/b978-0-323-95327-6.00028-2.

Changmai, M., Banerjee, P., Nahar, K., & Purkait, M. K. (2018). A novel adsorbent from carrot, tomato and polyethylene terephthalate waste as a potential adsorbent for Co (II) from aqueous solution: Kinetic and equilibrium studies. *Journal of Environmental Chemical Engineering, 6*(1), 246−257. Available from https://doi.org/10.1016/j.jece.2017.120.009, http://www.journals.elsevier.com/journal-of-environmental-chemical-engineering/.

Changmai, M., Das, P. P., Mondal, P., Pasawan, M., Sinha, A., Biswas, P., . . . Purkait, M. K. (2022). Hybrid electrocoagulation−microfiltration technique for treatment of nanofiltration rejected steel industry effluent. *International Journal of Environmental Analytical Chemistry, 102*(1), 62−83. Available from https://doi.org/10.1080/03067319.2020.1715381, http://www. tandf.co.uk/journals/titles/03067319.asp.

Chen, Z. Y., Gao, S., Zhou, R. B., Wang, R. D., & Zhou, F. (2022). Dual-crosslinked networks of superior stretchability and toughness polyacrylamide-carboxymethylcellulose hydrogel for delivery of alendronate. *Materials and Design, 217*. Available from https://www.journals. elsevier.com/materials-and-design, https://doi.org/10.1016/j.matdes.2022.110627.

Cheng, X., Zhu, Z., Liu, Y., Xue, Y., Gao, X., Wang, J., ... Wan, Q. (2020). Zeolitic imidazolate framework-8 encapsulating risedronate synergistically enhances osteogenic and antiresorptive properties for bone regeneration. *American Chemical Society, China ACS Biomaterials Science and Engineering, 6*(4), 2186−2197. Available from https://doi.org/10.1021/acsbiomaterials.0c00195, http://pubs.acs.org/journal/abseba.

Das, P. P., Mondal, P., Anweshan., Sinha, A., Biswas, P., Sarkar, S., & Purkait, M. K. (2021). Treatment of steel plant generated biological oxidation treated (BOT) wastewater by hybrid process. *Separation and Purification Technology, 258.* Available from http://www.journals.elsevier.com/separation-and-purification-technology/, https://doi.org/10.1016/j.seppur.2020.118013.

Das, P. P., Anweshan, A., & Purkait, M. K. (2021). Treatment of cold rolling mill (CRM) effluent of steel industry. *Separation and Purification Technology, 274.* Available from http://www.journals.elsevier.com/separation-and-purification-technology/, https://doi.org/10.1016/j.seppur.2021.119083.

Das, P. P., Sharma, M., & Purkait, M. K. (2022). Recent progress on electrocoagulation process for wastewater treatment: A review. *Separation and Purification Technology, 292.* Available from http://www.journals.elsevier.com/separation-and-purification-technology/, https://doi.org/10.1016/j.seppur.2022.121058.

Das, P. P., Sontakke, A. D., Samanta, N. S., & Purkait, M. K. (2023). *Emerging contaminants in wastewater: Eco-toxicity and sustainability assessment. Industrial wastewater reuse: Applications, prospects and challenges* (pp. 63−87). India: Springer Nature. Available from https://link.springer.com/book/10.1007/978-981-99-2489-9, https://doi.org/10.1007/978-981-99-2489-9_4.

Das, P. P., Sontakke, A. D., & Purkait, M. K. (2023). *Rice straw for biofuel production. Green approach to alternative fuel for a sustainable future* (pp. 153−166). India: Elsevier. Available from https://www.sciencedirect.com/book/9780128243183, https://doi.org/10.1016/B978-0-12-824318-3.00034-5.

Das, P. P., Samanta, N. S., Dhara, S., & Purkait, M. K. (2023). *Biofuel production from algal biomass. Green approach to alternative fuel for a sustainable future* (pp. 167−179). India: Elsevier. Available from https://www.sciencedirect.com/book/9780128243183, https://doi.org/10.1016/B978-0-12-824318-3.00009-6.

Das, P. P., & Mondal, P. (2023). *Membrane-assisted potable water reuses applications: benefits and drawbacks* (pp. 289−309). Elsevier BV. Available from https://doi.org/10.1016/b978-0-323-99344-9.00014-1.

Das, P. P., Deepti., & Purkait, M. K. (2023). *Industrial wastewater to biohydrogen production via potential bio-refinery route* (pp. 159−179). Springer Science and Business Media LLC. Available from https://doi.org/10.1007/978-3-031-20822-5_8.

Das, P. P., Anweshan., Mondal, P., Sinha, A., Biswas, P., Sarkar, S., & Purkait, M. K. (2021). Integrated ozonation assisted electrocoagulation process for the removal of cyanide from steel industry wastewater. *Chemosphere, 263,* 128370. Available from https://doi.org/10.1016/j.chemosphere.2020.128370.

Das, P. P., Mondal, P., & Purkait, M. K. (2022). *Recent advances in synthesis of iron nanoparticles via green route and their application in biofuel production* (pp. 79−104). Springer Science and Business Media LLC. Available from https://doi.org/10.1007/978-981-16-9356-4_4.

Das, P. P., Dhara, S., & Purkait, M. K. (2023). The Anaerobic Ammonium Oxidation Process: Inhibition, Challenges and Opportunities. In M. P. Shah (Ed.), *Ammonia Oxidizing Bacteria: Applications in Industrial Wastewater Treatment* (pp. 56−82). Royal Society of Chemistry. Available from https://doi.org/10.1039/BK9781837671960-00056.

Das, P. P., Dhara, S., & Purkait, M. K. (2024). Ozone-based oxidation processes for the removal of pharmaceutical products from wastewater. In M. P. Shah, & P. Ghosh (Eds.), *Development in Wastewater Treatment Research and Processes* (pp. 287−308). Elsevier. Available from https://doi.org/10.1016/B978-0-443-19207-4.00003-3.

Dhara, S., Das, P. P., Uppaluri, R., & Purkait, M. K. (2023). *Phosphorus recovery from municipal wastewater treatment plants. Development in wastewater treatment research and processes: Advances in industrial wastewater treatment technologies: Removal of contaminants and recovery of resources* (pp. 49−72). India: Elsevier. Available from https://www.sciencedirect.com/book/9780323956840, https://doi.org/10.1016/B978-0-323-95684-0.00014-2.

Dhara, S., Shekhar Samanta, N., Das, P. P., Uppaluri, R. V. S., & Purkait, M. K. (2023). Ravenna grass-extracted alkaline lignin-based polysulfone mixed matrix membrane (MMM) for aqueous Cr(VI) removal. *ACS Applied Polymer Materials*, *5*(8), 6399−6411. Available from https://doi.org/10.1021/acsapm.3c00999, http://pubs.acs.org/journal/aapmcd.

Dhara, S., Das, P. P., Uppaluri, R., & Purkait, M. K. (2023). *Biological approach for energy self-sufficiency of municipal wastewater treatment plants* (pp. 235−260). Elsevier BV. Available from https://doi.org/10.1016/b978-0-323-99348-7.00006-0.

Dhara, S., Samanta, N. S., Uppaluri, R., & Purkait, M. K. (2023). High-purity alkaline lignin extraction from *Saccharum ravannae* and optimization of lignin recovery through response surface methodology. *International Journal of Biological Macromolecules*, *234*, 123594. Available from https://doi.org/10.1016/j.ijbiomac.2023.123594.

Duarah, P., Haldar, D., Patel, A. K., Dong, C. D., Singhania, R. R., & Purkait, M. K. (2022). A review on global perspectives of sustainable development in bioenergy generation. *Bioresource Technology*, *348*. Available from http://www.elsevier.com/locate/biortech, https://doi.org/10.1016/j.biortech.2022.126791.

Duarah, P., Das, P. P., & Purkait, M. K. (2023). *Technological advancement in the synthesis and application of nanocatalysts. Advanced application of nanotechnology to industrial wastewater* (pp. 191−214). India: Springer Nature. Available from https://link.springer.com/book/10.1007/978-981-99-3292-4, 10.1007/978-981-99-3292-4_10.

Khan, I., Saeed, K., & Khan, I. (2019). Nanoparticles: Properties, applications and toxicities. *Arabian Journal of Chemistry*, *12*(7), 908−931. Available from https://doi.org/10.1016/j.arabjc.2017.050.011, http://colleges.ksu.edu.sa/Arabic%20Colleges/CollegeOfScience/ChemicalDept/AJC/default.aspx, http://www.sciencedirect.com/science/journal/18785352.

Kianfar, E., Hajimirzaee, S., mousavian, S., & Mehr, A. S. (2020). Zeolite-based catalysts for methanol to gasoline process: A review. *Microchemical Journal*, *156*, 104822. Available from https://doi.org/10.1016/j.microc.2020.104822.

Li, J., Corma, A., & Yu, J. (2015). Synthesis of new zeolite structures. *Chemical Society Reviews*, *44*(20), 7112−7127. Available from https://doi.org/10.1039/c5cs00023h, http://pubs.rsc.org/en/journals/journal/cs.

Manrique, C., Guzmán, A., Pérez-Pariente, J., Márquez-Álvarez, C., & Echavarría, A. (2016). Effect of synthesis conditions on zeolite beta properties and its performance in vacuum gas oil hydrocracking activity. *Microporous and Mesoporous Materials*, *234*, 347−360. Available from http://www.elsevier.com/inca/publications/store/6/0/0/7/6/0, https://doi.org/10.1016/j.micromeso.2016.07.017.

Mondal, P., Samanta, N. S., Meghnani, V., & Purkait, M. K. (2019). Selective glucose permeability in presence of various salts through tunable pore size of pH responsive PVDF-co-HFP membrane. *Separation and Purification Technology*, *221*, 249−260. Available from http://www.journals.elsevier.com/separation-and-purification-technology/, https://doi.org/10.1016/j.seppur.2019.04.001.

Mondal, P., Samanta, N. S., Kumar, A., & Purkait, M. K. (2020). Recovery of $H_2SO_4$ from wastewater in the presence of NaCl and $KHCO_3$ through pH responsive polysulfone membrane: Optimization approach. *Polymer Testing*, *86*. Available from https://www.journals. elsevier.com/polymer-testing, https://doi.org/10.1016/j.polymertesting.2020.106463.

Muciño, G. E. G., Romero, R., García-Orozco, I., Serrano, A. R., Jiménez, R. B., & Natividad, R. (2016). Deactivation study of $K_2O$/NaX and $Na_2O$/NaX catalysts for biodiesel production. *Catalysis Today*, *271*, 220−226. Available from http://www.sciencedirect.com/science/journal/09205861, https://doi.org/10.1016/j.cattod.2015.09.054.

Nosrati, R., Olad, A., & Nofouzi, K. (2015). A self-cleaning coating based on commercial grade polyacrylic latex modified by $TiO_2$/Ag-exchanged-zeolite-A nanocomposite. Elsevier B.V., Iran. *Applied Surface Science*, *346*, 543−553. Available from https://doi.org/10.1016/j. apsusc.2015.040.056, http://www.journals.elsevier.com/applied-surface-science/.

Pérez-Botella, E., Valencia, S., & Rey, F. (2022). Zeolites in adsorption processes: State of the art and future prospects. *Chemical Reviews*, *122*(24), 17647−17695. Available from https:// doi.org/10.1021/acs.chemrev.2c00140, http://pubs.acs.org/journal/chreay.

Purkait, M. K., DasGupta, S., & De, S. (2005). Micellar enhanced ultrafiltration of phenolic derivatives from their mixtures. *Journal of Colloid and Interface Science*, *285*(1), 395−402. Available from https://doi.org/10.1016/j.jcis.2004.110.036.

Samanta, N. S., Anweshan., & Purkait, M. K. (2023). *Techniques in removal of organics and emerging contaminants from wastewater for water reuse application. Development in wastewater treatment research and processes: Advances in industrial wastewater treatment technologies: Removal of contaminants and recovery of resources* (pp. 73−96). India: Elsevier. Available from https://www.sciencedirect.com/book/9780323956840, https://doi. org/10.1016/B978-0-323-95684-0.00009-9.

Samanta, N. S., Banerjee, S., Mondal, P., Anweshan., Bora, U., & Purkait, M. K. (2021). Preparation and characterization of zeolite from waste Linz-Donawitz (LD) process slag of steel industry for removal of Fe3 + from drinking water. *Advanced Powder Technology*, *32*(9), 3372−3387. Available from https://doi.org/10.1016/j.apt.2021.07.023, http://www.elsevier.com.

Samanta, N. S., Das, P. P., Mondal, P., Changmai, M., & Purkait, M. K. (2022). Critical review on the synthesis and advancement of industrial and biomass waste-based zeolites and their applications in gas adsorption and biomedical studies. *Journal of the Indian Chemical Society*, *99*(11). Available from https://doi.org/10.1016/j.jics.2022.100761, https://www. sciencedirect.com/journal/journal-of-the-indian-chemical-society.

Samanta, N. S., Das, P. P., Mondal, P., Bora, U., & Purkait, M. K. (2022). Physico-chemical and adsorption study of hydrothermally treated zeolite A and FAU-type zeolite X prepared from LD (Linz−Donawitz) slag of the steel industry. *International Journal of Environmental Analytical Chemistry*, *10290397*. Available from http://www.tandf.co.uk/journals/titles/ 03067319.asp, https://doi.org/10.1080/03067319.2022.2079082.

Samanta, N. S., Anweshan., Mondal, P., Bora, U., & Purkait, M. K. (2023). Synthesis of precipitated calcium carbonate from LD-slag using $CO_2$. *Materials Today Communications*, *36*. Available from http://www.journals.elsevier.com/materials-today-communications/, https:// doi.org/10.1016/j.mtcomm.2023.106588.

Samanta, N. S., Mondal, P., & Purkait, M. K. (2023). *Nanofiltration technique for the treatment of industrial wastewater. Advanced application of nanotechnology to industrial wastewater* (pp. 165−190). India: Springer Nature. Available from https://link.springer.com/book/ 10.1007/978-981-99-3292-4, https://doi.org/10.1007/978-981-99-3292-4_9.

Sandomierski, M., Adamska, K., Ratajczak, M., & Voelkel, A. (2022). Chitosan - zeolite scaffold as a potential biomaterial in the controlled release of drugs for osteoporosis.

*International Journal of Biological Macromolecules, 223,* 812−820. Available from http://www.elsevier.com/locate/ijbiomac, https://doi.org/10.1016/j.ijbiomac.2022.11.071.

Sandomierski, M., Zielińska, M., & Voelkel, A. (2020). Calcium zeolites as intelligent carriers in controlled release of bisphosphonates. *International Journal of Pharmaceutics, 578,* 119117. Available from https://doi.org/10.1016/j.ijpharm.2020.119117.

Setiawan, T., Alvin Setiadi, G., & Mukti, R. R. (2023). Deployable and retrievable 3D-printed zeolite−polymer composites for wastewater treatment: A review. *Industrial & Engineering Chemistry Research.* Available from https://doi.org/10.1021/acs.iecr.3c02317.

Sharma, M., Das, P. P., Sood, T., Chakraborty, A., & Purkait, M. K. (2021). Ameliorated polyvinylidene fluoride based proton exchange membrane impregnated with graphene oxide, and cellulose acetate obtained from sugarcane bagasse for application in microbial fuel cell. *Journal of Environmental Chemical Engineering, 9*(6). Available from http://www.journals.elsevier.com/journal-of-environmental-chemical-engineering/, https://doi.org/10.1016/j.jece.2021.106681.

Sharma, M., Das, P. P., Sood, T., Chakraborty, A., & Purkait, M. K. (2022). Reduced graphene oxide incorporated polyvinylidene fluoride/cellulose acetate proton exchange membrane for energy extraction using microbial fuel cells. *Journal of Electroanalytical Chemistry, 907.* Available from https://www.journals.elsevier.com/journal-of-electroanalytical-chemistry, https://doi.org/10.1016/j.jelechem.2021.115890.

Sharma, M., Das, P. P., Chakraborty, A., & Purkait, M. K. (2022). Clean energy from salinity gradients using pressure retarded osmosis and reverse electrodialysis: A review. *Sustainable Energy Technologies and Assessments, 49.* Available from http://www.journals.elsevier.com/sustainable-energy-technologies-and-assessments, https://doi.org/10.1016/j.seta.2021.101687.

Sharma, M., Das, P. P., & Purkait, M. K. (2023). *Energy storage properties of nanomaterials. Advances in smart nanomaterials and their applications* (pp. 337−350). India: Elsevier. Available from https://www.sciencedirect.com/book/9780323995467, https://doi.org/10.1016/B978-0-323-99546-7.00005-7.

Sharma, M., Samanta, N. S., Chakraborty, A., & Purkait, M. K. (2023). *Simultaneous treatment of industrial wastewater and resource recovery using microbial fuel cell* (pp. 621−637). Elsevier BV. Available from https://doi.org/10.1016/b978-0-323-95327-6.00002-6.

Sharma, M., Das, P. P., Chakraborty, A., & Purkait, M. K. (2023). *Extraction of clean energy from industrial wastewater using bioelectrochemical process* (pp. 601−620). Elsevier BV. Available from https://doi.org/10.1016/b978-0-323-95327-6.00003-8.

Shekhar Samanta, N., Das, P. P., Sharma, M., & Purkait, M. K. (2023). *Recycle of water treatment plant sludge and its utilization for wastewater treatment* (pp. 239−264). Elsevier BV. Available from https://doi.org/10.1016/b978-0-323-99344-9.00010-4.

Shekhar Samanta, N., Das, P. P., Dhara, S., & Purkait, M. K. (2023). An overview of precious metal recovery from steel industry slag: Recovery strategy and utilization. *Industrial and Engineering Chemistry Research, 62*(23), 9006−9031. Available from https://doi.org/10.1021/acs.iecr.3c00604, http://pubs.acs.org/journal/iecred.

Soares, A. P., do Espírito Santo, R. F., Line, S. R. P., Pinto, M. d. G. F., Santos, P. d. M., Toralles, M. B. P., & do Espírito Santo, A. R. (2016). Bisphosphonates: Pharmacokinetics, bioavailability, mechanisms of action, clinical applications in children, and effects on tooth development. *Environmental Toxicology and Pharmacology, 42,* 212−217. Available from http://www.elsevier.com/locate/etap, https://doi.org/10.1016/j.etap.2016.01.015.

Sontakke, A. D., Das, P. P., Mondal, P., & Purkait, M. K. (2022). Thin-film composite nanofiltration hollow fiber membranes toward textile industry effluent treatment and environmental remediation applications: Review. *Emergent Materials, 5*(5), 1409−1427. Available from https://doi.org/10.1007/s42247-021-00261-y, https://www.springer.com/journal/42247.

Sontakke, A. D., Deepti., Samanta, N. S., & Purkait, M. K. (2023). *Smart nanomaterials in the medical industry. Advances in smart nanomaterials and their applications* (pp. 23−50). India: Elsevier. Available from https://www.sciencedirect.com/book/9780323995467, https://doi.org/10.1016/B978-0-323-99546-7.00025-2.

Sowunmi, A. R., Folayan, C. O., Anafi, F. O., Ajayi, O. A., Omisanya, N. O., Obada, D. O., & Dodoo-Arhin, D. (2018). Dataset on the comparison of synthesized and commercial zeolites for potential solar adsorption refrigerating system. *Data in Brief, 20,* 90−95. Available from https://www.journals.elsevier.com/data-in-brief, https://doi.org/10.1016/j.dib.2018.07.040.

Taghizadeh, F., Ghaedi, M., Kamali, K., Sharifpour, E., Sahraie, R., & Purkait, M. K. (2013). Comparison of nickel and/or zinc selenide nanoparticle loaded on activated carbon as efficient adsorbents for kinetic and equilibrium study of removal of Arsenazo (III) dye. *Powder Technology, 245,* 217−226. Available from https://doi.org/10.1016/j.powtec.2013.040.020.

Tangy, A., Pulidindi, I. N., & Gedanken, A. (2016). SiO2 beads decorated with SrO nanoparticles for biodiesel production from waste cooking oil using microwave irradiation. *Energy & Fuels, 30*(4), 3151−3160. Available from https://doi.org/10.1021/acs.energyfuels.6b00256.

Tao, S., Chen, S.-q., Zhou, W.-t., Yu, F.-y., Bao, L., Qiu, G.-x., ... Yuan, H. (2020). A novel biocompatible, simvastatin-loaded, bone-targeting lipid nanocarrier for treating osteoporosis more effectively. *RSC Advances, 10*(35), 20445−20459. Available from https://doi.org/10.1039/d0ra00685h.

Volli, V., & Purkait, M. K. (2015). Selective preparation of zeolite X and A from flyash and its use as catalyst for biodiesel production. *Journal of Hazardous Materials, 297,* 101−111. Available from http://www.elsevier.com/locate/jhazmat, https://doi.org/10.1016/j.jhazmat.2015.04.066.

Wang, X., Miao, D., Liang, X., Liang, J., Zhang, C., Yang, J., ... Sun, H. (2017). Nanocapsules engineered from polyhedral ZIF-8 templates for bone-targeted hydrophobic drug delivery. *Biomaterials Science, 5*(4), 658−662. Available from https://doi.org/10.1039/c6bm00915h, http://www.rsc.org/publishing/journals/bm/.

Wu, Q., Luan, H., & Xiao, F. S. (2022). Targeted synthesis of zeolites from calculated interaction between zeolite structure and organic template. *National Science Review, 9*(9). Available from https://doi.org/10.1093/nsr/nwac023, http://nsr.oxfordjournals.org/.

Xie, W., & Huang, X. (2006). Synthesis of biodiesel from soybean oil using heterogeneous KF/ZnO catalyst. *Catalysis Letters, 107*(1−2), 53−59. Available from https://doi.org/10.1007/s10562-005-9731-0.

Chapter 10

# Advancement in zeolite regeneration techniques

## 10.1 Introduction

A significant number of individuals reside in densely populated informal settlements characterized by inadequate or nonexistent access to interconnected sewer systems or sewage treatment facilities (Mondal, Samanta, Kumar, & Kumar Purkait, 2020). The absence of adequate access to sanitary infrastructure that ensures safety and proper management contributes to the prevalence of open defecation. The enhancement of global accessibility to sanitation infrastructure is thus intricately linked to the advancement of public health, hygiene, and water availability for a population exceeding 4.2 billion individuals around the globe. Non-sewered sanitation systems (NSSSs) refer to a decentralized approach to wastewater treatment (Baum, Luh, & Bartram, 2013; Castro, Shyu, Xaba, Bair, & Yeh, 2021; Starkl, Brunner, Feil, & Hauser, 2015). These systems are designed to collect, convey, and treat wastewater directly at the source, eliminating the need for centralized sewer lines and the associated expenses and disturbances. NSSSs are increasingly being recognized as viable alternatives for enhancing wastewater management, particularly in developing nations that are undergoing rapid urbanization, facing constraints in water availability, and grappling with insufficient sanitation infrastructure. The effluent water quality parameters specified by ISO 30500 encompass several key aspects, including chemical oxygen demand (COD), total suspended solids (TSSs), pH levels, presence of pathogens, total nitrogen (TN) content, and total phosphorus (TP) content (Mondal, Samanta, Meghnani, & Purkait, 2019; N. S. Samanta, Mondal, & Purkait, 2023). Although some field-tested NSSSs have successfully satisfied the effluent standards for COD and TSS, continuously achieving the nutrient needs outlined in ISO 30500 and ensuring continued use has proven to be a hard task. The TN content of blackwater consists of organic nitrogen, nitrate ($NO_3-N$), nitrite ($NO_2-N$), and ammonium ($NH_4^+-N$) (Castro, Goodwill, Rogers, Henderson, & Butler, 2014; Sahondo et al., 2020). However, it is worth noting that ammonium ($NH_4^+-N$) generally constitutes a substantial proportion of the TN content. In the absence of dilution from graywater and other sources, the nutrient concentrations in blackwater (BW), which refers

Waste-based Zeolite. DOI: https://doi.org/10.1016/B978-0-443-22316-7.00010-3
**247**

to wastewater directly discharged from toilets, can reach levels around 10 times greater than those typically recorded at municipal wastewater treatment plants. The design of NSSS should prioritize robustness to effectively manage the significant variations in nutrient loading events resulting from intermittent usage and dormancy periods (Knerr, Rechenburg, Kistemann, & Schmitt, 2011). Simultaneously, it is crucial to ensure that the construction and operation of NSSS remain simple. In the context of $NH_4^+-N$ removal, both aerobic and anoxic biological nutrient removal (BNR) techniques have frequently been utilized in centralized treatment facilities. In the context of expansive environments, the utilization of BNR demonstrates notable efficacy. This is primarily attributed to the ability of operators to meticulously monitor and regulate crucial parameters, including dissolved oxygen (DO), pH, and oxidation/reduction potential (ORP) (Lackner et al., 2014; Vlaeminck et al., 2009). Moreover, the operational expenses associated with BNR implementation in such settings are typically not prohibitive. For small-scale NSSS, aerobic and anoxic BNR processes have limited use due to their high energy demand for operation, the requirement of high precision control, the availability of dependable small-scale sensors for system monitoring, and the intermittent nature of NSSS (Varigala et al., 2020). In addition, it is worth noting that these BNR procedures generally do not effectively retrieve nitrogen for further reuse in downstream applications. The technological strategies employed in addressing the obstacles encountered by decentralized treatment systems have evolved to prioritize high-rate physical, chemical, and electrochemical treatment processes that can be efficiently implemented in the field with restricted process control (Bunce, Ndam, Ofiteru, Moore, & Graham, 2018). Emerging technologies for nonbiological treatment procedures in the context of NSSSs include air stripping, breakpoint chlorination, ion exchange, and membrane separation. In their study, Cid, Qu, and Hoffmann (2018) employed an electrochemical method to eliminate nitrogen by generating free chlorine and oxidizing ammonia to produce chloramines (Cid et al., 2018). The researchers observed a significant reduction in Total Kjeldahl nitrogen (TKN) levels, from an initial concentration of 43 mg/L down to 5 mg TKN/L, within a time frame of 3 hours. Although the method effectively eliminates nitrogen from wastewater, it is characterized by high energy consumption and lacks the capability to recover nitrogen for subsequent utilization.

The mineral known as zeolite, which is an aluminosilicate with a microporous structure, has been the subject of much research (Dhara, Samanta, Uppaluri, & Purkait, 2023; Samanta et al., 2021; Sharma, Samanta, et al., 2023; Sontakke et al., 2023). This is mostly due to its notable ability to exchange cations, particularly $NH_4^+$, and its effective regenerative properties (Alshameri et al., 2014; Wijesinghe et al., 2016). The zeolite framework consists of interconnected aluminosilicate tetrahedra arranged in layered channels, resulting in the formation of porous cavities capable of trapping ions and molecules (Kithome, Paul, Lavkulich, & Bomke, 1998). Therefore zeolites can also be used for the treatment of water and wastewater (Das, Samanta, Dhara, Purkait, & Shah, 2023; Das, Sontakke, Samanta, & Purkait, 2023; P.Das, Mondal, Sillanpää, Khadir, & Gurung, 2023; Samanta, Das, Dhara, & Purkait, 2023). In addition to zeolites, there are several other methods used for the treatment of water and wastewater, such as electrocoagulation (Changmai et al., 2022; Das, Anweshan, & Purkait, 2021; Das, Anweshan, Mondal, et al., 2021; Das, Dhara, & Purkait, 2023; Das, Mondal, et al., 2021; Das, Sharma, & Purkait, 2022; Das, Sontakke, Purkait, & Shah, 2023), membrane (Chakraborty et al., 2023; Das, Sontakke, Purkait, & Shah, 2023; Dhara, Shekhar Samanta, Das, Uppaluri, & Purkait, 2023; Sharma, Das, Sood, Chakraborty, & Purkait, 2021; Sharma, Das, Chakraborty, & Purkait, 2022; Sharma, Das, Sood, Chakraborty, & Purkait, 2022; Sontakke, Das, Mondal, & Purkait, 2022), adsorption (Chakraborty, Gautam, Das, & Hazarika, 2019; Das et al., 2023; Samanta, Das, & Mondal, 2022; Samanta, Das, Mondal, Changmai, & Purkait, 2022; Samanta et al., 2023; Samanta et al. 2023; Sharma, Das, Kumar, & Purkait, 2023), nanotechnology (Anweshan et al., 2023; Das, Duarah, Purkait, & Jafari, 2023; Das, Mondal, & Purkait, 2022; Duarah, Das, & Purkait, 2023; Sharma, Das, Purkait, Husen, & Siddiqi, 2023), and biological treatment (Das et al., 2023; Das, Deepti, & Purkait, 2023; Das, Sontakke, Purkait, et al., 2023; Dhara, Das, Uppaluri, Purkait, & Shah, 2023; Dhara, Das, Uppaluri, Purkait, Sillanpää, et al., 2023; Sharma, Das, Chakraborty, et al., 2023). Currently, there are more than 40 distinct varieties of naturally occurring zeolites, among which clinoptilolite stands out as the most prevalent on a global scale. The utilization of zeolite clinoptilolite in NSSSs offers several benefits. Firstly, it exhibits a remarkable capability to effectively manage elevated levels of $NH_4^+$ concentrations. Additionally, it possesses a substantial adsorption capacity, ranging from 3 to 30 mg of $NH_4^+-N$ per gram of zeolite. Furthermore, it can be regenerated for repeated usage, as demonstrated by a study that achieved up to 20 successful regeneration cycles. The pricing of zeolite is subject to variation based on factors such as kind, processing methods, and geographical location. However, on average, the cost of zeolite often falls within the range of \$85−\$160 per ton (Stefanakis, Akratos, Gikas, & Tsihrintzis, 2009). This affordability renders zeolite a financially viable technology for the purpose of ammonia recovery. The chemical regeneration process of zeolite as a strategy for $NH_4^+-N$ removal presents a significant difficulty. This procedure normally necessitates the utilization of substantial amounts of NaCl and water to eliminate sorbed cations from the surface of the zeolite. Moreover, a significant obstacle lies in the absence of consistency in the available literature about the frequency at which zeolite can undergo regeneration and be reused (Brandes, 1978). This variability is primarily influenced by the specific type of zeolite employed and the specific composition of ions found in the wastewater. A significant portion of the current research on regeneration has been carried out at the laboratory level, constraining its applicability to larger-scale systems working in natural environmental circumstances (Guo, Zeng, Li, & Park, 2007; Katsou, Malamis, Tzanoudaki, & Loizidou, 2011).

Numerous methodologies have been employed in the regeneration of swapped zeolites to attain a sustainable removal of $NH_4^+-N$. The adsorption of $NH_4^+-N$ from zeolite was efficiently facilitated by the NaCl solution, as demonstrated by Rahmani and Mahvi in 2004 (Rahmani, Mahvi, Mesdaghinia, & Nasseri, 2004). Additionally, the regeneration efficiency was further improved under alkaline conditions, as reported by Du, Liu, Cao, and Wang (2005). Nevertheless, the process of regenerating NaCl necessitates periodic replacement as a consequence of the buildup of $NH_4^+-N$ in the regenerant solution. This accumulation leads to secondary pollution, which in turn requires additional treatment (Lei et al., 2009). The utilization of a sodium hydroxide (NaOH) solution for alkaline regeneration effectively prevented the formation of ammonium−nitrogen ($NH_4^+-N$) by air stripping. However, this method was limited by its considerable expenses in terms of regenerant and the need for controlling gaseous pollution. In contrast to chemical methodologies, the biological regeneration technique has been devised for the purpose of eliminating $NH_4^+-N$ from zeolite by means of nitrification and/or denitrification processes. This strategy offers distinct benefits in terms of its environmentally sustainable nature and cost-effectiveness (Das, Dhara, & Purkait, 2024). To achieve conventional biological regeneration, several factors must be considered. These include the need for external alkalinity to regulate pH levels, the presence of a carbon source to facilitate denitrification, the maintenance of appropriate temperature conditions, and the allocation of sufficient reaction time (Jung, Chung, Shin, & Son, 2004). In recent years, researchers have utilized hypochlorite ions, which can be produced through the addition of NaClO or through electrochemical means, to oxidize adsorbed $NH_4^+$ to $N_2$. This approach has proven to be successful in achieving efficient and prompt regeneration of zeolites while avoiding any secondary contamination. Further investigation is required to ascertain the stability and adaptability of hypochlorite regeneration. This entails doing a comprehensive study on regeneration kinetics and identifying optimal conditions. Additionally, it is crucial to examine the impact of hypochlorite on the composition and textural features of regenerated zeolite (RZ) over an extended period of operation (Gendel & Lahav, 2013; Lahav, Schwartz, Nativ, & Gendel, 2013).

The main emphasis of this chapter centers on the various methodologies employed for the regeneration of zeolite. This study extensively examines and summarizes many aspects of chemical regeneration, biological regeneration, and electrochemical regeneration procedures. Additionally, this chapter also encompasses the obstacles and future suggestions pertaining to the regeneration of zeolite derived from various sources.

## 10.2  Regeneration process

### 10.2.1  Chemical regeneration

#### 10.2.1.1  Zeolite regeneration

The zeolites underwent regeneration using three distinct regenerants: a solution containing 5 g/L NaCl, a solution of NaClO with a molar ratio of hypochlorite to nitrogen ($ClO^-/N$) of 2.25, and a solution of NaClO−NaCl with the same aforementioned dosage. A total of six experiments were performed for each regenerant using a 250 mL beaker containing dry zeolites submerged in 200 mL of regenerant for durations of 2.5, 5, 10, 20, 30, and 40 minutes, respectively. Subsequently, the solid−liquid mixture underwent sedimentation, and the resulting liquid phase was subjected to filtration to determine the concentrations of $NH_4^+−N$, nitrate nitrogen $NO_3^-(−N)$, and TN. The RZ sample was introduced into a 250 mL beaker containing 50 mL of hydrochloric acid (HCl) with a pH range of 0.5−0.6. The purpose of this step was to facilitate the release of residual $NH_4^+−N$ from the RZ sample. The mixture was stirred for a duration of 2 hours, as described in the study conducted by Huang et al. (2015) (Huang, Yang, et al., 2015). The concentration of $NH_4^+−N$ that was released was determined by measuring samples that had been filtered using a membrane with a pore size of 0.45 mm. The zeolite regeneration efficiency (ZRE) is computed based on the Eq. (10.1).

$$ZRE = \left[1 - \frac{V_a AN_a}{V_w(AN_i - AN_e)}\right] \times 100\% \qquad (10.1)$$

where $AN_i$ and $AN_e$ represent the concentrations of $NH_4^+−N$ in the ammoniacal solution before and following zeolite adsorption, measured in milligrams per liter (mg/L), respectively. The variable "$V_w$" represents the volume of the ammoniacal solution, measured in liters (Ates & Akgül, 2016). "$AN_a$" denotes the concentration of $NH_4^+−N$ in the HCl solution, expressed in milligrams per liter, and is used to determine the remaining $NH_4^+−N$ after regeneration. Lastly, "$V_a$" represents the volume of the HCl solution, also measured in liters. The calculation of nitrogen removal efficiency (NRE) is performed following the zeolite adsorption regeneration process.

$$NRE = \left[1 - \frac{V_a AN_a + V_r TN_r}{V_w(AN_i - AN_e)}\right] \times 100\% \qquad (10.2)$$

The variable $TN_r$ represents the concentration of TN in the regenerant solution following the process of regeneration, measured in milligrams per liter (mg/L). Meanwhile, $V_a$ denotes the volume of the regenerant solution, expressed in liters (L) (Huang, Huang, Zhang, Jiang, & Ding, 2015). The disparity between NRE and ZRE lies in the proportion of $NH_4^+$ (ammonium) that undergoes conversion to $NO_3^-$ (nitrate) during the process of regeneration (Zhang et al., 2017). Fig. 10.1 shows the effect of regenerants on ZRE and nitrogen species after regeneration.

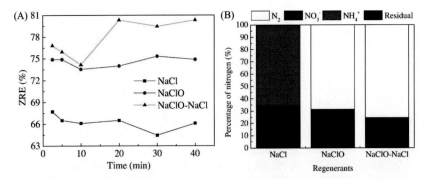

**FIGURE 10.1** Effects of regenerants on (A) ZRE and (B) nitrogen species after regeneration for natural zeolites . *From Zhang, W., Zhou, Z., An, Y., Du, S., Ruan, D., Zhao, C., Ren, N., & Tian, X. (2017). Chemosphere, 178, 565−572. https://doi.org/10.1016/j.chemosphere.2017.03.091.*

## 10.2.1.2 Optimization for NaClO−NaCl regeneration

The tests were conducted to investigate the impact of $ClO^-/N$, NaCl concentration, and pH on ZRE and NRE during the zeolite adsorption process coupled with NaClO−NaCl regeneration. Preliminary experiments were conducted to determine the suitable range of factors, including $ClO^-/N$ levels, NaCl concentration, and pH, by the implementation of single-factor testing at various levels (Vitzthum von Eckstaedt, Charles, Ho, & Cord-Ruwisch, 2016). The response surface methodology (RSM) technique, in conjunction with the Box-Behnken design, was employed to optimize the key parameters involved in the regeneration process of NaClO−NaCl. The experimental design employed in this study consisted of a three-level factorial trial, necessitating a total of 17 tests. The three components considered for this study were the $ClO^-/N$ ratio ranging from 1.8 to 3.0, the concentration of NaCl ranging from 0 to 20 g/L, and the pH ranging from 4 to 10. The response variables chosen for analysis were ZRE and NRE. In each of the regeneration tests, zeolites were subjected to immersion in a 200 mL solution containing NaClO and NaCl for a duration of 40 minutes. The establishment of second-order polynomial models for ZRE and NRE is given in Eqs. (10.3) and (10.4), respectively.

$$ZRE = 69.65 + 1.86A + 2.61B + 1.72C - 2.43AB + 1.37AC - 0.66BC$$

$$+ 0.29A^2 - 1.39B^2 + 2.23C^2 \qquad (10.3)$$

$$NRE = 57.74 + 0.19A + 1.50B + 2.19C - 1.81AB + 1.57AC - 1.02BC$$

$$+ 0.17A^2 - 1.35B^2 + 5.91C^2 \qquad (10.4)$$

Let A, B, and C represent the concentrations of $ClO^-/N$, NaCl, and pH, respectively. The acquired coefficients of determination $(R^2)$ for ZRE (0.984) and NRE (0.983), as well as the significant $P$-value at the 0.01 level

**TABLE 10.1** Significance of quadratic model coefficients and ANOVA for ZRE and NRE by NaClO−NaCl regeneration.

| Source | ZRE | | | NRE | | |
|---|---|---|---|---|---|---|
| | Mean square | F-value | P-value | Mean square | F-value | P-value |
| Model | 18.4 | 84.5 | 0.0019 | 23.4 | 79.5 | 0.0021 |
| A-$ClO^-/N$ | 27.5 | 126 | 0.0015 | 0.28 | 0.96 | 0.400 |
| B-NaCl | 54.3 | 250 | 0.0005 | 18.0 | 61.3 | 0.0043 |
| C-pH | 23.7 | 109 | 0.0019 | 38.5 | 131 | 0.0014 |
| AB | 23.6 | 109 | 0.0019 | 13.0 | 44.3 | 0.0069 |
| AC | 7.51 | 34.6 | 0.0098 | 9.86 | 33.5 | 0.0103 |
| BC | 1.76 | 8.08 | 0.0655 | 4.14 | 14.1 | 0.0331 |
| $A^2$ | 0.19 | 0.86 | 0.422 | 0.063 | 0.21 | 0.675 |
| $B^2$ | 4.42 | 20.4 | 0.0203 | 4.14 | 14.1 | 0.0331 |
| $C^2$ | 11.4 | 52.3 | 0.0055 | 79.9 | 272 | 0.0005 |

*Source*: From Zhang, W., Zhou, Z., An, Y., Du, S., Ruan, D., Zhao, C., Ren, N., & Tian, X. (2017). *Chemosphere, 178,* 565−572. https://doi.org/10.1016/j.chemosphere.2017.03.091.

derived from the analysis of variance (ANOVA) findings presented in Table 10.1, suggest that the proposed models exhibit a high level of accuracy and reliability. Both NRE and ZRE exhibited an upward trend when the concentration of NaCl in the $ClO^-/N$ range of 1.8−2.4 rose, ultimately reaching their peak values at the highest NaCl concentrations (Wang et al., 2006). The ANOVA findings presented in Table 10.1 indicate that the concentrations of NaCl had a statistically significant impact on both ZRE and NRE at a significance level of 0.01. Additionally, the influence of $ClO^-/N$ was found to be statistically significant on ZRE ($P < .01$) but not significant on NRE. The findings of the study demonstrate that the presence of a high concentration of $Na^+$ facilitates the exchange of $NH_4^+$ from zeolite. Additionally, the process of chlorination of $NH_4^+$ using hypochlorite further enhances the ion exchange process. However, when the $ClO^-/N$ ratio is beyond the threshold value (breakpoint), the production of $NO_3^-$ experiences a significant rise, leading to a gradual change in NRE for $ClO^-/N$ ratios greater than 2.4. The ANOVA findings presented in Table 10.1 indicate that the pH value exhibited a statistically significant impact on both ZRE and NRE, with a significance level of .01. Additionally, the interaction effect between the pH value and $ClO^-/N$ was found to be significant at the

.01 level for ZRE and at the .05 level for NRE (Moussavi, Talebi, Farrokhi, & Sabouti, 2011). Both zero-valent iron (ZVI) and nanozero-valent iron (nZVI) exhibited a drop in reactivity initially, followed by an increase as the pH climbed from 4 to 10. This observation suggests that an alkaline environment promotes the regeneration of sodium hypochlorite (NaClO) and sodium chloride (NaCl) through the use of ZVI and nZVI. The $NH_4^+$ ions undergo conversion to $NH_3$ at elevated pH levels, hence facilitating the extraction of more $NH_4^+$ ions. An alkaline environment increased the efficiency of ion exchange in a NaCl solution. Moreover, $NH_3$ molecules, whose proportion increases as the pH level rises, have a higher susceptibility to oxidation by NaClO compared to $NH_4^+$. From a synthetic perspective, it can be observed that the presence of high $ClO^-/N$ ratio, NaCl concentration, and pH levels all contribute to the advantageous regeneration of NaClO−NaCl. However, it should be noted that a high $ClO^-/N$ ratio leads to an increased conversion of $NH_4^+$ to $NO_3^-$ and a higher likelihood of $NH_4^+$−N buildup in the regenerant solution (Li et al., 2010; Wen, Ho, & Tang, 2006). Hence, the RSM-optimized condition characterized by a reduced $ClO^-/N$ ratio was selected for subsequent experiments. The ideal conditions for NaClO−NaCl regeneration were determined to be a $ClO^-/N$ ratio of 1.75, a NaCl concentration of 20 g/L, and a pH of 10.0. Under these conditions, the resulting zinc removal efficiency and nickel removal efficiency were found to be 73.6% and 65.2%, respectively. The optimal condition was experimentally evaluated three times, yielding ZRE and NRE values of 73.7 ± 0.4% and 64.9 ± 0.6%, respectively. The small relative deviations (0.14% and 0.46%) observed between the modeled and experimental outcomes provide evidence supporting the accuracy of the response models and the presence of an optimal point.

## 10.2.2 Biological regeneration

The stoichiometric reaction that represents the equilibrium exchange process between the $NH_4^+$ ion in the solution and the $Na^+$ ion linked to the zeolite can be stated as follows (Jung et al., 2004):

$$Z-Na^+ + NH_4^+ \rightarrow Z-NH_4^+ + Na^+ \qquad (10.5)$$

In this context, Z represents the chemical compound known as zeolite. The selectivity of different ions is contingent upon the dimensions, electron configuration, and electric properties of cations. Chemical regeneration of the zeolite can be achieved by introducing a substantial quantity of sodium chloride into the solution after it has become saturated with ammonium. Upon immersion of the zeolite saturated with ammonium into a solution, a minor quantity of ammonium ions will be displaced, as indicated by Eq. (10.2):

$$Z-NH_4^+ + Na^+ + Cl^- \rightarrow Z-Na^+ + NH_4^+ + Cl^- \tag{10.6}$$

$$[NH_4^+] = \frac{X_{NH_4^+}[Na^+]}{a\left(1 - X_{NH_4^+}\right)} \tag{10.7}$$

The concentration of ammonium ion ($[NH_4^+]$) and sodium ion ($[Na^+]$) are denoted by meq/L. The variable $X_{NH_4^+}$ represents the ionic fraction of ammonium within the zeolite phase, whereas the variable a denotes the selectivity coefficient.

Eq. (10.7) can be reformulated as:

$$\frac{a[NH_4^+]}{[Na^+] + a[NH_4^+]} \tag{10.8}$$

Dividing the right side of Eq. (10.8) by $NH_4^+$, we get:

$$\frac{a}{[Na^+]}]/[NH_4^+] + a \tag{10.9}$$

Eq. (10.9) can be summed up as:

$$X_{NH_4^+} \frac{a}{R + a} \tag{10.10}$$

Here, $R = [Na^+/NH_4^+]$

The regeneration process of zeolite that is saturated with ammonium can be elucidated by utilizing Eq. (10.6). The reduction in the $X_{NH_4^+}$ parameter indicates the release of ammonium ions from the zeolite, a process commonly referred to as regeneration. Two potential approaches could be employed for the process of regeneration. There is an increase in the concentration of sodium ions ($Na^+$) due to chemical regeneration. Another factor contributing to the decrease in the concentration of $NH_4^+$ is the process of biological regeneration through nitrification. The primary objective of this investigation was to examine the bioregeneration process of zeolite that had become saturated with ammonium. This was achieved by reducing the concentration of ammonium through nitrification in the liquid phase. Conversely, the process of zeolite regeneration is contingent upon the selectivity coefficient. Typically, the rate of regeneration tends to be higher when the selectivity coefficient is of lesser magnitude (Green, Mels, & Lahav, 1996; Lahav & Green, 1998; Semmens, Klieve, Schnobrich, & Tauxe, 1981).

## 10.2.3 Electrochemical process

Solutions of synthetic $(NH_4)_2SO_4$ were produced, containing ammonium−nitrogen concentrations of 100 mg/L. The process of ammonia adsorption was assessed within a 2000-mL beaker. Zeolites were added to a beaker in measured quantities of 100 g. Subsequently, a solution of $(NH_4)_2SO_4$ with a concentration of 100 mg of $NH_4^+-N$ per liter and a pH of approximately 6.5 was poured into the beaker, with a total volume of 1000 mL. The beaker was subjected to agitation at a temperature of 20°C for a duration of 30 minutes (Wu et al., 2008). The initial experiments have verified that a contact duration of 30 minutes is adequate to utilize 80% of the zeolite's adsorption capacity. This capacity remains consistent at an average value of 0.64 mg $NH_4^+-N/g$ zeolite. The residual ammonia concentrations in the solution were examined. The two phases were physically separated by the process of filtration (Feng, Sugiura, Shimada, & Maekawa, 2003). Subsequently, a quantity of 100 g of zeolites was introduced into the electrochemical cell, which was filled with a synthetic solution comprising 300 mL and a concentration of 0.5 g per liter of $Na_2SO_4$, along with varying quantities of NaCl. The regeneration trials were conducted under ambient conditions (Lin & Wu, 1997). The process of electrolysis was terminated after a duration of 2 hours had transpired. Following each regeneration cycle of the zeolite, the RZ was isolated from the regeneration solution to assess its adsorption capacity. This assessment was conducted by equilibrium tests or by comparing the adsorption capacity throughout a 30-minute time frame. The study involved the investigation of the impact of different operating parameters by manipulating one parameter while maintaining the others at a constant level. To examine the impact of varying sodium chloride (NaCl) dosages on the process of zeolite regeneration, different concentrations of NaCl at 1.0, 1.5, and 2.0 g/L (w/v) were introduced into the regeneration solutions. The study examined the impact of different current densities on a galvanostatic system, specifically at current densities of 20, 40, and 60 mA/cm$^2$ (Lei et al., 2009).

## 10.3 Challenges and future prospective

Zeolites, renowned for their distinctive porous structure and exceptional adsorption properties, find extensive utilization across diverse industrial sectors. Nevertheless, the process of regeneration presents a number of challenges, encompassing matters such as the preservation of structural integrity and the efficient utilization of energy resources. Several obstacles are associated with the process of zeolite regeneration.

1. Structural integrity: One of the primary difficulties encountered in the process of zeolite regeneration is in the preservation of the zeolite framework's structural integrity. The integrity of the porous structure, which plays a vital role in facilitating efficient adsorption, may be disrupted when subjected to regeneration procedures using elevated temperatures or aggressive chemical agents. This phenomenon may result in a diminishment of surface area, pore volume, and, eventually, a decline in adsorption capacity (Yu et al., 2015).

2. Energy consumption: The process of zeolite regeneration frequently necessitates significant energy inputs, particularly in cases when elevated temperatures are utilized. The environmental impact and operational expenses connected with zeolite-based procedures are a cause for concern due to the energy-intensive nature of certain regeneration techniques (Vanlangendonck, Corbisier, & Van Lierde, 2005).
3. Selectivity issues: The preservation of zeolite selectivity during the process of regeneration presents a notable obstacle. Zeolites are engineered to exhibit selective adsorption properties toward specific molecules, and the process of regeneration must effectively remove undesired impurities while preserving the zeolite's selectivity toward the desired target molecules (Klieve & Semmens, 1980).
4. Scale-up challenges: The expansion of zeolite regeneration techniques for industrial applications presents supplementary complexities. To achieve successful application on an industrial scale, it is imperative to overcome various challenges, including the consistency of regeneration across huge volumes, scalability of equipment, and economic feasibility (Booker, Cooney, & Priestley, 1996).
5. Environmental impact: The increasing concern lies in the environmental ramifications associated with zeolite regeneration processes, particularly those that include the utilization of toxic compounds. The identification of ecologically viable alternatives that may effectively replace zeolite-based processes is of utmost importance for achieving long-term sustainability (Guida, Potter, Jefferson, & Soares, 2020).

Given the significant role that zeolites play in multiple sectors, it is imperative to optimize the regeneration processes to ensure long-lasting and efficient utilization. The proposals presented in this discourse pertain to crucial domains encompassing technological advancement, interdisciplinary cooperation, and sustainable methodologies to influence the trajectory of zeolite regeneration in the future.

1. Technological innovation: Future research should prioritize the development and optimization of enhanced regeneration technologies. The precision and efficiency of zeolite regeneration can be enhanced through the implementation of several techniques, including in situ monitoring, artificial intelligence-driven process control, and the integration of advanced materials.
2. Energy-efficient regeneration methods: Emphasize on strategies for energy-efficient regeneration techniques. This encompasses the utilization of alternate energy sources, such as renewable energy, alongside the advancement of regeneration techniques that aim to reduce energy consumption while preserving the integrity and function of zeolites.
3. Industry-academia partnerships: Promote enhanced cooperation between academic institutions and industrial partners. This collaborative effort has

the potential to enhance the translation of successful laboratory-scale experiments into commercially feasible and scalable methods for regenerating zeolites in industrial settings.

4. Life cycle assessments: Perform thorough life cycle studies on zeolite regeneration procedures to determine their total environmental impact. This study aims to offer significant perspectives on the sustainability of various regeneration techniques and direct the sector toward adopting more environmentally responsible approaches.

5. Public awareness campaigns: Propose the implementation of public awareness campaigns aimed at emphasizing the significance of zeolite regeneration in the context of sustainable industrial operations. Enhanced consciousness has the potential to generate backing for research endeavors, regulatory structures, and industrial methodologies that prioritize the regeneration of zeolites with ecologically sustainable approaches.

## 10.4  Conclusion

The examination of chemical regeneration, biological regeneration, and electrochemical regeneration procedures for zeolites constitutes a thorough exploration of sustainable and efficient methodologies aimed at restoring the adsorption capacity of zeolitic materials. Every way of regeneration offers distinct benefits and considerations, thereby contributing to the advancement of varied and eco-friendly approaches for prolonging the longevity of materials based on zeolite. Chemical regeneration is a widely recognized and effective method that involves the use of diverse chemical agents to desorb and eliminate impurities from the surfaces of zeolite materials. The method's inherent versatility enables the customization of regeneration agents to target specific pollutants and accommodate varying process conditions. The efficacy of chemical regeneration is emphasized by its capacity to restore the performance of zeolite for several cycles, rendering it highly suitable for industrial applications that prioritize efficiency and durability. The implementation of biological regeneration in the context of zeolite restoration is a novel and environmentally conscious approach. Through the utilization of the metabolic processes of microorganisms, this methodology presents an inherent and environmentally friendly solution for the degradation of adsorbed pollutants within the confined spaces of zeolite pores. The incorporation of biological regeneration into wastewater treatment and environmental remediation processes is in line with the increasing focus on environmentally friendly and biologically mediated technology. The utilization of electrical currents to promote the desorption of pollutants, known as electrochemical regeneration, is considered a technologically advanced and energy-efficient strategy. The utilization of this approach exhibits the potential to mitigate resource consumption and diminish environmental impact in contrast to conventional methods of regeneration. The deliberate utilization of electrical currents offers a precise and focused method for rejuvenating zeolites, hence improving their suitability for applications that need high levels of efficiency and accuracy. The selection of the regeneration technique is contingent upon the particular application, the pollutants present, and the intended trade-off between efficacy, environmental consequences, and energy usage. The examination of various processes of regeneration enhances our comprehension of their capabilities and constraints, hence assisting in the identification of the most appropriate approach for certain circumstances.

The comprehensive examination of chemical, biological, and electrochemical regeneration methods for zeolites demonstrates a wide array of approaches that have the capacity to bring about transformative advancements in the realm of materials based on zeolites. With the growing significance of environmental issues and the need for efficient resource utilization, the outcomes of this study provide a foundation for the advancement of sustainable and energy-efficient regeneration methods that are specifically designed for zeolite materials. These techniques aim to maintain the ongoing significance and practicality of zeolite materials in various industrial and environmental contexts.

## 10.5  AI disclosure

During the preparation of this work, the author(s) used Quillbot to improve their English and for smooth writing. After using this tool/service, the author(s) reviewed and edited the content as needed and take(s) full responsibility for the content of the publication.

## References

Alshameri, A., Yan, C., Al-Ani, Y., Dawood, A. S., Ibrahim, A., Zhou, C., & Wang, H. (2014). An investigation into the adsorption removal of ammonium by salt activated Chinese (Hulaodu) natural zeolite: Kinetics, isotherms, and thermodynamics. *Journal of the Taiwan Institute of Chemical Engineers, 45*(2), 554−564. Available from https://doi.org/10.1016/j.jtice.2013.050.008, https://www.sciencedirect.com/science/article/pii/S1876107013001168.

Anweshan., Das, P. P., Dhara, S., Purkait, M. K., Husen, A., & Siddiqi, K. S. (2023). *Chapter 12 - Nanosensors in food science and technology micro and nano technologies* (pp. 247−272). Elsevier. Available from https://www.sciencedirect.com/science/article/pii/B978032399546700015X, 10.1016/B978-0-323-99546-7.00015-X.

Ates, A., & Akgül, G. (2016). Modification of natural zeolite with NaOH for removal of manganese in drinking water. *Powder Technology, 287*, 285−291. Available from https://doi.org/10.1016/j.powtec.2015.100.021, https://www.sciencedirect.com/science/article/pii/S0032591015301108.

Baum, R., Luh, J., & Bartram, J. (2013). Sanitation: A global estimate of sewerage connections without treatment and the resulting impact on MDG progress. *Environmental Science & Technology, 47*(4), 1994−2000. Available from https://doi.org/10.1021/es304284f.

Booker, N. A., Cooney, E. L., & Priestley, A. J. (1996). Ammonia removal from sewage using natural Australian zeolite. *Water Quality International '96, 34*(9), 17−24. Available from https://doi.org/10.1016/S0273-1223(96)00782-2, https://www.sciencedirect.com/science/article/pii/S0273122396007822.

Brandes, M. (1978). Characteristics of effluents from gray and black water septic tanks. *Journal (Water Pollution Control Federation)*, *50*(11), 2547–2559. Available from http://www.jstor.org/stable/25040185.

Bunce, J. T., Ndam, E., Ofiteru, I. D., Moore, A., & Graham, D. W. (2018). A review of phosphorus removal technologies and their applicability to small-scale domestic wastewater treatment systems. *Frontiers in Environmental Science*, *6*. Available from: https://www.frontiersin.org/articles/10.3389/fenvs.2018.00008.

Castro, C. J., Shyu, H. Y., Xaba, L., Bair, R., & Yeh, D. H. (2021). Performance and onsite regeneration of natural zeolite for ammonium removal in a field-scale non-sewered sanitation system. *Science of The Total Environment*, *776*145938. Available from https://doi.org/10.1016/j.scitotenv.2021.145938, https://www.sciencedirect.com/science/article/pii/S0048969721010056.

Castro, C. J., Goodwill, J. E., Rogers, B., Henderson, M., & Butler, C. S. (2014). Deployment of the microbial fuel cell latrine in Ghana for decentralized sanitation. *Journal of Water, Sanitation and Hygiene for Development*, *4*(4), 663–671. Available from https://doi.org/10.2166/washdev.20140.020, https://doi.org/10.2166/washdev.2014.020.

Chakraborty, S., Gautam, S. P., Das, P. P., & Hazarika, M. K. (2019). Instant controlled pressure drop (DIC) treatment for improving process performance and milled rice quality. *Journal of The Institution of Engineers (India): Series A*, *100*(4), 683–695. Available from https://doi.org/10.1007/s40030-019-00403-w, https://doi.org/10.1007/s40030-019-00403-w.

Chakraborty, S., Das, P. P., Mondal, P., Sillanpää, M., Khadir, A., & Gurun, K. (2023). 34 - Recent advances in membrane technology for the recovery and reuse of valuable resources (pp. 695–719). Elsevier. Available from https://www.sciencedirect.com/science/article/pii/B9780323953276000282, 10.1016/B978-0-323-95327-6.00028-2.

Changmai, M., Das, P. P., Mondal, P., Pasawan, M., Sinha, A., Biswas, P., ... Purkait, M. K. (2022). Hybrid electrocoagulation–microfiltration technique for treatment of nanofiltration rejected steel industry effluent. *International Journal of Environmental Analytical Chemistry*, *102*(1), 62–83. Available from https://doi.org/10.1080/03067319.2020.1715381, https://doi.org/10.1080/03067319.2020.1715381.

Cid, C. A., Qu, Y., & Hoffmann, M. R. (2018). Design and preliminary implementation of onsite electrochemical wastewater treatment and recycling toilets for the developing world. *Environmental Science.: Water Research & Technology*, *4*(10), 1439–1450. Available from https://doi.org/10.1039/C8EW00209F, http://doi.org/10.1039/C8EW00209F.

Das, P. P., Anweshan., & Purkait, M. K. (2021). Treatment of cold rolling mill (CRM) effluent of steel industry. *Separation and Purification Technology*, *274*119083. Available from https://doi.org/10.1016/j.seppur.2021.119083, https://www.sciencedirect.com/science/article/pii/S1383586621007930.

Das, P. P., Deepti., & Purkait, M. K. (2023). *Industrial wastewater to biohydrogen production via potential bio-refinery route* (pp. 159–179). Cham: Springer International Publishing. Available from https://doi.org/10.1007/978-3-031-20822-5_8, 10.1007/978-3-031-20822-5_8.

Das, P. P., Anweshan., Mondal, P., Sinha, A., Biswas, P., Sarkar, S., & Purkait, M. K. (2021). Integrated ozonation assisted electrocoagulation process for the removal of cyanide from steel industry wastewater. *Chemosphere*, *263*128370. Available from https://doi.org/10.1016/j.chemosphere.2020.128370, https://www.sciencedirect.com/science/article/pii/S0045653520325650.

Das, P. P., Mondal, P., Anweshan., Sinha, A., Biswas, P., Sarkar, S., & Purkait, M. K. (2021). Treatment of steel plant generated biological oxidation treated (BOT) wastewater by hybrid process. *Separation and Purification Technology, 258*118013. Available from https://doi.org/10.1016/j.seppur.2020.118013, https://www.sciencedirect.com/science/article/pii/S1383586620324862.

Das, P. P., Samanta, N. S., Dhara, S., Purkait, M. K., & Shah, M. P. (2023). *Chapter 13 - Biofuel production from algal biomass* (pp. 167–179). Elsevier. Available from https://www.sciencedirect.com/science/article/pii/B9780128243183000096 10.1016/B978-0-12-824318-3.00009-6.

Das, P. P., Sharma, M., & Purkait, M. K. (2022). Recent progress on electrocoagulation process for wastewater treatment: A review. *Separation and Purification Technology, 292*121058. Available from https://doi.org/10.1016/j.seppur.2022.121058, https://www.sciencedirect.com/science/article/pii/S1383586622006153.

Das, P. P., Mondal, P., & Purkait, M. K. (2022). *Recent advances in synthesis of iron nanoparticles via green route and their application in biofuel production* (pp. 79–104). Singapore: Springer Nature Singapore. Available from https://doi.org/10.1007/978-981-16-9356-4_4, 10.1007/978-981-16-9356-4_4.

Das, P. P., Dhara, S., & Purkait, M. K. (2023). *Hybrid electrocoagulation and ozonation techniques for industrial wastewater treatment* (pp. 107–128). Singapore: Springer Nature Singapore. Available from https://doi.org/10.1007/978-981-99-2560-5_6, 10.1007/978-981-99-2560-5_6.

Das, P. P., Mondal, P., Sillanpää, M., Khadir, A., & Gurung, K. (2023). *14 - Membrane-assisted potable water reuses applications: benefits and drawbacks*. Elsevier. Available from https://www.sciencedirect.com/science/article/pii/B9780323993449000141 10.1016/B978-0-323-99344-9.00014-1.

Das, P. P., Sontakke, A. D., Purkait, M. K., & Shah, M. P. (2023). *Chapter 2 - Electrocoagulation process for wastewater treatment: applications, challenges, and prospects* (pp. 23–48). Elsevier. Available from https://www.sciencedirect.com/science/article/pii/B9780323956840000154, 10.1016/B978-0-323-95684-0.00015-4.

Das, P. P., Duarah, P., Purkait, M. K., & Jafari, S. M. (2023). *5 - Fundamentals of food roasting process unit operations and processing equipment in the food industry* (pp. 103–130). Woodhead Publishing. Available from https://www.sciencedirect.com/science/article/pii/B9780128186183000057, 10.1016/B978-0-12-818618-3.00005-7.

Das, P. P., Sontakke, A. D., Purkait, M. K., & Shah, M. P. (2023). *Chapter 12 - Rice straw for biofuel production* (pp. 153–166). Elsevier. Available from https://www.sciencedirect.com/science/article/pii/B9780128243183000345, 10.1016/B978-0-12-824318-3.00034-5.

Dhara, S., Shekhar Samanta, N., Das, P. P., Uppaluri, R. V. S., & Purkait, M. K. (2023). Ravenna grass-extracted alkaline lignin-based polysulfone mixed matrix membrane (MMM) for aqueous Cr(VI) removal. *ACS Applied Polymer Materials, 5*(8), 6399–6411. Available from https://doi.org/10.1021/acsapm.3c00999, https://doi.org/10.1021/acsapm.3c00999.

Dhara, S., Samanta, N. S., Uppaluri, R., & Purkait, M. K. (2023). High-purity alkaline lignin extraction from Saccharum ravannae and optimization of lignin recovery through response surface methodology. *International Journal of Biological Macromolecules, 234*123594. Available from https://doi.org/10.1016/j.ijbiomac.2023.123594, https://www.sciencedirect.com/science/article/pii/S0141813023004877.

Das, P. P., Sontakke, A. D., Samanta, N. S., & Purkait, M. K. (2023). *Emerging contaminants in wastewater: Eco-toxicity and sustainability assessment* (pp. 63–87). Singapore: Springer Nature Singapore. Available from https://doi.org/10.1007/978-981-99-2489-9_4 10.1007/978-981-99-2489-9_4.

Das, Pranjal P., Dhara, S., & Purkait, Mihir K. (2024). Ozone-based oxidation processes for the removal of pharmaceutical products from wastewater. In Maulin P. Shah, & Pooja Ghosh (Eds.), *Development in Wastewater Treatment Research and Processes* (pp. 287−308). Elsevier. Available from https://doi.org/10.1016/B978-0-443-19207-4.00003-3.

Dhara, S., Das, P. P., Uppaluri, R., Purkait, M. K., & Shah, M. P. (2023). *Chapter 3 - Phosphorus recovery from municipal wastewater treatment plants* (pp. 49−72). Elsevier. Available from https://www.sciencedirect.com/science/article/pii/B9780323956840000142, 10.1016/B978-0-323-95684-0.00014-2.

Dhara, S., Das, P. P., Uppaluri, R., Purkait, M. K., Sillanpää, M., Khadir, A., & Gurung, K. (2023). *12 - Biological approach for energy self-sufficiency of municipal wastewater treatment plants* (pp. 235−260). Elsevier. Available from https://www.sciencedirect.com/science/article/pii/B9780323993487000060, 10.1016/B978-0-323-99348-7.00006-0.

Du, Q., Liu, S., Cao, Z., & Wang, Y. (2005). Ammonia removal from aqueous solution using natural Chinese clinoptilolite. *Separation and Purification Technology*, *44*(3), 229−234. Available from https://doi.org/10.1016/j.seppur.2004.040.011, https://www.sciencedirect.com/science/article/pii/S1383586605000493.

Duarah, P., Das, P. P., & Purkait, M. K. (2023). *Technological advancement in the synthesis and application of nanocatalysts* (pp. 191−214). Singapore: Springer Nature Singapore. Available from https://doi.org/10.1007/978-981-99-3292-4_10, 10.1007/978-981-99-3292-4_10.

Feng, C., Sugiura, N., Shimada, S., & Maekawa, T. (2003). Development of a high performance electrochemical wastewater treatment system. *Journal of Hazardous Materials*, *103*(1), 65−78. Available from https://doi.org/10.1016/S0304-3894(03)00222-X, https://www.sciencedirect.com/science/article/pii/S030438940300222X.

Gendel, Y., & Lahav, O. (2013). A novel approach for ammonia removal from fresh-water recirculated aquaculture systems, comprising ion exchange and electrochemical regeneration. *Aquacultural Engineering*, *52*, 27−38. Available from https://doi.org/10.1016/j.aquaeng.2012.070.005, https://www.sciencedirect.com/science/article/pii/S0144860912000623.

Green, M., Mels, A., & Lahav, O. (1996). Biological-ion exchange process for ammonium removal from secondary effluent. *Water Quality International '96 Part 1: Nutrient Removal*, *34*(1), 449−458. Available from https://doi.org/10.1016/0273-1223(96)00534-3, https://www.sciencedirect.com/science/article/pii/0273122396005343.

Guida, S., Potter, C., Jefferson, B., & Soares, A. (2020). Preparation and evaluation of zeolites for ammonium removal from municipal wastewater through ion exchange process. *Scientific Reports*, *10*(1)12426. Available from https://doi.org/10.1038/s41598-020-69348-6, https://doi.org/10.1038/s41598-020-69348-6.

Guo, X., Zeng, L., Li, X., & Park, H. -S. (2007). Removal of ammonium from RO permeate of anaerobically digested wastewater by natural zeolite. *Separation Science and Technology*, *42*(14), 3169−3185. Available from https://doi.org/10.1080/01496390701514949, https://doi.org/10.1080/01496390701514949.

Huang, H., Yang, L., Xue, Q., Liu, J., Hou, L., & Ding, L. (2015). Removal of ammonium from swine wastewater by zeolite combined with chlorination for regeneration. *Journal of Environmental Management*, *160*, 333−341. Available from https://doi.org/10.1016/j.jenvman.2015.060.039, https://www.sciencedirect.com/science/article/pii/S0301479715301316.

Huang, H., Huang, L., Zhang, Q., Jiang, Y., & Ding, L. (2015). Chlorination decomposition of struvite and recycling of its product for the removal of ammonium-nitrogen from landfill leachate. *Chemosphere*, *136*, 289−296. Available from https://doi.org/10.1016/j.chemosphere.2014.100.078, https://www.sciencedirect.com/science/article/pii/S0045653514012715.

Jung, J.-Y., Chung, Y.-C., Shin, H.-S., & Son, D.-H. (2004). Enhanced ammonia nitrogen removal using consistent biological regeneration and ammonium exchange of zeolite in modified SBR process. *Water Research, 38*(2), 347−354. Available from https://doi.org/10.1016/j.watres.2003.090.025. Available from: https://www.sciencedirect.com/science/article/pii/S0043135403005414.

Katsou,E., Malamis, S., Tzanoudaki, M., Haralambous, K.J., & Loizidou, M. (2011). Regeneration of natural zeolite polluted by lead and zinc in wastewater treatment systems. In *Selected papers presented at the 2nd International Conference on Research Frontiers in Chalcogen Cycle Science and Technology,* Delft, The Netherlands, May 31-June 1, 2010. 0304-3894 (Vol. 189, no. 3, pp., 773−786). Available from: https://www.sciencedirect.com/science/article/pii/S0304389410016419. doi: 10.1016/j.jhazmat.2010.12.061.

Kithome, M., Paul, J. W., Lavkulich, L. M., & Bomke, A. A. (1998). Kinetics of ammonium adsorption and desorption by the natural zeolite clinoptilolite. *Soil Science Society of America Journal, 62*(3), 622−629. Available from https://doi.org/10.2136/sssaj1998.03615995006200030011x, https://doi.org/10.2136/sssaj1998.03615995006200030011x.

Klieve, J. R., & Semmens, M. J. (1980). An evaluation of pretreated natural zeolites for ammonium removal. *Water Research, 14*(2), 161−168. Available from https://doi.org/10.1016/0043-1354(80)90232-8, https://www.sciencedirect.com/science/article/pii/0043135480902328.

Knerr, H., Rechenburg, A., Kistemann, T., & Schmitt, T. G. (2011). Performance of a MBR for the treatment of blackwater. *Water Science and Technology, 63*(6), 1247−1254. Available from https://doi.org/10.2166/wst.20110.367, https://doi.org/10.2166/wst.2011.367.

Lackner, S., Gilbert, E. M., Vlaeminck, S. E., Joss, A., Horn, H., & van Loosdrecht, M. C. M. (2014). Full-scale partial nitritation/anammox experiences − An application survey. *Water Research, 55,* 292−303. Available from https://doi.org/10.1016/j.watres.2014.020.032, https://www.sciencedirect.com/science/article/pii/S0043135414001481.

Lahav, O., & Green, M. (1998). Ammonium removal using ion exchange and biological regeneration. *Water Research, 32*(7), 2019−2028. Available from https://doi.org/10.1016/S0043-1354(97)00453-3, https://www.sciencedirect.com/science/article/pii/S0043135497004533.

Lahav, O., Schwartz, Y., Nativ, P., & Gendel, Y. (2013). Sustainable removal of ammonia from anaerobic-lagoon swine waste effluents using an electrochemically-regenerated ion exchange process. *Chemical Engineering Journal, 218,* 214−222. Available from https://doi.org/10.1016/j.cej.2012.120.043, https://www.sciencedirect.com/science/article/pii/S138589471201683X.

Lei, X., Li, M., Zhang, Z., Feng, C., Bai, W., & Sugiura, N. (2009). Electrochemical regeneration of zeolites and the removal of ammonia. *Journal of Hazardous Materials, 169*(1), 746−750. Available from https://doi.org/10.1016/j.jhazmat.2009.040.012, https://www.sciencedirect.com/science/article/pii/S0304389409005639.

Li, M., Feng, C., Zhang, Z., Lei, X., Chen, N., & Sugiura, N. (2010). Simultaneous regeneration of zeolites and removal of ammonia using an electrochemical method. *Microporous and Mesoporous Materials, 127*(3), 161−166. Available from https://doi.org/10.1016/j.micromeso.2009.070.009, https://www.sciencedirect.com/science/article/pii/S1387181109003357.

Lin, S. H., & Wu, C. L. (1997). Electrochemical nitrite and ammonia oxidation in sea water. *Journal of Environmental Science and Health. Part A: Environmental Science and Engineering and Toxicology., 32*(8), 2125−2138. Available from https://doi.org/10.1080/10934529709376672, https://doi.org/10.1080/10934529709376672.

Mondal, P., Samanta, N. S., Meghnani, V., & Purkait, M. K. (2019). Selective glucose permeability in presence of various salts through tunable pore size of pH responsive PVDF-co-HFP membrane. *Separation and Purification Technology, 221,* 249−260. Available from https://doi.org/10.1016/j.seppur.2019.040.001, https://www.sciencedirect.com/science/article/pii/S1383586618346252.

Mondal, P., Samanta, N. S., Kumar, A., & Kumar Purkait, M. (2020). Recovery of H2SO4 from wastewater in the presence of NaCl and KHCO3 through pH responsive polysulfone membrane: Optimization approach. *Polymer Testing*, *86*106463. Available from https://doi.org/10.1016/j.polymertesting.2020.106463, https://www.sciencedirect.com/science/article/pii/S0142941820301239.

Moussavi, G., Talebi, S., Farrokhi, M., & Sabouti, R. M. (2011). The investigation of mechanism, kinetic and isotherm of ammonia and humic acid co-adsorption onto natural zeolite. *Special Section: Symposium on Post-Combustion Carbon Dioxide Capture*, *171*(3), 1159–1169. Available from https://doi.org/10.1016/j.cej.2011.050.016, https://www.sciencedirect.com/science/article/pii/S1385894711005638.

Rahmani, A. R., Mahvi, A. H., Mesdaghinia, A. R., & Nasseri, S. (2004). Investigation of ammonia removal from polluted waters by Clinoptilolite zeolite. *International Journal of Environmental Science & Technology*, *1*(2), 125–133. Available from https://doi.org/10.1007/BF03325825, https://doi.org/10.1007/BF03325825.

Sahondo, T., Hennessy, S., Sindall, R. C., Chaudhari, H., Teleski, S., Lynch, B. J., ... Hawkins, B. T. (2020). Field testing of a household-scale onsite blackwater treatment system in South Africa. *Science of The Total Environment*, *703*135469. Available from https://doi.org/10.1016/j.scitotenv.2019.135469, https://www.sciencedirect.com/science/article/pii/S0048969719354622.

Samanta, N., Das, P. P., & Mondal, P. (2022). Physico-chemical and adsorption study of hydrothermally treated zeolite A and FAU-type zeolite X prepared from LD (Linz–Donawitz) slag of the steel industry. *International Journal of Environmental Analytical Chemistry*, *13*, 1–23. Available from https://doi.org/10.1080/03067319.2022.2079082.

Samanta, N. S., Banerjee, S., Mondal, P., Anweshan., Bora, U., & Purkait, M. K. (2021). Preparation and characterization of zeolite from waste Linz-Donawitz (LD) process slag of steel industry for removal of Fe3 + from drinking water. *Advanced Powder Technology*, *32*(9), 3372–3387. Available from https://doi.org/10.1016/j.apt.2021.070.023, https://www.sciencedirect.com/science/article/pii/S0921883121003605.

Samanta, N. S., Das, P. P., Dhara, S., & Purkait, M. K. (2023). An overview of precious metal recovery from steel industry slag: Recovery strategy and utilization. *Industrial & Engineering Chemistry Research*, *62*(23), 9006–9031. Available from https://doi.org/10.1021/acs.iecr.3c00604.

Samanta, N. S., Das, P. P., Mondal, P., Changmai, M., & Purkait, M. K. (2022). Critical review on the synthesis and advancement of industrial and biomass waste-based zeolites and their applications in gas adsorption and biomedical studies. *Journal of the Indian Chemical Society*, *99*(11)100761. Available from https://doi.org/10.1016/j.jics.2022.100761, https://www.sciencedirect.com/science/article/pii/S001945222200423X.

Samanta, N. S., Mondal, P., & Purkait, M. K. (2023). *Nanofiltration technique for the treatment of industrial wastewater* (pp. 165–190). Singapore: Springer Nature Singapore. Available from https://doi.org/10.1007/978-981-99-3292-4_9, 10.1007/978-981-99-3292-4_9.

Samanta, N. S., Das, P. P., Sharma, M., Purkait, M. K., Sillanpää, M., Khadir, A., & Gurung, K. (2023). *12 - Recycle of water treatment plant sludge and its utilization for wastewater treatment* (pp. 239–264). Elsevier. Available from https://www.sciencedirect.com/science/article/pii/B9780323993449000104, 10.1016/B978-0-323-99344-9.00010-4.

Semmens, M., Klieve, J., Schnobrich, D., & Tauxe, G. W. (1981). Modeling ammonium exchange and regeneration on clinoptilolite. *Water Research*, *15*(6), 655–666. Available from https://doi.org/10.1016/0043-1354(81)90157-3, https://www.sciencedirect.com/science/article/pii/0043135481901573.

Sharma, M., Das, P. P., Sood, T., Chakraborty, A., & Purkait, M. K. (2021). Ameliorated polyvinylidene fluoride based proton exchange membrane impregnated with graphene oxide, and cellulose acetate obtained from sugarcane bagasse for application in microbial fuel cell. *Journal of Environmental Chemical Engineering*, *9*(6)106681. Available from https://doi.org/10.1016/j.jece.2021.106681, https://www.sciencedirect.com/science/article/pii/S2213343721016584.

Sharma, M., Das, P. P., Sood, T., Chakraborty, A., & Purkait, M. K. (2022). Reduced graphene oxide incorporated polyvinylidene fluoride/cellulose acetate proton exchange membrane for energy extraction using microbial fuel cells. *Journal of Electroanalytical Chemistry*, *907*115890. Available from https://doi.org/10.1016/j.jelechem.2021.115890, https://www.sciencedirect.com/science/article/pii/S1572665721009164.

Sharma, M., Das, P. P., Chakraborty, A., & Purkait, M. K. (2022). Clean energy from salinity gradients using pressure retarded osmosis and reverse electrodialysis: A review. *Sustainable Energy Technologies and Assessments*, *49*101687. Available from https://doi.org/10.1016/j.seta.2021.101687, https://www.sciencedirect.com/science/article/pii/S2213138821007013.

Sharma, M., Samanta, N. S., Chakraborty, A., Purkait, M. K., Sillanpää, M., Khadir, A., & Gurung, K. (2023). *30 - Simultaneous treatment of industrial wastewater and resource recovery using microbial fuel cell* (pp. 621−637). Elsevier. Available from https://www.sciencedirect.com/science/article/pii/B9780323953276000026, 10.1016/B978-0-323-95327-6.00002-6.

Sharma, M., Das, P. P., Kumar, S., & Purkait, M. K. (2023). Polyurethane foams as packing and insulating materials. *American chemical society, polyurethanes: Preparation, properties, and applications volume 3: Emerging applications, 1454*. Available from https://doi.org/10.1021/bk-2023-1454.ch004, https://doi.org/10.1021/bk-2023-1454.ch004.

Sharma, M., Das, P. P., Chakraborty, A., Purkait, M. K., Sillanpää, M., Khadir, A., & Gurung, K. (2023). *29 - Extraction of clean energy from industrial wastewater using bioelectrochemical process* (pp. 601−620). Elsevier. Available from https://www.sciencedirect.com/science/article/pii/B9780323953276000038, 10.1016/B978-0-323-95327-6.00003-8.

Sharma, M., Das, P. P., Purkait, M. K., Husen, A., & Siddiqi, K. S. (2023). *Chapter 16 - Energy storage properties of nanomaterials micro and nano technologies* (pp. 337−350). Elsevier. Available from https://www.sciencedirect.com/science/article/pii/B9780323995467000057, 10.1016/B978-0-323-99546-7.00005-7.

Sontakke, A. D., Das, P. P., Mondal, P., & Purkait, M. K. (2022). Thin-film composite nanofiltration hollow fiber membranes toward textile industry effluent treatment and environmental remediation applications: review. *Emergent Materials*, *5*(5), 1409−1427. Available from https://doi.org/10.1007/s42247-021-00261-y, https://doi.org/10.1007/s42247-021-00261-y.

Sontakke, A. D., Deepti., Samanta, N. S., Purkait, M. K., Husen, A., & Siddiqi, K. S. (2023). *Chapter 2 - Smart nanomaterials in the medical industry micro and nano technologies* (pp. 23−50). Elsevier. Available from https://www.sciencedirect.com/science/article/pii/B9780323995467000252, 10.1016/B978-0-323-99546-7.00025-2.

Starkl, M., Brunner, N., Feil, M., & Hauser, A. (2015). Ensuring sustainability of non-networked sanitation technologies: An approach to standardization. *Environmental Science & Technology*, *49*(11), 6411−6418. Available from https://doi.org/10.1021/acs.est.5b00887, https://doi.org/10.1021/acs.est.5b00887.

Stefanakis, A. I., Akratos, C. S., Gikas, G. D., & Tsihrintzis, V. A. (2009). Effluent quality improvement of two pilot-scale, horizontal subsurface flow constructed wetlands using natural zeolite (clinoptilolite). *Microporous and Mesoporous Materials*, *124*(1), 131−143. Available from https://doi.org/10.1016/j.micromeso.2009.050.005, https://www.sciencedirect.com/science/article/pii/S1387181109002340.

Vanlangendonck, Y., Corbisier, D., & Van Lierde, A. (2005). Influence of operating conditions on the ammonia electro-oxidation rate in wastewaters from power plants (ELONITA™ technique). *Water Research*, *39*(13), 3028−3034. Available from https://doi.org/10.1016/j.watres.2005.050.013, https://www.sciencedirect.com/science/article/pii/S0043135405002551.

Varigala, S. K., Hegarty-Craver, M., Krishnaswamy, S., Madhavan, P., Basil, M., Rosario, P., . . . Luettgen, M. (2020). Field testing of an onsite sanitation system on apartment building blackwater using biological treatment and electrochemical disinfection. *Environmental Science.: Water Research Technology*, *6*(5), 1400−1411. Available from https://doi.org/10.1039/C9EW01106D, http://doi.org/10.1039/C9EW01106D.

Vitzthum von Eckstaedt, S., Charles, W., Ho, G., & Cord-Ruwisch, R. (2016). Novel process of bio-chemical ammonia removal from air streams using a water reflux system and zeolite as filter media. *Chemosphere*, *144*, 257−263. Available from https://doi.org/10.1016/j.chemosphere.2015.080.048, https://www.sciencedirect.com/science/article/pii/S0045653515300576.

Vlaeminck, S. E., Terada, A., Smets, B. F., Linden, D. Van der, Boon, N., Verstraete, W., & Carballa, M. (2009). Nitrogen removal from digested black water by one-stage partial nitritation and anammox. *Environmental Science & Technology*, *43*(13), 5035−5041. Available from https://doi.org/10.1021/es803284y, https://doi.org/10.1021/es803284y.

Wang, Y., Liu, S., Xu, Z., Han, T., Chuan, S., & Zhu, T. (2006). Ammonia removal from leachate solution using natural Chinese clinoptilolite. *Journal of Hazardous Materials*, *136*(3), 735−740. Available from https://doi.org/10.1016/j.jhazmat.2006.010.002, https://www.sciencedirect.com/science/article/pii/S0304389406000161.

Wen, D., Ho, Y.-S., & Tang, X. (2006). Comparative sorption kinetic studies of ammonium onto zeolite. *Journal of Hazardous Materials*, *133*(1), 252−256. Available from https://doi.org/10.1016/j.jhazmat.2005.100.020, https://www.sciencedirect.com/science/article/pii/S0304389405006394.

Wijesinghe, D. T. N., Dassanayake, K. B., Sommer, S. G., Jayasinghe, G. Y., Scales, P. J., & Chen, D. (2016). Ammonium removal from high-strength aqueous solutions by Australian zeolite. *Journal of Environmental Science and Health, Part A.*, *51*(8), 614−625. Available from https://doi.org/10.1080/10934529.2016.1159861, https://doi.org/10.1080/10934529.2016.1159861.

Wu, Z., An, Y., Wang, Z., Yang, S., Chen, H., Zhou, Z., & Mai, S. (2008). Study on zeolite enhanced contact−adsorption regeneration−stabilization process for nitrogen removal. *Journal of Hazardous Materials*, *156*(1), 317−326. Available from https://doi.org/10.1016/j.jhazmat.2007.120.029, https://www.sciencedirect.com/science/article/pii/S0304389407017748.

Yu, W., Yuan, P., Liu, D., Deng, L., Yuan, W., Tao, B., Cheng, H., & Chen, F. (2015). Facile preparation of hierarchically porous diatomite/MFI-type zeolite composites and their performance of benzene adsorption: The effects of NaOH etching pretreatment. *Journal of Hazardous Materials*, *285*, 173−181. Available from https://doi.org/10.1016/j.jhazmat.2014.110.034, https://www.sciencedirect.com/science/article/pii/S0304389414009418.

Zhang, W., Zhou, Z., An, Y., Du, S., Ruan, D., Zhao, C., . . . Tian, X. (2017). Optimization for zeolite regeneration and nitrogen removal performance of a hypochlorite-chloride regenerant. *Chemosphere*, *178*, 565−572. Available from https://doi.org/10.1016/j.chemosphere.2017.030.091, https://www.sciencedirect.com/science/article/pii/S0045653517304629.

# Index

Printed in the United States
by Baker & Taylor Publisher Services